Towards the "Perfect" Weather Warning

Editor
Brian Golding
Met Office
Exeter, UK

This book is an open access publication.
ISBN 978-3-030-98991-0   ISBN 978-3-030-98989-7   (eBook)
https://doi.org/10.1007/978-3-030-98989-7

Foreword

Enhancing international cooperation for developing countries to reduce disaster risk is a key target of the global blueprint to reduce disaster losses, the Sendai Framework for Disaster Risk Reduction 2015–2030.

In that spirit, I welcome this publication as a shining example of cooperation and a vindication of the initiative embarked on 10 years ago by WMO to create the High Impact Weather (HIWeather) project as part of its World Weather Research Programme. HIWeather is also an important contribution towards achieving another of the Sendai Framework's targets to substantially increase the availability of, and access to, multi-hazard early warning systems.

UNDRR and WMO are passionately partnering in our advocacy for improved access to early warning systems, especially in low- and middle-income countries. These countries bear a disproportionate burden in terms of mortality and economic losses as a result of the rising number and increasing intensity of extreme weather events in a warming world.

This publication is testament to the contribution that HIWeather has made over the last decade to convince policymakers that early warning systems should be far more than mechanisms to issue warnings of impending hazard events. An effective multi-hazard early warning system is one in which hazards are monitored, and forecasts and warnings issued. They must also increasingly take account of the systemic nature of disaster risk and generate risk scenarios for the areas and population likely to be affected by a forecast event.

Recently, we have seen an enormous escalation in the numbers of people evacuating in response to early warnings, saving many thousands of lives. Trust in the risk governance system in place and good communication are vital to these successes. The effectiveness of any early warning system is not a matter of whether warnings are issued but rather if the warnings lead to appropriate and timely action to save lives and reduce damage to critical infrastructure.

Thanks to WMO and its members, we know what works. The challenge before us is to make that available in places which need it most but are currently underserved.

While the majority of UN Member States that signed the Paris Agreement identify early warning systems as a "top priority", WMO's most recent Climate Services report shows that a high number of LDCs and SIDS do not have multi-hazard early warning systems in place. Many LDCs, notably in Africa, lack capacity to translate early warning into early action.

I have no doubt that this publication will boost efforts to fill capacity gaps in early warning systems and encourage greater investment in this area.

This book brings together expert contributions from around the globe and from many disciplines. It will be of very practical use to Sendai Framework focal points in government, national disaster management agencies and others engaged in developing and implementing national strategies for disaster risk reduction. It will be of particular benefit to countries with weak warning systems, and I note with gratitude that it is an open access publication thanks to the support of WMO and the HIWeather Trust Fund.

水島直美.

Mami Mizutori
Special Representative of the UN Secretary-General for Disaster Risk Reduction
Geneva, Switzerland

Preface and Acknowledgements

This book is about the contribution of early warnings to reducing damage, disruption and distress from natural hazards. Its theme is partnership – between producers and receivers of warnings, and between the many experts who contribute to creating a warning. We dedicate this book to everyone who has saved a life by issuing an effective early warning and in the hope that it will be of help to those who we rely on to do so in the future.

The background to writing this book is the 10-year High Impact Weather (HIWeather) project of the World Weather Research Programme of the World Meteorological Organization (WMO). The project is, itself, an example of a very successful partnership – across the wide variety of disciplines involved in warning production, between weather services and academia, and between countries. The work of HIWeather is described on its website at http://hiweather.net and can be followed by signing up to its newsletter.

The writing of this book has also been a partnership, with nearly 50 expert contributors brought together and their contributions integrated by the chapter coordinators. We have aimed for a coherent narrative across the warning chain in which the individual areas of expertise are brought together seamlessly. As editor, I provided the overall structure and areas of content, and edited the final text for consistency. The expert input is provided by the individual contributors listed in each chapter. You, the reader, are the final judge of whether we have successfully combined all of this diverse expertise.

The HIWeather project aims to raise the capability for early warnings to save lives and reduce damage across the world in pursuit of the aims of the Sendai Agreement of 2015. Our book has the same aim and is therefore targeted at emergency responders, weather services and governments in every country, but especially in those with poorly developed warning systems. To maximise its impact, we are delighted to be able to offer this as an open access publication that can be accessed everywhere without limitations of available budgets. We are very grateful to the World Meteorological Organization and to the contributors to the HIWeather Trust Fund for making this possible.

I am grateful to the Met Office for supporting my co-chairmanship of the HIWeather project, and particularly for providing the time for me to edit this book. As a National Weather Service, a primary role of the Met Office is the provision of weather warnings for the protection of life and property.

Towards the "Perfect" Weather Warning: Bridging Disciplinary Gaps Through Partnership and Communication

Overview This book is about making weather warnings more effective in saving lives, property, infrastructure and livelihoods, but the underlying theme of the book is partnership. The book represents the warning process as a pathway linking observations to weather forecasts to hazard forecasts to socio-economic impact forecasts to warning messages to the protective decision, via a set of five bridges that cross the divides between the relevant organisations and areas of expertise. Each bridge represents the communication, translation and interpretation of information as it passes from one area of expertise to another and ultimately to the decision maker. Without effective partnerships between the disciplines and/or organisations involved at each stage in the warning process, information is lost and distorted. Making the whole system work effectively also requires a partnership of those involved within a policy structure that brings together government, private business, civil society and the voluntary sector. As we explore the partnerships upon which each bridge is built, we look at the expertise and skills that each partner brings, at the challenges of communication between them, and at structures and methods of working that build effective partnerships. We have chosen to order the book according to the "first mile" paradigm in which the decision maker comes first, and then we work back up the production chain through the warning and forecast to the observations, emphasising the importance of co-design and co-production throughout the warning process.

Audience The target audience for the book will be professionals and trainee professionals with a role in the warning chain, i.e. in weather services, emergency management agencies, disaster risk reduction agencies, risk management sections of infrastructure agencies and relevant parts of government. With this in mind, we aim to focus on producing a succinct and clear exposition of the evidence and theory, with an emphasis on putting it into practice.

Brian Golding
Exeter, UK

Contents

Chapter 1
Introduction

Brian Golding

Abstract We outline the objectives of the book, setting them in the context of disaster risk management in a changing world and introducing the role of warnings in mitigating weather-related disasters. We describe the warning value chain, linking the needs of the decision-maker with forecasting capability; identify the blockages where information is lost, which we call the "valleys of death"; and introduce partnership as the core theme of the book. We then summarise the structure of the book and of each chapter, concluding by emphasising the importance of having a governance framework to monitor performance, inform troubleshooting and determine investment.

Keywords Sendai framework · Disaster risk management · Partnership · Governance

Disasters happen with appalling frequency in our world, resulting in death, injury, destruction, disruption and economic loss that can set back the development efforts of affected countries by decades. Disasters from natural hazards are growing at a rapid rate for several reasons. Population growth, migration and urbanisation are all leading to the poorest, most vulnerable, people occupying land that is more exposed to severe and frequent threats, while urbanisation and migration are also producing communities that are both more dependent on infrastructure and lacking in local knowledge of their environment. At the same time, climate change is increasing the severity and frequency of many of those threats. In 2015, the world community met together to sign the Sendai Framework for Disaster Risk Reduction (UNDRR 2015), aimed at building a safer world for everyone.

The signing of the Sendai Framework was a critical step towards mobilising resources to counter these trends towards more frequent, more costly disasters.

B. Golding (✉)
Met Office, Exeter, UK

WMO/WWRP HIWeather project, Geneva, Switzerland
e-mail: brian.golding@metoffice.gov.uk

© The Author(s) 2022
B. Golding (ed.), *Towards the "Perfect" Weather Warning*,
https://doi.org/10.1007/978-3-030-98989-7_1

Much of its content deals with planning and policies that will reduce exposure to hazards, especially in new developments, and reduce vulnerability in existing ones. However, threats will always occur that exceed the protection in place, and the Sendai Framework also promotes the critical role of early warnings in enabling people to survive and recover from disasters. Since the cost of disasters can severely set back progress in development, the implementation of more effective responses to weather-related hazards is a theme in many of the Sustainable Development Goals, also agreed by the world community in 2015.

The objective of this book is to save lives and livelihoods and to reduce injury, damage and disruption from weather-related hazards in all parts of the world by helping those who create policy and plan, design and operate warning systems to make use of the latest research on what makes a good warning, so that their warnings may be more effective in our rapidly changing world. The material for this book has been gathered as part of the World Meteorological Organisation's High Impact Weather (HIWeather) project under the auspices of the World Weather Research Programme (Golding et al. 2019).

While the frequency and severity of threats are growing for the reasons given above, the ability to avoid or reduce their impacts is also growing as a result of scientific research and technical innovation. Weather forecasting has seen spectacular advances in prediction accuracy in the last half century, with 5-day forecasts now more accurate than 1-day forecasts were then. These improvements have come as a result of technical achievements in computing and satellite observation as well as from the application of new science. The ability to gather and communicate information has always been at the heart of weather forecasting and warning, but the revolution in mobile communication of the past 20 years has enabled warning messages to reach a much greater number of people, even in remote areas of developing countries, so that there is a much greater awareness of the approach of weather-related hazards.

We have written this book for professionals and trainees who are involved in setting policy and planning, implementing and operating warning systems or parts of them. They may be in central or local government, in emergency management, in the management of businesses or utilities, in international aid agencies or in community response groups. The material is also a suitable introduction for those planning research on aspects of the warning chain especially those undertaking transdisciplinary research across the physical and social sciences.

In this book, we refer to anyone who acts on a warning as a *decision-maker*. They may be an individual acting to protect themselves by deciding whether to evacuate or to postpone a journey. They may be a manager responsible for the staff, customers and plant of a business. They may be an emergency manager responsible for the safety of a community. Or they may be a government minister responsible for the safety of a nation. Each has different levels of responsibility and will take different decisions in order to exercise the power they have been given.

This book is focused on the production and use of warnings. We distinguish a *forecast*, which produces information about the future state of the weather or some other aspect of the environment, without consideration of its use, from a *warning* which provides information about a threat so as to enable a response. The response will be different according to the lead time, the confidence level, the severity of the threat, the vulnerability of those threatened and other factors. For instance, a long range, low confidence warning may be used to initiate training or other "no regrets" preparatory activities, whereas a short range, high confidence warning may be used for more costly responses such as closing down a factory.

A decision-maker is a *user* of a warning but may also be a *producer* of a warning for someone else. For instance, an emergency manager in a city will receive a warning and initiate city-wide activities to protect citizens from the threat. At the same time, they may themselves issue a warning to citizens, advising them to take specific action. Similarly, a weather forecast centre in the path of a storm may need to respond to warnings issued by emergency services in order to protect its staff and to maintain operations.

We characterise the production of warnings as a *value chain* whose aim is to provide the information that enables the best decisions to be taken, both by individuals and by those with responsibility to protect others. In a perfect warning chain, the warning received by the end user would contain precise and accurate information that perfectly met their need, contributed by each of the many players in the chain. In real warning chains, information, and hence value, are always lost as well as gained at each link in the chain. In business, the term "valley of death" was coined as a metaphor of the failure of research to lead to successful innovation. NRC (2001) adopted this to represent the failure of research to translate into operational weather forecasting improvements. In Fig. 1.1, we use it more generally to represent the failure of the expert information generated in warning organisations to lead to the desired responses due to inadequate communication along the warning chain. The height of each mountain may be interpreted as the maturity of the expertise available for use in weather warnings. Successful communication of information from one contributor of expertise to the next is represented by spanning the valleys with bridges, whose height can represent the success of the communication between those contributors in avoiding the loss of information. Without a bridge, there is no communication, and the expertise of a particular contributor is completely lost. This representation of the warning process is, of course, a gross oversimplification of reality. Real warnings are created from a complex web of

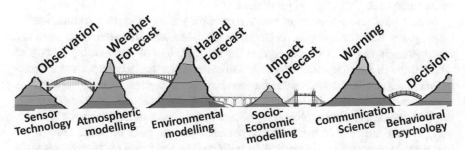

Fig. 1.1 The valleys of death concept of a warnings value chain. (© Crown Copyright 2020, Met Office)

interactions taking place continuously among a wide variety of people more or less involved in the core activities shown. At the same time, distinct activities may in some cases be combined in a single person. There are also professionals whose expertise lies in being one of these bridges. Nevertheless, the concept is a useful one that highlights the very broad range of disciplines involved – and the need for those disciplines to communicate with each other effectively.

This conceptual value chain can be read in either direction – there are no arrows on the diagram! Indeed, it is important that it is read in *both* directions. In designing a warning system, the starting point must be the decisions that need to be taken to protect life, property, infrastructure and livelihoods. These should be the basis for deciding what information is required and how it should be delivered – sometimes referred to as the "first mile paradigm", because the user is at the start of the process. These decisions will vary according to the hazard that is being responded to, the person making them and the environment in which that person sits. In some cases, they will be so distinctive that a specific tailored form of warning is needed for that person. In other cases, a generic form of warning may need to be designed to meet the common requirements of a variety of people. The need to take a decision demands specific information, which can be traced up the value chain to define the expertise and resources required to produce it. However, if we do this without considering the capabilities of the upstream contributors, the warning system will certainly fail. Not only are there limitations to what science and technology make possible, but there may be capabilities available that enable more effective decisions to be taken that hadn't been thought possible. Thus, design of a warning system is an iterative process, starting from the decision-maker, progressing up the value chain, then continually returning to the decision-maker in a process of mutual adjustment. On the other hand, when the warning system is in operation, the flow of information is predominantly down the value chain from producer to user – and this needs to happen quickly, accurately and reliably to maintain the value of the information. Nevertheless, the return flow of information is still crucial, providing updates from those involved "on the ground" so that the warning producers can maintain their situational awareness. Since this book is aimed particularly at those designing a new or improved warning system, we adopt the first mile paradigm for the ordering of the chapters – starting with the decision-maker and moving up the value chain to the information producers.

In this book *communication* and *partnership* are key words. Communication takes place between people, between institutions and between data systems. Limitations in any of these can inhibit the effectiveness of the warning system of which they are a part. Institutional communication is particularly important in creating an environment for successful partnership. At the highest level, government is responsible for creating the legislative framework that facilitates partnership working between organisations. However, while a good governance framework is necessary, it is not sufficient in itself. Shared personal knowledge of the aims, culture and language of the partners is critical to building the trust that enables outstanding performance when the threat is real.

Warnings of weather-related hazards have a long history going back to the foundations of national weather services in the late nineteenth century, largely in response to the implementation of the telegraph as the first telecommunication network, enabling instant communication of observations and warnings. The first applications were in maritime safety, at a time when most propulsion was by sail and shipwrecks were commonplace. With the growth of aviation in the early twentieth century, the extreme sensitivity of early flying machines required great care to avoid dangerous weather conditions. As a result, much of the modern structure of weather services was put in place to support the safety of aviation. Moving to the second half of the twentieth century, the massive growth in personal transport led to increased requirements for warnings of adverse road conditions. Weather prediction advanced rapidly through the application of satellite observing and computer prediction. At the same time, rapidly expanding populations in locations exposed to hazardous weather generated the need for a wider range of warnings on land, particularly of storms and floods. These developments exposed limitations in the ability of weather services to meet the needs of decision-makers. Whereas a sailing ship's captain would know what the impact of a gale force wind would be and an aircraft pilot would know the safe visibility for landing at his destination airport, users of this wider range of weather warnings were less likely to understand how the predicted hazard would affect them. More recently, the ability to get warning information to people has advanced rapidly with the availability of dedicated radio and television services and the mobile phone. As the range of warnings has extended into aspects of the weather that are inherently less certain, it has become more important to communicate the confidence that the hazard will occur at the location and severity predicted. Taken together these challenges have led to the development of a range of new warning paradigms incorporating impact and risk as well as hazard. While not yet adopted universally, these newer types of warning will become the norm over the next few years and their core capabilities are described here, drawing on the latest capabilities in probabilistic weather and hazard prediction at ever finer geographical scales. With an increasing diversity of responsibility in responding to warnings, the study of how people receive and react to a warning has become a critical input to warning design, and educational programmes have been developed to grow familiarity in the warnings and the desired responses. However, weather services and emergency managers have often lagged behind commercial business in applying the science of behavioural psychology to help ensure positive rather than negative responses to messages. This is now changing, as more weather services set up social science groups to advise them on warning design. As a result of these changes, an ever-increasing range of expertise is being brought to bear on warning production. While some weather services continue to employ the full range of expertise in a single organisation, it is increasingly recognised that the achievement of a critical mass of expertise in these newer disciplines may be best achieved through the building of partnerships between complementary organisations.

The justification for implementing warnings as part of a risk reduction strategy is that (a) the cost of a warning system is much less than most other risk reduction options and (b) the benefits to society far outweigh the costs. When considering the

benefits, there is often a focus on economic benefits, and these are certainly important. In the most-developed countries, economic costs of weather-related hazards
are large and increasing, due to growth in hazardous events due to climate change,
growth in economic vulnerability due to increased wealth and growth in exposure
due to the spread of populations into more hazard-prone areas. The relatively small
number of deaths and injuries remains important, however, as the cost of these to
society is high. The impact of indirect hazard impacts on people's health, well-being
and productivity is hidden but of increasing interest to researchers and potentially a
significant additional cost. By contrast those affected in the least-developed countries have much less wealth to destroy, and hazard impacts are primarily measured
in fatalities and injuries. Thankfully these are reducing, though they remain far too
high. However, the smallness of the economic losses hides the fact that a person
who loses an uninsured house is destitute regardless of whether their house was
worth $10 or $10 million, and such losses need to be measured against an appropriate comparator such as a country's GDP, to understand their significance.

This book is distinctive in the literature on warnings, in that it concentrates on the
partnerships connecting experts across the "valleys of death" (Fig. 1.2). The need
to build partnerships is not unique to effective warning systems. Consideration of
how they work in other contexts can provide useful pointers to those attempting to
make progress in warnings. There is a whole literature on building successful business partnerships (Rosen 2007; Morten 2009; Swientozielskyj 2016), which is
increasingly being applied more widely (de Bruijn and Tucker 2002, Bang and
Frith, 2017, Stibbe and Prescott 2020, Bucher et al. 2020, WISER 2020). The

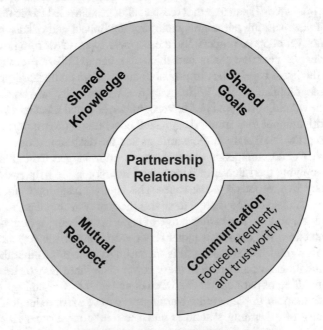

Fig. 1.2 Aspects of a working partnership. (Source – Rob Honch, ECCC)

lessons learned there are equally relevant to the production of warnings (Golnaraghi 2012). Every partnership has a formal and an informal aspect. The formal provides the legal basis that ensures the informal can proceed without the danger of being derailed by accidental ignorance or misunderstanding. Non-disclosure agreements, memoranda of understanding, contracts, etc. all have their place, enabling the respective organisations to be open with each other. Part of the formal process must also be regular reviews that challenge the justification and effectiveness of the partnership and, if necessary, dissolve it without delay. However, the real work lies in building the relationship that these represent, and that takes time, patience, lots of listening, strong leadership and hard work. A successful partnership requires that each partner understands the culture and can speak the language of their counterparts. They must know their organisational viewpoints and aims. Ultimately, a partnership works if the partners trust each other, something that can only be achieved by actively working together over a long period of time. It can easily be interrupted by changes of personnel, so must be actively maintained, both when it is going well and when it is not. It is in the pressurised situation of an incipient disaster that a failure of trust is most likely to surface – with potentially fatal results.

In the main part of this book, each chapter describes one of the bridges in our conceptual value chain (Fig. 1.3). It first describes the expertise and methods of those at the decision-maker's end of the bridge, highlighting the information they need to make their contribution. It then moves across the bridge to investigate the constraints and limitations of the provider of that information. Then the nature of the bridge itself is addressed, identifying the characteristics that inhibit communication and showing how these may be overcome through the building of strong partnerships. Each chapter includes some examples of real partnerships that have addressed these issues, the challenges they encountered and the outcomes that were achieved. Finally, the key points for success are summarised for reference.

Before proceeding to the core chapters of the book, which are organised according to the "first mile paradigm", described above, Chap. 2 describes an effective risk management framework in the context of societal drivers of risk and the responses in the United Nations 2030 agenda, introducing the main components, the roles of the main actors and the need for evaluation. We introduce some key definitions and emphasise the contribution of early warnings within this framework.

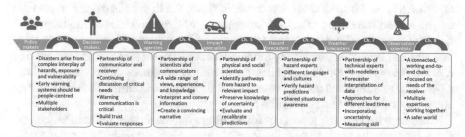

Fig. 1.3 Contents of the book in summary

In Chap. 3 we explore the challenges of achieving a level of risk perception, in each decision-maker, that is commensurate with the most cost-effective action while being consistent with the warning producer's capabilities. Firstly, we look at the evidence for how people respond to warnings and how the nature and delivery of the warning affect that. Then we look at the aims of the person providing the warning, the constraints within which they must act and the judgement process by which they decide when the level of confidence needed for a warning is reached. Then we address the connection between the two, provided by the delivery of the warning, and how a partnership between warner and receiver can produce a more effective response.

Chapter 4 looks at the range of actors who produce warnings in the public and private sectors, the sources of information they draw on to comprehend the nature of the hazard, its impacts and the implications for those exposed and the process of drawing that information together to produce a warning. We consider the wide range of experts who provide the tools to assess the impacts of the predicted hazards and the challenges and limitations of these tools and the information they produce. Then we look at the diverse ways in which these tools need to take account of the way their outputs will feed into warnings and the nature of the partnerships that can facilitate this.

Chapter 5 focuses on translation of the hazard into its impact which is at the heart of current efforts, within the WMO HIWeather project and elsewhere, to improve the effectiveness of warnings by incorporating impact information into the warning process. At the same time, it presents some of the most difficult and demanding challenges in contrasting methodology and language. In general, the hazards we are concerned with can be described well by repeatable processes that may be couched in mathematical language. By contrast, their impacts depend on the social and economic characteristics of the communities that they affect; repeatable characteristics can only be discerned statistically and often cannot be related to any mathematically describable process. While the hazard can usually be defined quite precisely – if not always accurately – descriptions of its impact may depend substantially on the perceptions of the observer. As a result, the experts on each side of this bridge may have quite different and conflicting views as to what information it is appropriate to exchange. Here we explore the needs of the impact scientist first, remembering that relevant impacts are those of interest to the end user. A key challenge is in obtaining historical information on impacts, especially where the raw data are confidential, and then of matching suitable hazard data to them. We then consider the constraints on the hazard forecaster, who may have access to large volumes of model predictions but cannot easily relate them to the times and locations of those being impacted and who has limited knowledge of model accuracy in hazardous situations. Creating a bridge between these two requires an open and pragmatic approach from both sides, with relationships built up over time, through joint working, so that the different ways of thinking can be absorbed.

Chapter 6 looks at the aspects of forecasting systems required to achieve consistency between the prediction of the state of the atmosphere and of related environmental hazards. We first look at the different approaches to hazard prediction and then consider the limitations in predicting their meteorological drivers. We note that different modelling structures are adopted in different hazard forecasting disciplines and consider how these relate to the user requirements for those hazards. We identify the benefits of seamless approaches to hazard prediction and the challenges of achieving them in a multi-institution situation.

Chapter 7 addresses the problem of monitoring and predicting the weather. We look at how atmospheric modellers use observations to initialise their forecasts. Effective use of data in models places specific requirements on the observations. We then consider the application of basic physics and engineering in producing sensors and the observing platforms that carry them. There is a long history of close working between sensor and platform designers and meteorologists that has produced spectacular advances in forecast accuracy. However, the latest high-resolution forecasting models require data that cannot be obtained with conventional approaches to either in situ or remotely sensed observing. At the same time, new capabilities in manufacturing and communication have made available a vast amount of relatively lower-quality observations that will require new collaboration models to bring into effective use.

Finally, in Chap. 8, we take a step back and consider the warning chain as a whole system whose aim is to avoid loss by delivering the needs of those taking decisions in response to the warnings. We emphasise that, within the chain, every actor is both a user of information coming from upstream actors and a provider of information to downstream actors and that the effective definition, communication and use of that information depends on partnerships.

An effective warning system saves lives and cost, builds trust and makes people more confident of their safety. Building such a system is worthy of the time and effort that it will take.

References

Bang, D. and C. D. Frith, 2017. Making better decisions in groups. *R. Soc. Open Sci.,* **4.** https://doi.org/10.1098/rsos.170193

de Bruijn, T. and A. Tukker, Eds., 2002. *Partnership and Leadership: Building Alliances for a Sustainable Future,* Springer

Bucher, A., A. Collins, B. H. Taylor, D. Pan, E. Visman, J. Norris, J. C. Gill, J. Rees, M. Pelling, M. T. Bayona, S. Cassidy and V. Murray, 2020. New Partnerships for Co-delivery of the 2030 Agenda for Sustainable Development. *Int. J. Disaster Risk Sci.,* **11,** 680–685. https://doi.org/10.1007/s13753-020-00293-8

Golding, B., E. Ebert, M. Mittermaier, A. Scolobig, S. Panchuk, C. Ross and D. Johnston, 2019. A value chain approach to optimising early warning systems. Contributing paper to *Global Assessment Report on Disaster Risk Reduction 2019.* UNDRR. 30pp. https://www.preventionweb.net/publications/view/65828

Golnaraghi, M., (Ed.), 2012. *Institutional Partnerships in Multi-Hazard Early Warning Systems: A Compilation of Seven National Good Practices and Guiding Principles.* Springer. https://doi.org/10.1007/978-3-642-25373-7

Morten, H., 2009. *Collaboration: How Leaders Avoid the Traps, Build Common Ground, and Reap Big Results*. Harvard Business Review Press. ISBN 10: 1422115151

NRC, 2001. *From Research to Operations in Weather Satellites and Numerical Weather Prediction: Crossing the Valley of Death*. National Academy Press, 96pp

Rosen, E., 2007. *The culture of collaboration: maximizing time, talent and tools to create value in the global economy*. San Francisco, Red Ape Pub. 304pp

Swientozielskyj, S., 2016. *Business Partnering: A Practical Handbook*. Taylor & Francis. ISBN:9781317440734

Stibbe, D. and D. Prescott, 2020. *The SDG partnership guidebook: A practical guide to building high impact multi-stakeholder partnerships for the Sustainable Development Goals*. The Partnering Initiative and UNDESA. https://sustainabledevelopment.un.org/content/documents/2698SDG_Partnership_Guidebook_1.01_web.pdf

UNDRR, 2015. Sendai Framework for Disaster Risk Reduction 2015-2030. Geneva, Switzerland: United Nations Office for Disaster Risk Reduction, 37. https://www.preventionweb.net/files/43291_sendaiframeworkfordrren.pdf. (Accessed 2/9/2021)

Weather and Climate Information Services for Africa (WISER), 2020. The Power of Partnership, https://www.metoffice.gov.uk/binaries/content/assets/metofficegovuk/pdf/business/international/wiser/wiser0223_highway_partnership_impact_article_0620.pdf. (Accessed 3/2/2021)

Chapter 2
Early Warning Systems and Their Role in Disaster Risk Reduction

Robert Šakić Trogrlić, Marc van den Homberg, Mirianna Budimir, Colin McQuistan, Alison Sneddon, and Brian Golding

Abstract In this chapter, we introduce early warning systems (EWS) in the context of disaster risk reduction, including the main components of an EWS, the roles of the main actors and the need for robust evaluation. Management of disaster risks requires that the nature and distribution of risk are understood, including the hazards, and the exposure, vulnerability and capacity of communities at risk. A variety of policy options can be used to reduce and manage risks, and we emphasise the contribution of early warnings, presenting an eight-component framework of people-centred early warning systems which highlights the importance of an integrated and all-society approach. We identify the need for decisions to be evidence-based, for performance monitoring and for dealing with errors and false information. We conclude by identifying gaps in current early warning systems, including in the social components of warning systems and in dealing with multi-hazards, and obstacles to progress, including issues in funding, data availability, and stakeholder engagement.

Keywords Governance · Risk management · People-centred · Hazard · Exposure · Vulnerability · Impact · Early Warning Systems · Risk

R. Šakić Trogrlić (✉)
International Institute for Applied Systems Analysis (IIASA), Laxenburg, Austria

Research done while with Practical Action, Rugby, UK
e-mail: trogrlic@iiasa.ac.at

M. van den Homberg
510 An Initiative of the Netherlands Red Cross, The Hague, The Netherlands

M. Budimir · C. McQuistan · A. Sneddon
Practical Action, Rugby, UK

B. Golding
Met Office, Exeter, UK

WMO/WWRP HIWeather project, Geneva, Switzerland

2.1   Introduction

Despite decades of progress in our understanding of disaster risks, how they should be dealt with and international agreements to build resilience of people and nations, hardly a week passes without devastating news of natural hazards causing havoc in both developed and developing countries. While the world was busy taming the beast of COVID-19, Tropical Cyclone Amphan unleashed its power over India and Bangladesh in May 2020, killing 72 people and causing over 13 billion dollars of damage in West Bengal (Sarkar 2021), with total loss and damage still unknown. In July 2020, heavy rainfall in Nepal triggered flooding and landslides, leaving a death toll and shattering the livelihoods of many.

Examples like these are countless. Although all impacts of natural hazards on people, economies and environment cannot be completely avoided, they can be substantially reduced. One of our 'best bets' is to implement early warning systems (EWS), as they nurture learning and understanding of natural hazards, provide us with warning information and give time to take early action, so as to avoid unnecessary consequences. Despite some progress in enhancing EWS globally, the recent report on the state of climate services (WMO 2020a) shows that, in the 73 countries considered, one-third of people are not covered by early warnings, and just 40% have multi-hazard EWS.

The world's climate is changing, and those changes also manifest themselves in a changing risk from weather-related hazards in every country. The intensity and frequency of hazards will change with climate change. This implies that, in some countries, there will be additional hazards for which EWS are required (e.g., EWS for heat waves in locations where this was previously not necessary) while others may become less significant. At the same time, socio-economic development in each country is changing the exposure to hazards and the vulnerability of their populations. Mitigating the increases in risk arising from these changes and further adaptation are crucial for sustainable development of societies. In this chapter we shall:

- Introduce the key concepts of hazard, exposure, vulnerability and risk.
- Outline measures that can be taken to reduce disaster risk.
- Situate EWS in the landscape of available options to reduce disaster risks.
- Elaborate on the main components of an EWS, presenting an eight-component framework of people-centred EWS which highlights the importance of an integrated and all-society approach.
- Identify gaps in current capability, especially in the social components of EWS and in dealing with multi-hazards, and obstacles to progress, including issues in funding and stakeholder engagement.

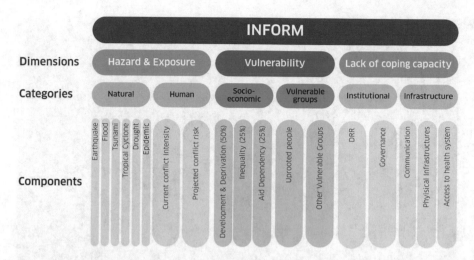

Fig. 2.1 The different risk dimensions, categories and components of INFORM. The final selection of the components and underlying indicators is country-specific. (Based on Marin-Ferrer et al. 2017)

2.2 Disaster Risks and Impacts

Risk from natural hazards arises from a combination of dimensions: natural hazard, exposure of people or assets to that hazard and the vulnerabilities and coping capacities of each person or asset to that hazard. Several multilateral organisations such as UNFCCC/IPCC, UNDRR and the Inter-Agency Standing Committee (IASC) Reference Group on Risk, Early Warning, and Preparedness, together with the European Commission, have put forward definitions of risk and its dimensions. For instance, IASC and the EU Joint Research Centre have developed the global open-source INFORM Risk Index that can be used to calculate risk at the national or sub-national level (Marin-Ferrer et al. 2017), specifically for humanitarian crises and disasters. Box 2.1 outlines the definitions by the United Nations Office for Disaster Risk Reduction. These definitions are widely accepted in the DRR community of practitioners. We note that the INFORM Risk Index relates closely to definitions proposed by UNDRR but also defines a methodology to calculate a composite risk index based on different risk dimensions, categories, components and indicators (Fig. 2.1).

2.2.1 Hazard

The hazard dimension is – in comparison with vulnerability and exposure – relatively well characterised, at least for single hazards. UNDRR (2020b) categorises hazards into biological, environmental, geological, hydrometeorological, technological and societal. Here we are primarily concerned with hydrometeorological

Box 2.1 Risk Dimension Definitions Based on the UNDRR Definitions (UNDRR 2016)

Vulnerability

The conditions determined by physical, social, economic and environmental factors or processes which increase the susceptibility of an individual, a community, assets or systems to the impacts of hazards.

Coping capacity

The ability of people, organisations and systems, using available skills and resources, to manage adverse conditions, risk or disasters.

Exposure

The situation of people, infrastructure, housing, production capacities and other tangible human assets located in a hazard-prone area or lying in the path of a specific hazard. Measures of exposure can include the number of people or types of assets in an area.

Hazard

A process, phenomenon or human activity that may cause loss of life, injury or other health impacts, property damage, social and economic disruption or environmental degradation. Hazards may be natural, anthropogenic or socio-natural in origin. Natural hazards are predominantly associated with natural processes and phenomena. Anthropogenic, or human-induced, hazards are induced entirely or predominantly by human activities and choices. Several hazards are socio-natural, in that they are associated with a combination of natural and anthropogenic factors, including environmental degradation and climate change. Hazards may be single, sequential or combined in their origin and effects. Each hazard is characterised by its location, intensity or magnitude, frequency and probability. Biological hazards are also defined by their infectiousness, toxicity, etc.

hazards. Hazards are dynamic in nature due to both climate variability and climate change. Forecasts of hazards occurring can range from climate change projections to decadal, seasonal, sub-seasonal and short-term forecasts. Early warning systems use seasonal up to short-term forecasts, a progression in which precision and confidence should grow as the length of the forecast decreases. The lead time for which useful information can be provided varies widely, from seasonal timescales for droughts to just a few seconds for an earthquake.

Apart from these different temporal dimensions of hazard forecasts, the spatial dimension is also very important. Spatial maps of the frequency of hazardous conditions are required for the planning and implementation of preparedness and response interventions as well as for longer-term interventions such as land-use zoning. These are typically based on observation or modelling of past conditions but should be adjusted using projections of future change (both human change and climate change). For example, observed flood depths can be combined to create a flood extent map representative of a historical flood, and hydrological and hydrodynamic models can be used to create hypothetical flood extent maps for different levels of probability.

There are slow- and sudden-onset hazards. Sudden-onset hazards refer to hazardous events that emerge quickly or unexpectedly, such as river and flash floods, wildfires or extreme winds. Slow-onset hazards occur gradually over time, such as droughts or sea-level rise. Some hazards can show intermediate-onset behaviour, such as disruptive winter weather. To add to the complexity, disasters are often consecutive. This means that the impacts of two or more disaster events overlap both spatially and temporally before recovery from the first event is considered to be complete. Multiple hazard events can be classified as compound events or cascading events (Ruiter et al. 2020), covering both the interaction of discrete natural hazards (Gill & Malamud 2014) and the interaction of natural hazards with shocks and stresses in social, cultural, political, economic, health and technological systems.

While the risks associated with multi-hazard events are recognised, and approaches for managing them are increasingly advocated as part of DRR policies and practice (UNDRR 2015), these risks are not well defined. Key challenges and gaps must be addressed to enable informed assessments of the likelihood of multi-hazard events and their impacts.

Hazards have different levels of intensity. Whereas scientists may describe a phenomenon using a physically continuous scale of intensity, for hazard warnings it is often more helpful to use discrete classes of intensity that are associated with degrees of impact, e.g. the Richter scale for earthquakes or the Fujita scale for tornadoes.

Several methodologies, including the INFORM Risk Index, merge aspects of the hazard and exposure dimensions into one risk dimension to reflect the probability of physical exposure associated with a specific hazard. For floods and drought, this identifies exposed cropland (e.g. in a floodplain or in a drought-prone area) and affected communities. An example is how UNEP, on their Global Risk Data Platform, calculates physical exposure to floods (UNEP 2021). To determine hazard exposure, hazard frequency data are combined with exposed population datasets. Long-term frequency data can be used to generate return periods, commonly used to communicate the probability of an event exceeding a certain magnitude happening in a given year. The ThinkHazard! tool of Global Facility for Disaster Reduction and Recovery (GFDRR) provides the likelihood of multiple natural hazards affecting a certain area, drawing from published hazard data, provided by a range of private, academic and public organisations (GFDRR 2021). Table 2.1 presents a non-exhaustive overview of hazard data providers.

Table 2.1 Non-exhaustive overview of hazard data providers

Primary hazard data providers	Data repositories
Communities	By hazard:
Local knowledge	FloodScan
Citizen science	FloodList
Government	Global Precipitation data sets (Sun et al. 2018)
National Meteorological and	Dartmouth Flood Observatory
Hydrological Services	Smithsonian Institution Volcanism Programme
UN	Global Historical Tsunami Database (NOAA)
World Meteorological Organization	Cyclones: International Best Track Archive for
(WMO)	Climate Stewardship (IBTrACS)
World Health Organization (WHO)	Earthquake database (USGS)
United Nations Environment	WHO Epidemic
Programme (UNEP)	For multiple hazards:
Global Facility for Disaster	UNEP Grid
Reduction and Recovery (GFDRR)	GFDRR ThinkHazard!
	Global geospatial earth observation-related data on
	drought and floods (Lindersson et al. 2020)

2.2.2 Vulnerability and Coping Capacity

Vulnerability and exposure are distinct but closely linked. Exposure is a necessary but not sufficient determinant of risk. It is possible to be exposed but not vulnerable (e.g. by living in a floodplain but having adequate means to modify building structure and behaviour to mitigate potential loss). Similarly, vulnerability to a hazard does not lead to impact until the vulnerable asset is exposed to the hazard. While vulnerability is defined with respect to a specific hazard, socio-economic factors, such as poverty and the lack of social networks and social support mechanisms, will aggravate or affect vulnerability levels irrespective of the type of hazard. Unfortunately, in many developing countries, this kind of socio-economic data is not available at a sufficiently granular level or gets lost in the way data are aggregated. Furthermore, this is a very dynamic landscape, for example, areas facing rapid urbanisation can be growing at a rate of 6 to 8% each year, and data can quickly become obsolete.

Although vulnerability data are often treated as static, there is growing evidence of the need to allow for its dynamic nature. For example, vulnerability of a household can change over short-term timescales, such as during the response and recovery phases of a disaster, perhaps due to loss of its income for a period. Vulnerability is also dynamic across different scales. For instance, a region's vulnerability can change due to deforestation or urbanisation.

The hazard-specific part of vulnerability may be described by vulnerability functions (also known as hazard damage curves), often used to describe physical vulnerability. These functions describe an exposed asset's response to the forces associated with a hazard, for instance, the reaction of a building to shaking of the earth during an earthquake, to wind during a tropical cyclone or to water depth in a flood. Vulnerability functions are often either proprietary or very generic, but they are critical for realistic assessment of potential loss. Once developed, they may be usable and adaptable to other areas with similar exposure profiles. Unfortunately, there are few openly available, high-quality vulnerability functions, such as the ones available from the open-source software CAPRA (Comprehensive Approach to Probabilistic Risk Assessment) platform (Cardona et al. 2010).

Coping capacity is an important component of disaster risk. It is usually conceptualised as short-term measures employed by individuals and communities in light of extreme events (Wamsler and Brink 2014), but it can also be considered at a country level (such as in the INFORM national risk index). Wisner et al. (2004) presented a range of coping strategies employed before, during and after an event. They identified preventative strategies, impact-minimising strategies, storing food and saleable assets, diversifying production and income sources, developing social support networks and post-event coping strategies. In some definitions, coping capacity is part of vulnerability, while in others, such as the aforementioned INFORM index, it is considered a separate risk dimension. Capacities should not be seen as opposite to vulnerability on a single spectrum, since vulnerable people might also possess a vast array of capacities (Gaillard et al. 2019).

2.2.3   Exposure of People and Assets

A hazard causes losses only when vulnerable people and assets are exposed to it. Exposure is thus the key that determines whether a hazard causes loss and whether vulnerabilities are tested. Exposure is a dynamic quantity changing on all timescales. On an annual timescale, a growing city has an increasing spatial extent, an increasing population and new buildings; a developing country has new infrastructure. At shorter timescales, people move around for summer holidays or festivals, and there are the daily movements of children to school, workers into and out of cities and travellers on roads, railways and aircraft. To adequately account for exposure in risk assessment, extensive data are needed in a form that enables it to be easily combined with hazard and vulnerability data.

In many countries, developing an exposure dataset is one of the biggest hurdles for completing a risk assessment. Low-resolution exposure data can be derived from existing and open global datasets, but they are not sufficient for detailed risk assessments that would be needed at a project or EWS level. Basic census data, asset inventories, city plans and topographic maps exist in most countries but are often out of date and are not always accessible to those who need them for reducing and managing disaster risks. Very few countries have dynamic exposure data suitable for use in early warnings. However, individual disaster risk managers and weather service personnel will use personal knowledge of major gatherings of people, for instance, in preparing their warnings and in promulgating them beyond the standard address lists. Exposure is strongly correlated with socio-economic indicators, as also used for vulnerability. Where full inventories do not exist, such indicators can serve as proxies to estimate the sectorial use of building stock and determine the exposure of productive assets used by communities for their livelihoods (often agriculture-based, such as exposed cropland).

Catastrophe risk modelling is used by banks, insurance companies, governments and industries to protect their assets. For insurance companies, assessing losses from disaster scenarios is central to ensuring their ability to pay out. Governments have obligations to reconstruct public assets and infrastructure after a disaster. Both have mostly focused on getting adequate physical exposure data. However, governments also have an implicit obligation to offer their populations emergency assistance (such as food and shelter) and to finance recovery/reconstruction activities (e.g. provision of support to poorer households, measures to support the recovery of the private sector) (Alton & Mahul 2017). Implicit liabilities are harder to quantify, and even if quantified, are usually of less absolute financial value for the poorer segment of society (*ibid.*). As Hallegatte et al. (2016) state: 'A flood or earthquake can be disastrous for poor people but have a negligible impact on a country's aggregate wealth or production if it affects people who own almost nothing and have very low incomes'. Consequently, these implicit liabilities are less well covered by Disaster Risk Finance and Insurance. It is of paramount importance for ensuring the well-being of all citizens in a country that disaster risk management interventions are properly designed. Overall, there is less understanding and quantification of the assets that are important to vulnerable and hazard-prone communities (Box 2.1).

2.2.4 Impacts

If risks are left unmanaged, disasters result in a vast array of impacts on people, societies, economies and environment. Impacts from natural hazards include negative, neutral and positive consequences. For instance, floods damage crops, property and infrastructure, but fill reservoirs. Damage to property from a storm may be followed by increased economic activity and rebuilding with healthier and safer homes. A disruption that causes loss to one business may provide an opportunity for other businesses to benefit. The terms 'loss' and 'damage'[1] are typically applied to the negative impacts of a disaster. The ultimate measure of the effectiveness of any disaster risk reduction measure is to assess the reduction in loss and damage. While they are often applied interchangeably, they may be used to differentiate between economic loss and physical damage (e.g. Koks 2016). Alternatively, some analysts distinguish between irreversible loss, e.g. fatalities from heat-related disasters, and recoverable damage, e.g. damages to buildings (Boyd et al. 2017; Mechler et al. 2019). Impacts may also be categorised as tangible or intangible and as direct or indirect. Tangible impacts can be expressed in monetary terms (e.g. disruption to businesses, costs of infrastructure destroyed), whereas intangible impacts cannot be easily expressed in monetary terms (e.g. casualties, impacts to mental health of individuals). Direct impacts can be directly associated with the action of the hazard

---

[1] Loss and damage is also one of the pillars of climate action in the Paris Agreement and refers to climate impacts which are beyond adaptation.

event where it strikes, whereas indirect impacts can be the result of cascading events and may be remote, e.g. interruptions of supply chains. It is very important for the design of risk reduction and management interventions to have a catalogue that has systematically and uniquely matched hazard information to the loss and damage associated with each historical disaster event. The World Meteorological Organization (WMO) has started an initiative to standardise how to catalogue high-impact events and their associated impact (WMO 2018a). Different approaches, methodologies and tools are used to collect the impact data. Damage and Needs Assessments (DNAs) are usually done at different intervals right after a disaster hits into the recovery phase. These DNAs, if government led, are consolidated into institutional databases where the data are accessible to the public usually at an aggregated level. Most governments have their own procedures for rapid and initial damage assessments. In addition, there are DNA methods that draw upon the capacity and expertise of national and international actors, such as the Damage, Loss, and Needs Assessment (DALA) and the Post Disaster Needs Assessment (PDNA). The PDNA is an inclusive, government-led and government-owned process, where the European Union, World Bank and United Nations provide technical support and facilitation as determined and requested by the government of the affected country for the recovery phase. DALA is a World Bank methodology used mostly for the immediate needs of a country. Table 2.2 provides a non-exhaustive overview of impact data providers, repositories and data collection methods.

Table 2.2 Non-exhaustive overview of impact data providers, repositories and data collection methods

Primary impact data providers	Data repositories per provider
Government	Reinsurance
Environment	Munich RE's NatCatSERVICE
Social welfare	Swiss Re SIGMA Explore
Health	Research centre
Public works	Centre for research on the epidemiology of
Energy	disasters EM-DAT
Water	Karlsruhe Institute of Technology CATDAT
Civil Protection Agencies	UNDRR
National Disaster Management Authorities	Preventionweb
Government international (OFDA, NOAA)	Sendai Desinventar
Humanitarian sector	UN OCHA
UN OCHA and other UN agencies	Reliefweb
NGOs	Humanitarian Data Exchange (HDX)
IFRC	IFRC
Affected communities	Disaster Response Emergency Fund appeals,
Local knowledge	plans and updates
Insurance and reinsurance companies	Country-specific, often National Disaster
MünichRe, SwissRe, LCW, AON	Management Authorities, e.g.
Media	United States: SHELDUS
Newspapers	Philippines: DSWD Dromic
Social media	
TV	
Community radio	

2.3   What Are Available Options to Deal with Disaster Risks?

When it comes to managing disasters and disaster risks, three approaches are often referred to: (i) disaster management (DM), (ii) disaster risk management (DRM) and (iii) disaster risk reduction (DRR). DM refers to the organisation, planning and application of measures preparing for, responding to and recovering from disasters (UNDRR 2016, p.14). DRM refers to the application of disaster risk reduction policies and strategies to prevent new risk, reduce existing disaster risk and manage residual risk, contributing to the strengthening of resilience and reduction of disaster losses (UNDRR 2016, p.15). DRR is aimed at preventing new and reducing existing disaster risk and managing residual risk, all of which contribute to strengthening resilience and, therefore, the achievement of sustainable development (UNDRR 2016, p. 16).

The evolution and application of these approaches mirror the shifts in thinking from hazards towards vulnerability and from top-down to bottom-up approaches (Paul et al. 2018). For instance, it is often emphasised that DM focused more on responding to and recovering from disasters (Jones et al. 2015), whereas DRM and DRR take a more comprehensive approach, including elements of prevention, mitigation and preparedness (Ouriachi-Peralta & Fakhruddin 2014).

An approach to managing and reducing disaster risk is often represented in the form of a disaster cycle, composed of four components:

1. Mitigation[2] encompasses strategies and practices aimed at reducing the likelihood or consequence of a hazard, e.g. levees, land-zoning and building practices (Coppola 2011).
2. Preparation/preparedness refers to strategies and measures for preparing for and reducing the impacts of disasters, e.g. early warning information, contingency planning and evacuation drills (Buckle 2012); more recently preparedness also includes initiatives around early warning early action and forecast-based financing.
3. Response encompasses strategies to reduce negative disaster impacts and avoid further possible implications, e.g. evacuation of people and property (WMO/EHA 2002).
4. Recovery involves aspects such as relief, reconstruction and rehabilitation (Wisner et al. 2012); usually, it refers to 'normalising' and returning to the pre-disaster situation (Coppola 2011), although contemporary thinking encourages the concept of 'building back better' (UNDRR 2015).

Although its prominence still prevails, especially among practitioners, the cycle is not without critics. In reality, these phases will never be so distinct and compartmentalised (Twigg 2015); they are rather in a constant interplay and continuum

---

[2] Mitigation as used in this chapter differs from mitigation as used in climate change discourse (i.e. used to refer to the cut in greenhouse gas emissions).

(Coppola 2011). This interplay is even more visible for slow-onset than for sudden-onset disasters.

Risk must be viewed in the context of the society in which it occurs. Every aspect of society is open to risk, and every member has a responsibility to respond to certain aspects of risk. Individuals may also have a responsibility on behalf of others as a result of their position in businesses and governmental or non-governmental bodies. Hence reducing disaster risks involves a wide range of both public and private actors. Private actors are individuals, households or communities that take action; for instance, communities are generally the first to respond to a disaster. Public actors are governmental institutions such as the National Meteorological and Hydrological Services (NMHS), disaster management authorities and government ministries responsible for water development and infrastructural works.

Disaster risk governance (DRG) refers to how public authorities, civil servants, media, private sector and civil society are organised at community, national and regional levels to manage and reduce disaster and climate-related risks (UNDP 2020). It is an essential part of DRG that all actors, from private individuals to businesses to the most senior government officials, understand the risks that they are exposed to and the level of responsibility they have for managing those risks. In many countries, domestic laws and policies define these levels of responsibility, e.g. the Philippine Disaster Risk Reduction and Management Act of 2010 (Republic of the Philippines 2010). In addition, international disaster response laws, rules and principles encompass a wide range of both global and regional international law and norms and bilateral treaties and agreements. Where a country has a federal structure, the law will state the conditions under which the provincial government should seek federal assistance. If a disaster caused by a natural hazard surpasses the capacity of a state to respond, the Inter-Agency Standing Committee can decide to initiate a humanitarian system-wide response (IASC 2020). In this case, the sovereign state can ask for and agree to international support. Actors operating at global, national and local levels require intra- and inter-organisational coordination.

A key aspect of DRG is the creation of a shared understanding, backed up by legislation, funding, management and enforcement, of where responsibility for assessing and managing risk lies. Responsibilities typically cascade from government ministries with responsibility for strategic risks to the whole country, to city councils holding the risk for their municipality, to infrastructure operators (often private businesses) having responsibility for risks to their systems and consequent risks to people using them, to businesses needing to protect themselves financially and their customers if their goods or services are interrupted, down to each individual having responsibility for actions to protect themselves. The higher up this chain the responsibility lies, the greater portion of risk is held and the greater the penalty of failure. Along with this shared understanding goes the requirement on each responsible actor to have a risk assessment and a risk management plan for their area of responsibility and to ensure that this is consistent with the plans of their stakeholders – whether higher up the chain, lower down or at the same level.

Often, measures for delivering DRR are classified as structural or non-structural (see, e.g. UNDRR 2016). Structural measures refer to engineering approaches resulting in physical infrastructure (e.g. flood walls), while non-structural measures refer to strategies involving policies, laws and 'soft approaches' (e.g. training, education, awareness-raising). Structural measures are more tailored towards hazard reduction, whereas non-structural measures aim to decrease vulnerability and exposure (Harries & Penning-Rowsell 2011) and increase coping capacity.

A large spectrum of actions can be taken, as part of risk reduction, to reduce, retain, transfer or absorb risk (UNFCCC 2012). Table 2.3 gives examples of Disaster Risk Reduction (DRR) actions and Climate Change Adaptation (CCA) actions and shows where early warning systems (EWS) fit in. At one end of the spectrum are actions that can be taken to protect against infrequent events with minor impacts. While these may be inconvenient, they do not justify major investments, so are best dealt with by early warnings that enable people to prepare for and avoid them and insurance to cover repair costs. Frequent events are best avoided altogether, either by land use planning, e.g. avoiding building on the floodplain; by use of natural protective features, e.g. coastal mangroves and salt marshes; by protective engineering, e.g. river levees, strengthened building codes; or by a combination of these measures. The most difficult to deal with are rare hazards with major impacts. Rarity and scale make engineering solutions unviable. Protection of life demands plans for large-scale evacuation to safe locations, backup for essential services and release of resources for rapid recovery. Insurance is a valuable contributor to recovery for moderately rare events, but for the most extreme, only governments have the necessary resources, supported where necessary by international financial mechanisms.

Table 2.3 Overview of public and private actions that can be taken to reduce, retain, transfer or absorb risk, adapted from van den Homberg and McQuistan, 2019. (DRR, disaster risk reduction; CCA, climate change adaptation; L&D, loss and damage)

Adjustment	Spectrum and *timing*	Private action; Tech level: examples	Public action; Tech level: examples
Incremental	DRR: Preparedness *Short-term* *Ex ante*	*Basic:* Fisherman put fish net around fish pond after receiving early warning	*Basic:* NGO locating relief items closer to the predicted to be affected area. Increase response capacity of communities
	DRR: Risk reduction CCA: Medium-term for next year's floods *Ex ante*	*Basic:* Household raises plinths/floors and diversifies their crops	*Intermediate to advanced:* A NMHS improves their hydro-meteorological modelling so that forecasts with better lead times and spatial resolution become available. Government-led irrigation system, building of dykes
	Humanitarian aid. *Directly after floods* *Ex post*	*Basic/none:* Support from within the community	*Intermediate:* Post-disaster public and donor assistance, such as relief items or cash transfers to households and money to governments for reconstruction of, e.g. roads and embankments

(continued)

Table 2.3 (continued)

Adjustment	Spectrum and *timing*	Private action; Tech level: examples	Public action; Tech level: examples
Fundamental	DRR and CCA (larger scale or intensity). *Long-term over several years* *Ex ante*	*Intermediate:* Access Interactive Voice Response service to get meteorological and agricultural advice	*Intermediate:* Improving access to information through digital inclusion, e.g. making sure early warning services are available in first language of beneficiaries, voice SMS early warning service, nationwide coverage of mobile networks, lower taxation on mobile users
	DRR and CCA (new to a particular region or resource system). *Medium to long-term* *Ex ante*	*Advanced:* Citizens participate in crowdsourcing of water levels	*Advanced:* Dam operator changes its way of releasing water by using advanced forecasting models. Forecast-based financing. A Rice Research Institute develops flood-tolerant rice
		Intermediate: Take a micro-insurance	*Intermediate:* Micro-insurance can be supported by mobile technology and/ or public-private partnerships to ensure commercial viability
	DRR and CCA (transform places) *Long term* *Ex ante*	*Intermediate:* Citizens contribute to constructing bio-dykes or ecological corridors	*Intermediate:* Large dams no longer being built, but several smaller ones. Green infrastructure such as bio-dykes; ecological corridors. Use of floodplains instead of building dykes
	L&D Curative: redress and rehabilitation *Short term* *Ex post*	*None:* Involuntary migration or staying put	*Intermediate:* Financial compensation for loss and damage that can be attributed to climate change. Active remembrance (e.g. through museum exhibitions, school curricula). Counselling

Given limited budgets and technological capacities, especially in developing countries, trade-offs and choices have to be made. A straightforward comparison of permanent and temporary or long-term and short-term risk reduction measures is problematic as multiple decision-makers with different mandates and political agendas are involved. For example, government agencies dealing with water development and irrigation are responsible for permanent and structural measures (e.g. building dikes), whereas disaster management and humanitarian agencies take decisions regarding temporary, EWS-informed and non-structural responses (Bischiniotis et al. 2020). It is likely that each agency will apply different evaluation protocols (Mechler 2016). On the one hand, economic valuations such as cost-benefit analysis are typically used to justify large-scale infrastructure expenditures, which often introduces a bias towards wealthier areas with more assets to lose (Hallegatte et al. 2016). On the other hand, EWS-based early actions are typically evaluated in terms of their reduction of human losses and livelihood impacts (Gros et al. 2019, Rai et al. 2020).

A systemic approach to risk management is essential to ensure that policy options and corresponding actions are sustainable in the long-term rather than short-term sticking plasters. It is important to move from silo approaches per individual hazard to multi-hazard approaches. Based on several sources of data, the Red Cross Climate Centre calculated that, of 132 unique extreme weather-related disasters occurring in 2020, of which 92 have overlapped with the COVID-19 pandemic (Walton & van Aalst 2020), 51.6 million people globally were directly affected by an overlap of floods, droughts, storms and the COVID-19 pandemic. Current methods for risk assessment and risk management need to evolve to capture (better) the systemic nature of risk. One can think of tools such as vulnerability and capacity assessments, contingency planning and visualisation techniques (Gill & Malamud 2014). Galasso et al. (2021) propose an approach to risk-based design of new urban settlements in which quantitative predictions of the impact of potential hazard scenarios form the foundation for a policy discussion between stakeholders. The challenges in this transformation to govern systemic risk are related to finding the optimal complexity. How detailed should the approach be, given limited resources and given limited data availability?

One possible way to speed up the transition from managing individual risks and disasters to managing compound and consecutive risks and disasters is to draw on insights from development aid. For example, we already know that poverty tends to increase in both developed and developing economies after a disaster such as a flood or storm (Karim and Noy 2016). Therefore, in the move towards systemic risk reduction, a core component should be a strong social programme to increase people's resilience even in the absence of explicit disaster-related triggers (Deryugina 2017). Adaptive and shock-responsive social protection systems have the potential to help people manage covariate risks comprehensively, including anticipating them, absorbing their impacts and managing future risks (Ulrichs et al. 2019). Examples from different social protection programmes in Latin America, South Asia and parts of Africa have shown that social protection can play an important role in reducing deprivation, increasing food security and avoiding negative risk coping strategies, among others. Moreover, some preliminary experience with adaptive social protection programmes in the Sahel (Daron et al. 2020) has shown the capacity to protect poor households from climate and other shocks before they occur, given their potential to scale up and be flexible, thus contributing to a long-term risk management strategy. Understanding the various cascading risks that increase vulnerability during different life phases can be useful in designing comprehensive social protection systems that are better prepared to handle multiple vulnerabilities and compound risks.

2.4   The Role of Early Warnings Systems in Disaster Risk Reduction

In the previous section, we showed that decision-makers and disaster and emergency managers have a large array of options to reduce disaster risks and their impacts on societies, economies and environment. Yet the dominant approach to dealing with disasters has been skewed towards responding and providing relief after they have happened. Over the years, there has been a shift in policy and practice with an increased understanding that preventing and preparing for disasters yields numerous benefits and contributes to resilient communities and societies. One of the central instruments in being more prepared is the provision of early warning systems (EWS), which we now explore in more detail.

2.4.1   The Emergence of Early Warning Systems

The emergence of EWS in international DRR policy and practice can be tracked through global agreements for disaster risk reduction and beyond. In 1994, during the World Conference on Natural Disaster Reduction held in Yokahoma, Japan, the State Members of the United Nations agreed on the Yokahoma Strategy and Plan of Action for a Safer World (IDNDR 1994). As one of the ten guiding principles, countries agreed that 'early warnings of impending disasters and their effective dissemination using telecommunications, including broadcast services, are key factors to successful disaster prevention and preparedness' (IDNDR 1994, p.6). The Yokahoma Strategy drew attention to a need for establishing and/or strengthening EWS and called for assistance in developing EWS for countries most vulnerable to natural hazards.

However, only limited progress in delivering integrated EWS at scale has been delivered. A greater attention to EWS in international arenas was given only after the devastating impacts of the Indian Ocean Tsunami in 2004 (WMO 2015a). This was reflected in the Hyogo Framework for Action 2005–2015, a global footprint for disaster risk reduction, recognised as a major shift towards focus on prevention and preparedness as opposed to response and recovery (Tozier de la Poterie & Baudoin 2015). As one of its five priorities for action, the Hyogo Framework lays out a need to identify, assess and monitor disaster risks and enhance early warning. In its general considerations, the Hyogo Framework states that EWS are 'essential investments that protect and save lives, property and livelihoods, contribute to sustainability of development, and are far more cost-effective in strengthening coping mechanism than is primary reliance on post-disaster response and recovery' (UNDRR 2005; p.5). Importantly, the Hyogo Framework emphasised a need for people-centred EWS, systems that will account for differentiated vulnerabilities, offer guidance on how to act on warning information and support action by decision-makers. Although the Hyogo Framework raised the profile of EWS worldwide, substantive gaps remained.

The successor of the Hyogo Framework is the Sendai Framework for Disaster Risk Reduction 2015–2030, a global agreement serving as guidance for countries to reduce their disaster risks at the time of writing of this chapter. Unlike the Hyogo Framework, the Sendai Framework does not identify EWS as one of its priority areas, but rather identifies it as one of seven global targets. Target (g) calls for countries to 'substantially increase the availability of and access to multi-hazard early warning systems and disaster risk information and assessments to people by 2030' (UNDRR 2015; p.12). Although multi-hazard early warning systems (MHEWS) was not a new concept, the Sendai Framework is the first global DRR policy blueprint that emphasises the importance of a multi-hazard approach in relation to early warnings. Given that the Sendai Framework is still relatively new and reporting on the targets is not yet fully developed, it remains to be seen to what extent the Sendai Framework has enhanced the delivery of EWS in both developed and developing countries.

In addition to global agreements for DRR, EWS are an important part of global climate action and the sustainable development goals (SDGs), as they are central for reducing vulnerability and enhancing resilience of people and nations. The Paris Agreement, a global document providing a framework for climate action, refers to EWS in Article 7 (on adaptation) and Article 8 (on averting, minimising and addressing loss and damage associated with the adverse effects of climate change) (UNFCCC 2015). One example of the intertwined nature of DRR and climate action at the global level is the establishment of the Climate Risk Early Warning Systems (CREWS) initiative during Conference of Parties 21 (COP21). CREWS is a financial mechanism, implemented by the Global Facility for Disaster Reduction and Recovery, the World Meteorological Organisation and the United Nations Office for Disaster Risk Reduction. It provides funding for least developed countries and small island developing states to implement risk-informed early warning services for weather-related hazards. In their 2019 Annual Report, CREWS states they supported 44 countries and over 10 million people in gaining access to better early warning services (WMO 2020b).

2.4.2   Early Warning Systems: Definition and Components

As explained by Kelman and Glantz (2014), there is no universal definition of EWS, as this is dependent on the context, scale and hazard in question. For the purpose of this chapter, we adopt the latest definition by UNDRR (2016) stating that an EWS is 'an integrated system of hazard monitoring, forecasting and prediction, disaster risk assessment, communication and preparedness activities, systems and processes that enables individuals, communities, governments, businesses and others to reduce disaster risks in advance of hazardous events'.

The World Meteorological Organisation and United Nations Office for Disaster Risk Reduction have published a widely used and internationally recognised checklist for multi-hazard and people-centred early warning systems, outlining four main

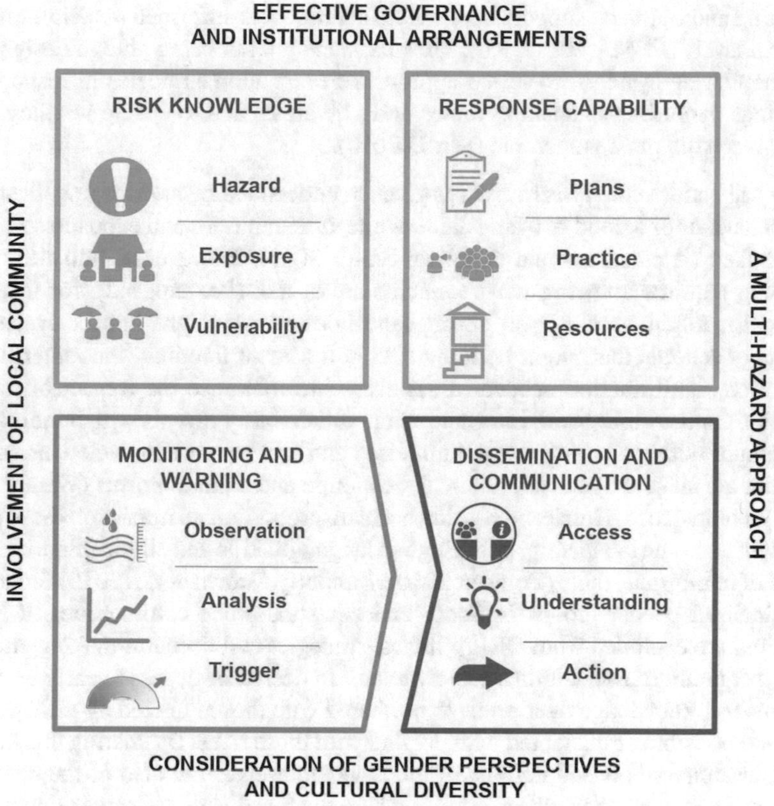

EFFECTIVE GOVERNANCE
AND INSTITUTIONAL ARRANGEMENTS

Fig. 2.2 Components of an early warning system. (Adopted from Brown et al. 2019)

elements and four overarching components of any early warning system (UNISDR 2006, WMO 2018b), as presented in Fig. 2.2. For an early warning system to be truly effective, all eight components must be considered and addressed in a holistic approach to ensure accurate, timely, reliable and understandable information reaches everyone in the right way for them to take action.

Risk Knowledge As through EWS there is an effort to reduce risks and prepare for hazards in a specific spatial area (e.g. community, city, region), it is imperative to know the nature of risk in the area. Risk assessments can help to identify the areas prone to hazard occurrences, the location and nature of vulnerable groups and critical infrastructure and assets in exposed locations. For instance, the Zurich Flood Alliance used a combination of digital mapping techniques, based on the OpenStreetMap (OSM) and community-based participatory methods, to map flood risks in Nepal, Peru and Mexico as a basis for risk reduction strategies (Practical Action 2018). In the United States, since 2009, The Federal Emergency Management Agency (FEMA) introduced the Risk Mapping, Assessment and Planning (Risk MAP) programme which provides risk assessment tools, flood mapping products, planning and outreach support in order to facilitate risk-informed decision-making at local levels (FEMA 2021). Risk information is vital in being able to design hazard monitoring systems, to set up appropriate evacuation strategies in response to warnings (including evacuation routes and safe areas) and to ensure warning messages reach the most vulnerable (WMO 2010).

Usually, more emphasis in EWS is given to understanding hazards (e.g. the physical behaviour of a flood or a landslide), while vulnerabilities and exposure are often overlooked (Alcántara-Ayala & Oliver-Smith 2019). However, a holistic understanding requires knowing what elements are at risk (i.e. exposed), for instance, roads that might be damaged during landslides which might impact evacuation efforts or schools that might be inundated in a case of flooding, thus interrupting education – information needed to act early and minimise the impact of natural hazards. On the other hand, information on vulnerability reveals which individuals and groups within a society are marginalised and will be more impacted, as vulnerabilities are shaped by social, political, economic and cultural norms (Wisner et al. 2004). For instance, Hurricane Katrina had a disproportionate impact on those poorest, with no home ownership, poor English language skills and ethnic minorities and those of immigrant status (Zoraster 2010). Similarly, Brown et al. (2019) found that marginalised gender groups in flood- and landslide-prone communities of Nepal and Peru are excluded from DRR policies, strategies and decision-making and that their marginalised role within a society results in decreased access to early warning.

This risk knowledge then needs to be shared with those affected by the risks and those responsible or mandated with dealing with those risks. By sharing this knowledge, awareness is raised not just of the risks themselves but also of the need and advocacy to reduce those risks – this is where the benefits of and engagement in an early warning system come in. Collaboration between stakeholders and sharing of knowledge, information and data are needed so that all are aware of the risks and the opportunities to take action to reduce those risks (WMO 2015a).

Monitoring and Warning Scientific understanding of the natural processes that generate hazards, together with past experience and monitoring of current conditions, enables the likelihood of their occurrence to be forecasted in advance (WMO 2010). The accuracy and reliability of these forecasts at different lead times before a hazard occurs can vary widely and are affected by a range of factors including hazard type, how suddenly the hazard occurs, how good the previous observational data and current monitoring are, how well the underlying processes are understood and how complex and replicable the hazard to be modelled is (WMO 2015a). There is a stark difference in forecasting capabilities for different hazards. While a tornado can only be forecast with certainty a few minutes ahead, the storm that spawned it, along with other severe weather hazards, can often be forecasted a few hours ahead. In contrast, development and movement of the weather system containing this storm and others may be predictable several days in advance. Prevalence of the general conditions favouring such storms may be identifiable months in advance. Using previous observations of hazards and their environmental impacts, and aligning these with capabilities for response, warning levels can be developed, whereby when a level of confidence is reached that a threshold of specific environmental conditions will be passed, a warning is issued. Warning systems vary widely according to the hazard, the nature of its impact and the organisation of response capabilities. Examples include the National Fire Danger Rating System in the United States, the Heat Health warning system in Hong Kong, the typhoon warning system in Japan, the storm surge warning system in the Netherlands, the National Severe Weather Warning Service in the United Kingdom and many others.

In order to ensure that warnings properly reflect the evidence, it is important that the evidence is made available for scrutiny, at least for the more severe events but ideally on a routine basis. A record of all warnings issued must be retained, together with the evidence used to justify each warning. Any event for which a warning was not issued, or was issued very late, should also be retained for scrutiny and analysis so that lessons may be learned for application. Similarly, warnings that were not followed by a hazardous event need to be retained, even if they were issued at very low probability, so that the reliability of the likelihood estimates can be assessed over the long term.

Dissemination and Communication Dissemination and communication refer to processes and procedures for distributing the warning and preparedness information in an understandable format to those with responsibility for taking action and to those at risk including the most vulnerable (Brown et al. 2019, UNDP 2018). In literature on EWS, dissemination and communication components are often clustered together. In short, dissemination is how the information reaches the end-users, while communication refers to its content. Appropriate, tailored communication of warning information is critical to ensure people get the right information in the right way to act in advance of hazards (WMO 2010). To ensure that warnings reach all those who need them, the needs of users must first be identified, and a suitably wide range of dissemination media selected to ensure that all are reached, including the most vulnerable and marginalised.

Communication of the warning information also needs to be carefully considered. The raw forecast information analysed by technical specialists in, for example, an NMHS is not appropriate to be communicated beyond this specialist expert group because it requires specific knowledge and skill to understand and interpret. Therefore, warning information needs to be re-packaged and tailored for different users. For instance, experiences of the Super Typhoon Haiyan in the Philippines in 2013 revealed a discrepancy between expert and lay people's understanding of what 'storm-surge' means, leading to higher impacts, as technical jargon got lost in translation and interpretation (Santos 2013). Evaluation of the effectiveness of warning communications is needed to assess whether the information, including the level of risk, was understood by users, whether it was felt to be useful, appropriate to needs and actionable.

Dealing with uncertainties in the forecasts is a challenge, and how to communicate this beyond producers of the warning varies among early warning systems. The majority of research and advice in communicating uncertainty in an operational context for natural hazard early warning encourages transparency and emphasises the importance of education and the development of trust (Morss et al. 2008, WMO 2008).

There have been substantial advancements in how warning and preparedness information is communicated and disseminated, especially with the advancement of technology (e.g. e-mails, text messages, radio broadcasts and mobile applications). For instance, Cumiskey et al. (2015) found that farmers in the low-lying, flash flood-prone district of Sunamganj located in the North-Eastern part of Bangladesh prefer mobile services for receiving flood warnings. This is a new opportunity as up to 85% of people have access to mobile phones. However, failures in EWS still mainly occur due to poor communication and dissemination practices (Basher 2006). This is especially true in developing countries, where many still lack access to the technologies for receiving warning information.

Response Capability Response capability refers to a community's knowledge of their risks, ability to act on warnings and familiarity with what they should do when a warning is issued (e.g. where and how to evacuate). It is important that, given accurate, timely and understandable warning information is available and communicated to appropriate users in advance of a disaster, people and institutions are able to respond and take action. A holistic early warning system not only provides warning information but also enables action to be taken based on those warnings. It should be noted here that 'response capability' refers to the capability of responding *to* the early warning information *before* the hazard event, as well as being prepared to respond effectively *after* the hazard event occurs. Response capability is rooted in resources, skills and networks that stakeholders have (Marchezini et al. 2018). It includes having clear lines of authorities and decision-making processes, organising drills and practice scenarios and clear protocols and procedures developed from national to community levels (WMO 2015a).

The capacity of users to take action before a disaster occurs, based on warning information, needs to be built in longer-term planning and preparedness activities. Preparedness plans based on an understanding of local and national knowledge and capacities are needed. Also, plans of how to respond to warnings have to be developed (WMO 2010). Those plans need to be practiced to develop familiarity through training and education (WMO 2002). People also need to have sufficient resources to respond, such as a safe location to go to, a safe route to that location and any other resources to enable them to take action.

Wherever possible, barriers to being able to take action need to be identified in advance, and measures taken to address them within the planning stages. People make decisions based on their perception of multiple risks (not just the natural hazard risk), their capacities and other circumstances. For instance, Elder et al. (2007) showed that, among other reasons, African American communities in New Orleans decided not to evacuate during Hurricane Katrina due to financial constraints, neighbourhood crime, perceived racism and inequities. Continually reviewing the effectiveness of response and any challenges experienced during disasters, and adapting plans in an iterative manner, ensures response plans are up to date (WMO 2010).

Effective Governance and Institutional Arrangements Effective governance of an EWS and robust institutional arrangements are key features to ensure that it operates smoothly. Mandates, responsibilities and long-term funding are required at national level for government institutions to be able to set up and operate a sustainable EWS (WMO 2010, 2018b). A legislatively underpinned commitment and consistent efforts at national level are needed over a long time period in order to address and develop all aspects. Clear standards must be set to ensure that warnings are issued when required and in a timely manner (e.g. at least 6 hours ahead of the event); that they describe the hazard, its location, timing and impact adequately; that they conform to a specified format (e.g. the Common Alerting Protocol); that a defined level of quality control is applied (e.g. a second person checks the warning before issue); that a record is kept of all warnings issued; and that the outcome is recorded. These standards should be monitored and statistics of conformance reported to stakeholders.

Where government commitment is lacking (e.g. through lack of funds), non-governmental organisations (NGOs) or the private sector have sought to fulfil this role, especially in developing countries. For example, Šakić Trogrlić et al. (2018) described how in Southern Malawi NGOs are supporting the government through delivery of community-based early warning systems for flooding. Similarly, in Nepal the Department of Hydrology and Meteorology based in Kathmandu is officially mandated as responsible for the flood early warning system monitoring and warning components. However, there are gaps in the responsibility for dissemination, communication and response capacity within the government institutional mandates. NGOs in Nepal have rushed to fill this void, setting up and operating local early warning services. In the past, these systems (national and local) have operated mostly independently of each other, providing potentially duplicative or conflicting sources of early warning information for local people (Meechaiya et al. 2019). This is not a sustainable solution, often reaching small segments of the population and prone to termination when funding ceases.

UNEP (2015) states that early warning information is a basic human right as climate change and disasters both contribute to human rights violation. As such the mandate for developing and sustaining an early warning system must rest on government bodies. EWS are considered a public task; they have the economic characteristics of public goods that make them difficult to be privately funded and therefore depend heavily on public funding for their proper functioning. Especially in developing countries where taxation systems do not ensure enough public funding, this poses challenges to the financial sustainability of its provision (Deltares 2015). Not every aspect of an EWS must be government operated, nor must a single organisation operate the whole system. Stakeholders involved in producing and using early warning systems range across institutional disciplines and operate at a variety of spatial scales. Effective governance should encourage communication and coordination between stakeholders.

Importance of governance is especially relevant in the context of cross-boundary aspects of an EWS, as hazards do not follow administrative borders. They can affect neighbouring countries simultaneously, and an aspect of approach to DRR in one country (e.g. land use change, construction of infrastructure) can affect the timing, severity and occurrence of a hazard in a neighbouring country. Therefore, a joint and coordinated approach including well-established governance structures (e.g. policy and planning frameworks and institutional design) and processes (e.g. public engagement and behaviour change, research and partnership with policy and practice) to cross-border EWS is crucial. This can include development of clear guidelines and protocols, sharing of historical and real-time data between the countries, sharing of modelling outputs and risk maps across boundaries, joint monitoring and operations, clear division of responsibilities and sustainable financial mechanisms.

Governance also includes the regular maintenance, evaluation and improvements of all elements of the system and of the system as a whole. A successful EWS meets the objectives it was designed for. With time, it will grow and evolve to meet new objectives. However, capabilities and needs are continually changing, so it is unlikely that any EWS will still be optimally meeting the needs of the community after 5 years of operation. Therefore, a key aspect of EWS management is regular review, based on a robust evaluation. Such reviews must address whether the EWS is successfully delivering the information required by users, whether it is still using the most appropriate technologies, whether it is still reaching those at risk in a timely manner and whether it is using the best information available.

It is essential that all aspects of an EWS are maintained, monitored and exercised, including through automated quality-control, structured reporting by trusted partners; monitoring of response through real-time media including social media, telephone and email; and post-event surveying – preferably including direct interviews. EWS managers need to be aware of the ethical dimensions of their systems, ensuring that users are not disadvantaged by reason of their personal characteristics, but also ensuring that their interactions with the EWS system do not, in themselves, have a negative effect. This requires that particular care is taken over confidentiality of feedback information that might, for instance, be used by credit or insurance agencies, in pressurised selling, or even by criminal groups. A particular challenge arising from the growth of social media is the need to counter false information. This requires constant monitoring of social media and rapid response with corrective information before false information is repeated. Where necessary, action should be taken to remove sources of false information.

Involvement of Local Community Early warning systems are only effective if they inherently and actively put people at the centre – ensuring all elements of the early warning system consider and prioritise those at risk from natural hazards (WMO 2010). Local authorities, non-governmental organisations and communities need to be involved in all aspects of early warning so that the system is designed to be appropriate for community needs and capacities. This way, the responses to warning information will be designed to protect people, households and communities from disasters.

A local, 'bottom-up', 'end-mile' or 'first-mile' approach to early warning, with active participation of local communities, including marginalised groups, enables engagement in and contribution to the system, ensuring reduced vulnerability and leveraging and strengthening of local capacities. Community-based early warning systems are good examples of involving local communities. For instance, Practical Action has been working with communities in Nepal since 2008 in setting up local flood early warning systems, with extensive involvement of local communities across the four components of EWS (Rai et al. 2020). Examples of community involvement in local-level early warning systems can also be found in high income countries. For instance, in Scotland, private developers Scottish Flood Forum and Scottish Environmental Protection Agency jointly implemented the RiverTrack. It is an affordable river level monitoring system providing real-time river levels to displays located in local homes and businesses, thus allowing for continuous monitoring.

The involvement of communities can also be framed in the context of citizen science, where the level of participation can increase from citizens as merely sensors, citizens as basic interpreters, citizens that directly participate in the EWS problem definition and data collection up to fully collaborative science (Paul et al. 2018). In many areas of the world, local communities also have rich local and indigenous knowledge on early warning (Acharya & Prakash 2019, Šakić Trogrlić et al. 2019), and there is an increasing focus on understanding how this type of information can be blended and integrated with scientific knowledge in EWS.

The Importance of Gender and Cultural Diversity Vulnerability to the impact of disasters is increased by gender inequality, gender norms and social marginalisation (Brown et al. 2019, UNISDR 2009). Women and marginalised groups including gender minorities are often excluded from DRR policies, strategies and decision-making (Brown et al. 2019, UNISDR 2009). In contexts of gender inequality, people of different genders access, process, interpret and respond to information in different ways, due to the social and cultural organisation of gender relations and the gendered division of labour (UNISDR 2009). For instance, Tyler and Fairbrother (2018) while researching a role of gender in decision-making at household level on wildfire evacuation found that men and women have differing conceptions on when they should evacuate: while women would prefer earlier evacuation, men prefer later evacuation. However, it is challenging for women to voice their concerns as men are culturally viewed as more authoritative voices in wildfire discussions. Fordham (2001) explored a gender perspective on early warning in DRR. She found that during the 1991 Cyclone in Bangladesh women were less likely to receive the warning; even when they did, cultural norms forbade their movement in public. Cultural diversity and marginalisation affect all elements of an early warning system (Brown et al. 2019). Marginalised people are often those most overlooked by early warning systems. People may be marginalised on the basis of age, sex, disability, race, ethnicity, religion, migration status, socio-economic status, place of residence, sexual orientation and gender identity. These groups require special consideration, focused attention, proactive engagement and sensitive or transformative approaches to ensure no one is left behind. The key consideration should be equity of outcome rather than equality of treatment. Cultural diversity and marginalisation affect all elements of an early warning system (Brown et al. 2019). Early warning systems need to take account of cultural differences in the perception of authority, of the cause of hazards, of the nature of prediction and in the availability and use of communication channels, among other factors.

A Multi-Hazard Approach People are at risk from multiple hazards with each having different likelihoods of occurring. For instance, they might live in a multi-hazardous location prone to both hydrometeorological and geophysical hazards, and different hazards can also interact (e.g., an earthquake triggering a landslide). As such, if we are taking a people-centric approach to early warning, we should develop an early warning system or early warning systems that address all hazards affecting the population in a certain location.

Where possible, early warning systems should link hazard-specific systems together to ensure people are provided with early warning for all hazards they are at risk from (WMO 2018b). Such a multi-hazard early warning system would provide a holistic understanding of forecasted hazards that may occur and their complex, interrelated relationships, such as whether these hazards occur alone, simultaneously, cascadingly or cumulatively (UNDRR 2020a).

For instance, the United Kingdom's Natural Hazard Partnership[3] (Hemingway & Gunawan 2018) publishes the Daily Hazard Assessment, an overview of 21 natural hazards that could affect the United Kingdom over the next 5 days. The hazards covered are air (e.g. aero allergens and air pollutions, hail, rain, lightning), land (e.g. avalanches, earthquakes, landslides), water (e.g. surface water flooding, drought) and space (e.g. space weather, near Earth and space objects). While multi-hazard early warning systems that are truly integrated across hazards are rare, a multi-hazard approach to early warning is achievable, for example: building new hazard early warning systems upon existing systems; coordinating across responsible institutions to share data, forecasts or outputs; and/or developing consistent, coordinated or combined communication materials.

2.4.3 Early Warning Systems as Preparedness and Risk Reduction

As Kelman and Glantz (2014) note, a common misunderstanding in relation to EWS is that they exist only to be activated once a hazard occurs. However, the aim of an EWS is not just to facilitate institutional, community or individual response to an impending hazard, but to (ideally) introduce a long-term risk reduction behaviour as well as instigate anticipatory action. To ensure EWS lead to both long- and

---

[3] http://www.naturalhazardspartnership.org.uk/

short-term risk reduction behaviour before a disaster arrives, the EWS should be integrated in the community's everyday life, as opposed to being only used when a disaster is imminent.

If designed, implemented and operated in its entirety, taking into account all of the parts described in the previous section (i.e. all eight components), EWS present an opportunity to reduce disaster risks and foster a 'culture' of preparedness. For instance, mapping of disaster risks conducted as a part of EWS can inform spatial development and serve as a basis for policies that would delineate disaster-prone areas and introduce some of the available measures (e.g. limited development, introduction of insurance schemes, disaster prevention infrastructure), in turn reducing risks in these areas. Similarly, paying attention to differentiated vulnerabilities of individuals and members of communities provides an opportunity to design actions that would both decrease their vulnerabilities in the long term (e.g. designing inclusive decision-making processes and increasing access to services) and improve their capacity in terms of EWS (e.g. designing communication practices for people with hearing impairment or evacuation protocols for people with physical disabilities).

Depending on the type of a hazard and the lead time of the warning that is possible, EWS offers a window of opportunity for early actions. Warning information is useless if not followed by appropriate actions that will minimise impacts by reducing risks or increase preparedness for a better response. For instance, this requires moving away from warnings that tell what the weather will be, to warnings that tell what the weather will do. WMO has developed guidelines for how National Meteorological and Hydrological Services can implement 'impact-based forecasting', i.e. providing a forecast of the potential consequences of a hydrometeorological event, in terms of its effects on, e.g. people and infrastructure. It can also be sector specific, such as for agriculture, tourism or humanitarian aid. These types of forecasts and warnings are designed to provide detailed information on who or what is exposed and vulnerable to the particular hazard. For impact forecast and warning services, exposure is explicitly considered along with hazard and vulnerability (WMO 2015b). This requires NMHS to transform and collaborate with other sector-specific government agencies, private sector and humanitarian agencies to be able to provide such impact-based forecasts. It also requires changes in mandates of NMHS as well as other government agencies. If impact-based forecasts are provided, this also brings responsibilities to act on this information. Several agencies involved in humanitarian response such as International Federation of Red Cross, World Food Programme, Food and Agriculture Organization and UN OCHA are working in parallel on mechanisms to release funding based on impact-based forecasts.

In 2008, the Red Cross Red Crescent movement introduced Forecast-based Financing (FbF) for early action and preparedness for response. FbF enables access to a Disaster Response Emergency Fund, a funding source habitually only available for humanitarian response, via an Early Action Protocol (EAP). The EAP is triggered when an impact-based forecast—i.e., the expected (humanitarian) impact as a result of the expected weather—reaches a predefined danger level (IFRC 2018). An

EAP outlines the potential high-risk-prone areas where the FbF mechanism could be activated, the prioritised risks to be tackled by early actions, the number of households to be reached against an expected activation budget, the forecast sources of information, the expected lead time for activation and the agencies responsible for implementation and coordination. Around ten EAPs for mostly sudden-onset disasters and one for slow-onset disasters have been established and approved since the first one in 2018. Early actions are determined in collaboration with to-be-affected communities and need to comply with a number of criteria (IFRC 2018) in order to be able to be executed and to be cost-effective. Very often one of the early actions is the transfer of cash to the to-be-affected communities. Most early action protocols are based on hazards for which the lead time of the warning allows for sufficient implementation time.

However, even if there are only a few hours available to have certainty of a hazardous event (e.g. a flash flood), if EWS is implemented as a preparedness (e.g. clear responsibilities of roles, defined evacuation routes and identified shelters) and risks are reduced (e.g. people trained in alternative livelihood options, existing insurance schemes), impacts could be minimised.

As described by IFRC (2008), in an example of a cyclone, there are multiple preparedness and risk reduction actions available in different timeframes, each with different requirements for dissemination. Given that climate change projections indicate an increased likelihood of intense tropical cyclones (an early warning for years in advance), risk reduction actions could be introducing strict building codes and promoting cyclone-proof housing, while preparedness actions could be raising awareness of cyclone risks and training communities for disaster response. On a seasonal timescale, forecasts of above-average cyclone activity are available, providing an opportunity to revisit contingency plans, replenish stocks and conduct emergency drills. Early warning information of likely development of cyclones in a particular stretch of the ocean can now provide weeks of advance warning, prompting awareness of the potential for storm warnings. Days before the cyclone makes landfall, when forecasts are quite accurate in identifying locations to be hit, evacuation can be prepared, warnings can be sent to communities at risk while housing can be cyclone-proofed. For example, machine learning models trained on historical cyclone events in the Philippines are being used to predict 3 days ahead whether more than 10% of houses in a municipality will be damaged. If this threshold is surpassed, early actions are taken in the form of household strengthening and early harvesting of rice or abaca trees (Wagenaar et al. 2021). Then, just hours before the event, final warnings provide the trigger for evacuation to storm shelters.

2.5 Gaps in Early Warning Systems

Early warning systems for natural hazards have come a long way, facilitated by advances in technology (e.g. monitoring, forecasting and dissemination technology) and science (understanding of the processes involved), by increased policy support (both at global and national levels) and by growth in understanding what integrated and people-centred EWS are (i.e. the components described above). However, there are still large gaps that warrant further research, investments, policy change and practice. In Table 2.4, we summarise the main gaps according to the eight components of EWS.

Substantive gaps remain across all components of EWS. Gaps in the 'technical' aspects of EWS (e.g. quality of monitoring equipment, forecasting capability, dissemination channels) are a hindrance to effective EWS in many parts of the world, especially in developing countries. For instance, observing networks are often inadequate, particularly across Africa, where in 2019 just 26% of stations reported

Table 2.4 Common gaps in EWS

Components of early warning system	Gaps identified in the literature
Risk information	A predominant focus on hazard with a lack of understanding of vulnerability and exposure Lack of integration of risk information in decision-making Data gaps – especially in developing countries Difficult access to data for risk information – particularly open access/sharing across disciplines or organisations
Monitoring and warning	Uncertainty in forecasting and climate change influencing forecasting capability Varying skills of forecast information: accuracy, reliability, resolution Lead time Spatial and temporal resolution Varying quality of historical data records limits prediction skill Lack of validation/evaluation of forecast skill Lack of monitoring infrastructure, technical and human capacity, especially in developing countries Lack of sustainability of monitoring and forecasting systems Inadequate monitoring Prediction capabilities for rapid-onset hazards (e.g. flash floods and landslides) and lack of systems for some hazards (e.g. dust and sandstorms, flash floods)
Dissemination and communication	Dominance of experts at the expense of user-focused communication Top-down dissemination takes time, reducing lead time Lack of feedback mechanisms between users and producers Lack of access to warning information, especially for the most vulnerable groups Inadequate communication systems to provide timely, accurate and meaningful warning information to those at risk Underdeveloped dissemination infrastructure in developing countries Lack of impact-based warning information Inadequately standardised nomenclature, protocols and standards Ineffective engagement of media and private sector Fragmented monitoring responsibilities Communication content/message not adapted for specific user needs/capabilities

(continued)

Table 2.4 (continued)

Components of early warning system	Gaps identified in the literature
Response capability	Weak public response to warnings
	Lack of risk awareness and understanding – lack of outreach/education and practice
	Lack of post-event reviews and poor incorporation of lessons learned
	Unclear authorities and decision-making processes hindering the response
	Lack of simulation exercises and evacuation drills
	Lack of inducing long-term risk reduction behaviour
	Lack of adequate safe spaces, concerns over safe spaces, lack of safe routes
	Barriers to taking action even if would want to, e.g. caring responsibilities or insufficient lead time
	Concerns over leaving assets/possessions (guarding and staying put)
	Behavioural reasons for not responding (e.g. risk perception based on previous experience of hazards and staying put)
Effective governance and Institutional arrangements	Inadequate multi-agency and institutional collaboration and clarity of roles and responsibilities
	Lack of funding (i.e. disaster finance still heavily focused on response)
	Weak budgetary and political support in some countries
	Inadequate coordination between local, national and regional levels
	Gaps in legal, institutional and coordination frameworks, especially in developing countries
	Political failures to take action (e.g. timing, lack of resources, fear of litigation)
	Weak integration of EWS in national plans
	Inadequate recognition of links between disaster risk reduction, climate change adaptation and sustainable development
	Insufficient coordination among actors responsible for EWS
Multi-hazard approach	Most countries report warning systems for single hazards (i.e. lack of multi-hazard EWS)
	Very few countries have all hazards covered. And rarely are they integrated (sharing data, risk analysis, interactions, one-communication channel/method, synthesised SOPs for response)
Involvement of local community	Lack of engagement of those at risk is the design and operation of EWS
	Practical challenges of community engagement (e.g. physical distance, funding, timeframes)
	Lack of using participatory approaches
	Lack of inclusion of local, traditional and indigenous knowledge
Gender perspectives and cultural diversity	Gender incorporation in EWS rarely considered
	Lack of consideration of cultural diversity and linguistic barriers
	Marginalised people not included or considered in a meaningful way in assessment of risk and unable to participate meaningfully in DRR/DRM/EWS preparedness plans, etc.

Based on Basher 2006, Grasso 2014, UNDP 2018, WMO 2015b, Zommers and Singh 2014 according to the WMO requirements (WMO 2020a). Good monitoring and forecasting depend on high-quality data. Yet, data quality and preservation of long-term records remain a challenge. Moreover, hazard data remain the focus of most EWS, with data on vulnerabilities and exposure sidelined. This results in an inability to provide impact-based and tailored warning information.

The 'social' component of EWS also remains marginalised in comparison to the technical aspects. Despite a rhetoric of importance of community involvement, consideration of gender and marginalised groups and differentiated vulnerabilities, these often remain box-ticking exercises, given inadequate attention. EWS are a long way from being considered as social processes, and a 'culture' of preparedness is rarely achieved in practice. For instance, inadequate attention is given to public awareness and training on how to respond to warning information, while systems in place continue to favour relief over early action. Furthermore, in many parts of the world, information fails to reach those at the sharpest end of natural hazards. Research on transboundary EWS in Bangladesh, India and Nepal showed, for example, that access to EWS technology is not distributed fairly (van den Homberg & McQuistan 2019). Overall, there is an insufficient capacity worldwide to translate early warning into early action (WMO 2020a).

Good governance remains a significant challenge in many parts of the world. Early warning systems remain unfunded and politically unfavoured, with inadequate policies and institutional structures in place. For instance, gaps remain in legal frameworks for EWS. A recent review of the role of national laws in managing flood risk by Mehryar and Surminski (2020), focusing on 139 national laws from 33 countries, found that national laws have a prevailing focus on the response and recovery strategies while placing less emphasis on proactive risk reduction and preparedness, including EWS. Taking legal responsibility for warnings and their dissemination remains one of the key issues in operationalisation of a flood EWS (Parker 2017). Responsibilities for different aspects of early warning largely remain scattered across departments and institutions, resulting in an uncoordinated and unsustainable approach. There is a plethora of reporting frameworks for the Global Agreements (i.e. Sustainable Development Goals, Sendai Framework and Paris Agreement), with indicators that relate to (parts of) EWS. However, these are often high-level, based on (too optimistic) self-reporting and not harmonised. As a result, there is also a lack of high spatial and temporal resolution data on whether early warnings are received, understood and acted upon.

As mentioned previously, despite multi-hazard frameworks being a target of EWS, they remain underdeveloped and rarely, if ever, achieved in practice. With a global push for multi-hazard EWS, it remains worrying that in many countries, EWS are inadequate or non-existent even for single hazards.

In addition to gaps across the eight components of an EWS, there are other significant gaps. For instance, in evaluating the performance of an EWS (Sättele et al. 2016). As suggested by WMO (2018a, 2018b), the checklist developed around the four core components of an EWS (i.e. risk knowledge, monitoring and warning, response capability, dissemination and communication) offers a series of practical actions and initiatives which should be considered when evaluating EWS. An EWS needs to be continuously reviewed and assessed in order to incorporate the learnings, adapt needed improvements and create an effective EWS. This is across all areas including (among others) evaluating forecast skill, data collection/monitoring accuracy and logistics, lead time, effectiveness of access to and understanding of warning information and people's abilities to act based on warnings. Furthermore, there are significant differences between countries in the availability of skills for EWS. For instance, in developing country contexts where resources are limited, the government departments responsible for EWS are often extremely restricted, both in terms of number of staff available to the department and in terms of the range of skills hired. Naturally, physical science skills are the most urgent types of skills needed in, for example, NMHS, but there are a range of skills and specialties required for a fully operational EWS (e.g., skills in social sciences, science communication, public relations). Without them, robust monitoring and warning thresholds may be developed, but they will not be effective in enabling early action. In contexts where these perceived 'softer' skills are not recruited or resourced within the EWS-mandated government department, it leaves gaps either (1) where those mandates are perceived as beyond the institution's capacities and therefore not attempted or (2) where NMHS staff are required to act beyond their training, experience, skills and knowledge in areas outside their expertise.

2.6 Summary

- Disaster risks arise from a complex interplay between physical hazards and the exposure of vulnerable people, assets and systems to them. Understanding disaster risk, and its distribution in time and space, is fundamental for management and reduction of these risks.
- We have presented the ingredients of disaster risks and available options to deal with them, with a specific focus on the role of early warning systems. We presented an eight-component framework of people-centred EWS, highlighting the importance of an integrated and all-society approach.
- If designed, implemented and operated in its entirety, such an EWS can reduce disaster risks, foster preparedness and early action and build resilience of populations at risk. In order to realise these benefits, warnings must be received, understood and acted on: they must be useful, usable and used.
- Successful operation of an EWS requires assured long-term funding and involves a vast array of stakeholders, including local communities, government departments at different levels, private sector, media and regional players.
- Equal importance should be given to the social components (e.g. community involvement, communication) as to the technical aspects of an EWS.
- EWS need to account for the occurrence of multiple hazards.
- Realising the full potential of EWS requires systematic changes in the current status quo, including (but not limited to) increased funding and prioritisation of EWS, improvements in horizontal and vertical governance arrangements, development of new technologies with corresponding capacity development and enhanced involvement of communities at risk.

References

Acharya A. and A. Prakash, 2019. When the river talks to its people: Local knowledge-based flood forecasting in Gandak River basin, India. *Environmental Development,* **31,** 55–67. DOI: https://doi.org/10.1016/J.ENVDEV.2018.12.003

Alcántara-Ayala I. and A. Oliver-Smith, 2019. Early Warning Systems: Lost in Translation or Late by Definition? A FORIN Approach. *Int. J. Disaster Risk Science,* **10(3)**, 317–331. DOI: https://doi.org/10.1007/s13753-019-00231-3

Alton M. L. and O. Mahul, 2017. *Assessing Financial Protection against Disasters: A guidance note on conducting a disaster risk finance diagnostic.* Washington, DC: World Bank Group, 60. https://documents1.worldbank.org/curated/en/102981499799989765/pdf/117370-REVISED-PUBLIC-DRFIFinanceProtectionHighRes.pdf (Accessed 2/9/2021)

Basher R., 2006. Global early warning systems for natural hazards: systematic and people-centred. *Phil. Trans. Roy. S., A: Mathematical, Physical and Engineering Sciences.* Royal Society **364(1845)**, 2167–2182. DOI: https://doi.org/10.1098/rsta.2006.1819

Bischiniotis K., H. de Moel, M. van den Homberg, A. Couasnon, J. Aerts, N. G. Guimarães, E. Zsoter and B. van den Hurk, 2020. A framework for comparing permanent and forecast-based flood risk-reduction strategies. *Sci. Tot. Environ.,* **720,** 137572. DOI: https://doi.org/10.1016/j.scitotenv.2020.137572

Boyd E., R. A. James, R. G. Jones, H. R. Young and F. E. L. Otto, 2017. A typology of loss and damage perspectives. *Nature Climate Change,* **7(10)**, 723–729. DOI: https://doi.org/10.1038/nclimate3389

Brown S., M. Budimir, A. Sneddon, D. Lau, P. Shakya and S. Upadhyay, 2019. *Gender Transformative Early Warning Systems: Experiences from Nepal and Peru.* Rugby, Warwickshire, United Kingdom: Practical Action. Available at: https://floodresilience.net/resources/item/gender-transformative-early-warning-systems-experiences-from-nepal-and-peru/ (Accessed 01/03/21).

Buckle P., 2012. Preparedness, warning and evacuation. In: Wisner B., J. C. Gaillard and I. Kelman (eds) *The Routledge Handbook of Hazards and Disaster Risk Reduction.* Abingdon: Routledge, 481–492. ISBN 9780415523257

Cardona O. D., M. G. Ordaz, E. Reinoso, L. E. Yamín and A. H. Barbat, 2010. CAPRA - Comprehensive Approach to Probabilistic Risk Assessment: International Initiative for Risk Management Effectiveness. *14th European Conference on Earthquake Engineering.* paper presented at the European Conference on Earthquake Engineering. Ohrid, Macedonia, 10. https://www.researchgate.net/profile/Alex-Barbat/publication/259598259_CAPRA_-_Comprehensive_Approach_to_Probabilistic_Risk_Assessment_International_Initiative_for_Risk_Management_Effectiveness/links/0c96052cd8a91e18b3000000/CAPRA-Comprehensive-Approach-to-Probabilistic-Risk-Assessment-International-Initiative-for-Risk-Management-Effectiveness.pdf (Accessed 2/9/2021)

Coppola D. P., 2011. *Introduction to international disaster management* (2nd edition). Elsevier. ISBN 978-0-12-382174-4. https://doi.org/10.1016/C2009-0-64027-7

Cumiskey L., M. Werner, K. Meijer, S. H. M. Fakhruddin and A. Hassan, 2015. Improving the social performance of flash flood early warnings using mobile services. *Int. J. Disaster Resilience in the Built Environment* **6(1)**, 57–72. DOI: https://doi.org/10.1108/IJDRBE-08-2014-0062

Daron J., M. Allen, M. Bailey, L. Ciampi, R. Cornforth, C. Costella, N. Fournier, R. Graham, K. Hall, C. Kane, I. Lele, C. Petty, N. Pinder, J. Pirret, J. Stacey and H. Ticehurst, 2020. Integrating seasonal climate forecasts into adaptive social protection in the Sahel. *Climate and Development,* 1–8. DOI: https://doi.org/10.1080/17565529.2020.1825920

Deltares, 2015. *Mobile services for flood early warning in Bangladesh: final report.* Delft, The Netherlands: Deltares. https://www.deltares.nl/app/uploads/2015/11/Deltares-Mobile-Services-for-Early-Warning-in-Bangladesh-Final-Report_web.pdf. (Accessed 2/9/2021)

Deryugina T., 2017. The Fiscal Cost of Hurricanes: Disaster Aid versus Social Insurance. *Amer. Economic J.: Economic Policy,* **9(3)**, 168–198. DOI: https://doi.org/10.1257/pol.20140296

Elder K., S. Xirasagar, N. Miller, S. A. Bowen, S. Glover and C. Piper, 2007. African Americans' Decisions Not to Evacuate New Orleans Before Hurricane Katrina: A Qualitative Study. *Amer. J. Pub. Health. Amer. Pub. Health Assoc.* **97,** S124–S129. DOI: https://doi.org/10.2105/AJPH.2006.100867e

FEMA, 2021. *Risk Mapping, Assessment and Planning (Risk MAP) | . Risk Mapping, Assessment and Planning (Risk MAP).* https://www.fema.gov/flood-maps/tools-resources/risk-map (Accessed 03/03/21).

Fordham M., 2001. *Challenging boundaries: a gender perspective on early warning in disaster and environmental management.* United Nations Division for the Advancement of Women. https://www.unisdr.org/files/8264_EP52001Oct261.pdf (Accessed 2/9/2021)

Gaillard J. C., J. R. D. Cadag and M. M. F. Rampengan, 2019. People's capacities in facing hazards and disasters: an overview. *Nat. Hazards,* **95(3)**, 863–876. DOI: https://doi.org/10.1007/s11069-018-3519-1

Galasso, C., J. McCloskey, M. Pelling, M. Hope, C. J. Bean, G. Cremen, R. Guragain, U. Hancilar, J. Menoscal, K. Mwang'a, J. Phillips, D. Rush and H. Sinclair, 2021. Editorial. Risk-based, Pro-poor Urban Design and Planning for Tomorrow's Cities. *Int. J. Disaster Risk Reduction,* **58**, 102158. https://doi.org/10.1016/j.ijdrr.2021.102158

GFDRR, 2021. ThinkHazard. Available at: https://thinkhazard.org/en/ (Accessed 28/02/21).

Gill J. C. and B. D. Malamud, 2014. Reviewing and visualizing the interactions of natural hazards. *Rev. Geophys.* **52(4)**, 680–722. DOI: https://doi.org/10.1002/2013RG000445

Grasso V. F., 2014. The State of Early Warning Systems. In: Singh A and Zommers Z (eds) *Reducing Disaster: Early Warning Systems for Climate Change.* Dordrecht: Springer Netherlands, 109–125. https://doi.org/10.1007/978-94-017-8598-3_6

Gros C., M. Bailey, S. Schwager, A. Hassan, R. Zingg, M. M. Uddin, M. Shahjahan, H. Islam, S. Lux, C. Jaime and E. Coughlan de Perez, 2019. Household-level effects of providing forecast-based cash in anticipation of extreme weather events: Quasi-experimental evidence from humanitarian interventions in the 2017 floods in Bangladesh. *Int. J. Disaster Risk Reduction,* **41,** 101275. https://doi.org/10.1016/j.ijdrr.2019.101275

Hallegatte S., A. Vogt-Schilb, M. Bangalore and J. Rozenberg, 2016. *Unbreakable: Building the Resilience of the Poor in the Face of Natural Disasters.* The World Bank. http://elibrary.worldbank.org/doi/book/10.1596/978-1-4648-1003-9 (Accessed 06/02/19).

Harries T. and E. Penning-Rowsell, 2011. Victim pressure, institutional inertia and climate change adaptation: The case of flood risk. *Global Environmental Change,* **21(1)**, 188–197. https://doi.org/10.1016/j.gloenvcha.2010.09.002

Hemingway R. and O. Gunawan, 2018. The Natural Hazards Partnership: A public-sector collaboration across the UK for natural hazard disaster risk reduction. *Int. J. Disaster Risk Reduction,* **27,** 499–511. https://doi.org/10.1016/j.ijdrr.2017.11.014

van den Homberg M. and C. McQuistan, 2019. Technology for Climate Justice: A Reporting Framework for Loss and Damage as Part of Key Global Agreements. In: Mechler R., L. M. Bouwer, T. Schinko, S. Surminski and J. Linnerooth-Bayer (eds) *Loss and Damage from Climate Change: Concepts, Methods and Policy Options.* Cham: Springer International Publishing, pp513–545. https://doi.org/10.1007/978-3-319-72026-5_22 (Accessed 05/02/19).

IASC, 2020. *What does the IASC humanitarian system-wide level 3 emergency response mean in practice?* New York, USA: Inter-Agency Standing Committee. https://interagencystandingcommittee.org/system/files/l3_what_iasc_humanitarian_system-wide_response_means_final.pdf (Accessed 2/9/2021)

IDNDR, 1994. *Yokahoma strategy and plan for action for a safer world.* Yokahoma, Japan: United Nations. https://www.ifrc.org/Docs/idrl/I248EN.pdf (Accessed 2/9/2021)

IFRC, 2008. *Early Warning > Early Action.* Geneva, Switzerland: IFRC. https://media.ifrc.org/ifrc/document/early-warning-early-action/ (Accessed 03/03/21).

IFRC, 2018. *Forecast-based Financing Practitioners Manual.* Geneva, Switzerland: IFRC. https://manual.forecast-based-financing.org/. (Accessed 2/9/2021)

Jones S., B. Manyena and S. Walsh, 2015. Chapter 4 - Disaster Risk Governance: Evolution and Influences. In: Shroder JF, Collins AE, Jones S, Manyena B and Jayawickrama J (eds) *Hazards, Risks and Disasters in Society*. Boston: Academic Press, 45–61. http://www.sciencedirect.com/science/article/pii/B9780123964519000044 (Accessed 03/04/19).

Karim A. and I. Noy, 2016. Poverty and Natural Disasters: A Regression Meta-Analysis. *Rev. Economics Institutions*, **7(2)**, 26. DOI:10.5202/REI.V7I2.222

Kelman I. and M. H. Glantz, 2014. Early Warning Systems Defined. In: Singh A and Zommers Z (eds) *Reducing Disaster: Early Warning Systems For Climate Change*. Dordrecht: Springer Netherlands, 89–108. https://doi.org/10.1007/978-94-017-8598-3_5

Lindersson S., L. Brandimarte, J. Mård and G. D. Baldassarre, 2020. A review of freely accessible global datasets for the study of floods, droughts and their interactions with human societies. *WIREs Water*, **7(3)**: e1424. DOI: https://doi.org/10.1002/wat2.1424

Marchezini V., F. E. A. Horita, P. M. Matsuo, R. Trajber, M. A. Trejo-Rangel and D. Olivato, 2018. A Review of Studies on Participatory Early Warning Systems (P-EWS): Pathways to Support Citizen Science Initiatives. *Frontiers in Earth Science*. Frontiers 6. https://www.frontiersin.org/articles/10.3389/feart.2018.00184/full (Accessed 20/11/20).

Marin-Ferrer M., K. Poljansek and L. Vernaccini, 2017. *Index for risk management - INFORM: concept and methodology, version 2017*. Joint Research Centre (European Commission). http://op.europa.eu/en/publication-detail/-/publication/b1ef756c-5fbc-11e7-954d-01aa75ed71a1/language-en/format-PDF (Accessed 28/02/21).

Mechler R., 2016. Reviewing estimates of the economic efficiency of disaster risk management: opportunities and limitations of using risk-based cost–benefit analysis. DOI https://doi.org/10.1007/s11069-016-2170-y

Mechler R., E. Calliari, L. M. Bouwer, T. Schinko, S. Surminski, J. Linnerooth-Bayer, J. Aerts, W. Botzen, E. Boyd, N. D. Deckard, J. S. Fuglestvedt, M. González-Eguino, M. Haasnoot, J. Handmer, M. Haque, A. Heslin, S. Hochrainer-Stigler, C. Huggel, S. Huq, R. James, R. G. Jones, S. Juhola, A. Keating, S. Kienberger, S. Kreft, O. Kuik, M. Landauer, F. Laurien, J. Lawrence, A. Lopez, W. Liu, P. Magnuszewski, A. Markandya, B. Mayer, I. McCallum, C. McQuistan, L. Meyer, K. Mintz-Woo, A. Montero-Colbert, J. Mysiak, J. Nalau, I. Noy, R. Oakes, F. E. L. Otto, M. Pervin, E. Roberts, L. Schäfer, P. Scussolini, O. Serdeczny, A. de Sherbinin, F. Simlinger, A. Sitati, S. Sultana, H. R. Young, K. van der Geest, M. van den Homberg, I. Wallimann-Helmer, K. Warner and Z. Zommers, 2019. Science for Loss and Damage. Findings and Propositions. In: Mechler R., L. M. Bouwer, T. Schinko, S. Surminski and J. Linnerooth-Bayer (eds) *Loss and Damage from Climate Change: Concepts, Methods and Policy Options*. Cham: Springer International Publishing, 3–37. https://doi.org/10.1007/978-3-319-72026-5_1

Meechaiya C., E. Wilkinson, E. Lovell, S. Brown and M. Budimir, 2019. *The governance of Nepal's flood early warning system: opportunities under federalism*. London: Overseas Development Institute. https://odi.org/en/publications/the-governance-of-nepals-flood-early-warning-system-opportunities-under-federalism/ (Accessed 2/9/2021)

Mehryar S. and S. Surminski, 2020. *The role of national laws in managing flood risk and increasing future flood resilience*. London, UK: Grantham Research Institute on Climate Change and Environment. https://www.lse.ac.uk/granthaminstitute/publication/the-role-of-national-laws-in-managing-flood-risk-and-increasing-future-flood-resilience/ (Accessed 03/03/21).

Morss R. E., J. L. Demuth and J. K. Lazo, 2008. Communicating Uncertainty in Weather Forecasts: A Survey of the U.S. Public. *Wea. Forecast*, **23(5)**, 974–991. https://doi.org/10.1175/2008WAF2007088.1

Ouriachi-Peralta T. and S. H. M. Fakhruddin, 2014. Integrating local knowledge in disaster risk reduction: a case study for Indonesia. *Asian J. Environ. Disaster Management*, **6**. DOI:https://doi.org/10.3850/S1793924014000297

Parker D. J., 2017. Flood Warning Systems and Their Performance. In *Oxford Research Encyclopedia of Natural Hazard Science*. https://oxfordre.com/naturalhazardscience/view/10.1093/acrefore/9780199389407.001.0001/acrefore-9780199389407-e-84 (Accessed 03/03/21).

Paul J. D., W. Buytaert, S. Allen, J. A. Ballesteros-Cánovas, J. Bhusal, K. Cieslik, J. Clark, S. Dugar, D. M. Hannah, M. Stoffel, A. Dewulf, M. R. Dhital, W. Liu, J. L. Nayaval, B. Neupane, A. Schiller, P. J. Smith and R. Supper, 2018. Citizen science for hydrological risk reduction and resilience building. *Wiley Interdisciplinary Reviews: Water,* **5(1)**: e1262. DOI:https://doi.org/10.1002/wat2.1262

Practical Action, 2018. *Participatory digital mapping: building community resilience in Nepal, Peru, and Mexico.* Rugby, Warwickshire, United Kingdom: Practical Action. https://infohub.practicalaction.org/bitstream/handle/11283/620791/Digital+Mapping_web.pdf?sequence=1 (Accessed 03/03/21).

Rai R. K., M. J. C. van den Homberg, G. P. Ghimire and C. McQuistan, 2020. Cost-benefit analysis of flood early warning system in the Karnali River Basin of Nepal. *Int. J. Disaster Risk Reduction,* **47,** 101534. DOI:https://doi.org/10.1016/j.ijdrr.2020.101534

Republic of the Philippines, 2010. *Philippine Disaster Risk Reduction and Management Act of 2010.* https://www.ifrc.org/PageFiles/100077/Philippines_2009_Philippine%20Disaster%20Risk%20Reduction%20and%20Management%20Act%20of%202010.pdf. (Accessed 2/9/2021)

Ruiter M. C. De, A. Couasnon, M. J. C. Van Den Homberg, J. E. Daniell, J. C. Gill and P. J. Ward, 2020. Why We Can No Longer Ignore Consecutive Disasters. *Earth's Future,* **8(3)**: e2019EF001425. https://doi.org/10.1029/2019EF001425

Šakić Trogrlić R., G. B. Wright, A. J. Adeloye, M. J. Duncan and F. Mwale, 2018. Taking stock of community-based flood risk management in Malawi: different stakeholders, different perspectives. *Environ. Hazards,* **17(2),** 107–127. https://doi.org/10.1016/j.pdisas.2021.100171

Šakić Trogrlić R., G. B. Wright, M. J. Duncan, M. J. C. van den Homberg, A. J. Adeloye, F. D. Mwale and J. Mwafulirwa, 2019. Characterising Local Knowledge across the Flood Risk Management Cycle: A Case Study of Southern Malawi. *Sustainability,* **11(6),** 1681. DOI:https://doi.org/10.3390/su11061681

Santos L. A., 2013. *Storm surge: Lost in translation and interpretation.* Devex. https://www.devex.com/news/sponsored/storm-surge-lost-in-translation-and-interpretation-82311 (Accessed 01/03/21).

Sarkar S., 2021. Rapid assessment of cyclone damage using NPP-VIIRS DNB and ancillary data. *Nat. Hazards,* **106(1)**, 579–593. DOI:https://doi.org/10.1007/s11069-020-04477-9

Sättele M., M. Bründl and D. Straub, 2016. Quantifying the effectiveness of early warning systems for natural hazards. *Nat. Hazards Earth Syst. Sci.,* **16(1),** 149–166. DOI:https://doi.org/10.5194/nhess-16-149-2016

Sun Q., C. Miao, Q. Duan, H. Ashouri, S. Sorooshian and K.-L. Hsu, 2018. A Review of Global Precipitation Data Sets: Data Sources, Estimation, and Intercomparisons. *Rev. Geophys.,* **56(1),** 79–107. https://doi.org/10.1002/2017RG000574

Tozier de la Poterie A. and M.-A. Baudoin, 2015. From Yokohama to Sendai: Approaches to Participation in International Disaster Risk Reduction Frameworks. *Int. J. Disaster Risk Science,* **6(2),** 128–139. DOI https://doi.org/10.1007/s13753-015-0053-6

Twigg J., 2015. *Disaster Risk Reduction* (2nd edition). London, UK: Overseas Development Institute. https://odihpn.org/wp-content/uploads/2011/06/GPR-9-web-string-1.pdf (Accessed 2/9/2021)

Tyler M. and P. Fairbrother, 2018. Gender, households, and decision-making for wildfire safety. *Disasters,* **42(4),** 697–718. DOI:https://doi.org/10.1111/disa.12285

Ulrichs M., R. Slater and C. Costella, 2019. Building resilience to climate risks through social protection: from individualised models to systemic transformation. *Disasters,* **43(S3),** S368–S387. DOI:https://doi.org/10.1111/disa.12339

UNDP, 2018. *Five approaches to build functional early warning systems.* United Nations Development Programme. https://www.eurasia.undp.org/content/dam/rbec/docs/UNDP%20Brochure%20Early%20Warning%20Systems.pdf (Accessed 2-9-2021)

UNDP, 2020. *Issue Brief: disaster risk governance.* https://www.undp.org/content/dam/undp/library/crisis%20prevention/disaster/Issue_brief_disaster_risk_reduction_governance_11012013.pdf (Accessed 03/05/21).

UNDRR, 2005. *Hyogo Framework for Action 2005-2015: Building the resilience of nations and communities to disasters.* Geneva, Switzerland: United Nations Office for Disaster Risk Reduction, 25. https://www.unisdr.org/2005/wcdr/intergover/official-doc/L-docs/Hyogo-framework-for-action-english.pdf (Accessed 2/9/2021)

UNDRR, 2015. *Sendai Framework for Disaster Risk Reduction 2015-2030.* Geneva, Switzerland: United Nations Office for Disaster Risk Reduction, 37. https://www.preventionweb.net/files/43291_sendaiframeworkfordrren.pdf. (Accessed 2/9/2021)

UNDRR, 2016. *Report of the open-ended intergovernmental expert working group on indicators and terminology relating to disaster risk reduction.* New York: United Nations. https://www.unisdr.org/we/inform/publications/51748 (Accessed 08/05/19).

UNDRR, 2020a. *Early warning system.* http://www.undrr.org/terminology/early-warning-system (Accessed 03/03/21).

UNDRR, 2020b. *Hazard definition & classification review.* https://www.undrr.org/publication/hazard-definition-and-classification-review (Accessed 08/06/21)

UNEP, 2015. *Early warning as a human right: building resilience to climate-related hazards.* UNEP. https://wedocs.unep.org/xmlui/handle/20.500.11822/7429 (accessed 03/03/21).

UNEP, 2021. *Global Risk Data Platform.* https://preview.grid.unep.ch/index.php?preview=data&events=floods&evcat=3&lang=eng (Accessed 28/02/21).

UNFCCC, 2012. *A literature review on the topics in the context of thematic area 2 of the work programme on loss and damage: a range of approaches to address loss and damage associated with the adverse effects of climate change.* UNFCCC. https://unfccc.int/resource/docs/2012/sbi/eng/inf14.pdf. (Accessed 2/9/2021)

UNFCCC, 2015. *Paris Agreement.* Paris, France: United Nations Framework Convention on Climate Change. https://unfccc.int/sites/default/files/english_paris_agreement.pdf (Accessed 2-9-2021)

UNISDR, 2006. *Developing early warning systems, a checklist: third international conference on early warning (EWC III),* 27-29 March 2006, Bonn, Germany. Geneva, Switzerland: UNISDR. http://www.undrr.org/publication/developing-early-warning-systems-checklist-third-international-conference-early-warning (Accessed 01/03/21).

UNISDR, 2009. *Making disaster risk reduction gender-sensitive: Policy and practical guidelines.* Geneva, Switzerland: UNISDR. https://www.unisdr.org/files/9922_MakingDisasterRiskReductionGenderSe.pdf (Accessed 2/9/2021)

Wagenaar D., T. Hermawan, M. J. C. van den Homberg, J. C. J. H. Aerts, H. Kreibich, H. de Moel and L. M. Bouwer, 2021. Improved Transferability of Data-Driven Damage Models Through Sample Selection Bias Correction. *Risk Analysis,* **41(1)**, 37–55. DOI: https://doi.org/10.1111/risa.13575

Walton D. and M. van Aalst, 2020. *Climate-related extreme weather events and COVID-19.* IFRC Climate Centre. https://climatecentre.org/downloads/files/Extreme%20weather%20events%20and%20COVID%2019%20IFRC2020(1).pdf. (Accessed 2/9/2021)

Wamsler C. and E. Brink, 2014. Moving beyond short-term coping and adaptation. *Environ. Urbanization* **26(1),** 86–111. DOI:https://doi.org/10.1177/0956247813516061

Wisner B., J. C. Gaillard and I. Kelman, 2012. *The Routledge handbook of hazards and disaster risk reduction.* (1st edition). London: Routledge. ISBN 9780415523257

Wisner B., P. Blaikie, T. Cannon and I. Davis, 2004. *At risk: natural hazards, people's vulnerability and disasters* (2nd edition). London: Routledge. ISBN 978-0415252164

WMO, 2002. *Guidelines on Improving Public Understanding of and Response to Warnings.* Geneva: WMO. https://library.wmo.int/doc_num.php?explnum_id=9222 (Accessed 2/9/2021)

WMO, 2008. *Guidelines on communicating forecast uncertainty.* Geneva, Switzerland: World Meteorological Organisation. https://library.wmo.int/doc_num.php?explnum_id=4687 (Accessed 2/9/2021)

WMO, 2010. *Guidelines on early warning systems and application of nowcasting and warning operations*. Geneva: WMO. https://library.wmo.int/doc_num.php?explnum_id=9456 (Accessed 2/9/2021)

WMO, 2015a. *Synthesis of the Status and Trends with the Development of Early Warning Systems. A Contribution to the Global Assessment Report 2015 (GAR15)*. Geneva, Switzerland: WMO. https://www.preventionweb.net/english/hyogo/gar/2015/en/bgdocs/WMO,%202014a. pdf (Accessed 2/9/2021)

WMO, 2015b. *WMO Guidelines on multi-hazard impact-based forecast and warning services*. Geneva, Switzerland: World Meteorological Organisation. https://library.wmo.int/doc_num. php?explnum_id=7901. (Accessed 2/9/2021)

WMO, 2018a. *Cataloguing high-impact events and associated loss and damage*. https://unfccc. int/sites/default/files/resource/Excom%207%20submission%20from%20WMO%20(002).pdf (Accessed 2/9/2021)

WMO, 2018b. *Multi-hazard Early Warning Systems: A Checklist, Outcome of the first Multi-hazard Early Warning Conference, 22-23 May 2017, Cancun, Mexico*. Geneva, Switzerland: World Meteorological Organization, 20. https://etrp.wmo.int/pluginfile.php/21553/mod_page/ content/18/MultihazardChecklist.pdf (Accessed 2/9/2021)

WMO, 2020a. *2020 State of Climate Services*. Geneva: WMO. https://library.wmo.int/doc_num. php?explnum_id=10385 (Accessed 2/9/2021)

WMO, 2020b. *CREWS Report Series Annual Report*. Geneva: WMO. https://library.wmo.int/ doc_num.php?explnum_id=10226 (Accessed 2/9/2021)

WMO/EHA, 2002. *Disasters and emergencies: definitions*. Addis Ababa, Ethiopia: WMO/ EHA. http://apps.who.int/disasters/repo/7656.pdf.

Zommers Z. and A. Singh, 2014. Introduction. In: Singh A and Zommers Z (eds) *Reducing Disaster: Early Warning Systems for Climate Change*. Dordrecht: Springer Netherlands, 1–19. https://doi.org/10.1007/978-94-017-8598-3_1

Zoraster R. M., 2010 Vulnerable Populations: Hurricane Katrina as a Case Study. *Prehospital and Disaster Medicine, 25(1),* 74–78. https://doi.org/10.1017/S1049023X00007718

Chapter 3
Connecting Warning with Decision and Action: A Partnership of Communicators and Users

Anna Scolobig, Sally Potter, Thomas Kox, Rainer Kaltenberger, Philippe Weyrich, Julia Chasco, Brian Golding, Douglas Hilderbrand, Nadine Fleischhut, Dharam Uprety, and Bikram Rana

---

A. Scolobig (✉)
University of Geneva, Geneva, Switzerland

International Institute for Applied Systems Analysis, Laxenburg, Austria

WMO/WWRP HIWeather project, Geneva, Switzerland
e-mail: anna.scolobig@unige.ch

S. Potter
GNS Science, Lower Hutt, New Zealand

WMO/WWRP HIWeather project, Geneva, Switzerland

T. Kox
Ludwig Maximilian University of Munich, Munich, Germany

WMO/WWRP HIWeather project, Geneva, Switzerland

R. Kaltenberger
Zentralanstalt für Meteorologie und Geodynamik (ZAMG), Vienna, Austria

WMO/WWRP HIWeather project, Geneva, Switzerland

P. Weyrich
Swiss Federal Institute of Technology (ETH), Zurich, Switzerland

J. Chasco
World Meteorological Organisation, Geneva, Switzerland

WMO/WWRP HIWeather project, Geneva, Switzerland

B. Golding
Met Office, Exeter, UK

WMO/WWRP HIWeather project, Geneva, Switzerland

D. Hilderbrand
National Weather Service, Silver Spring, MD, USA

N. Fleischhut
Max Planck Institute for Human Development, Berlin, Germany

D. Uprety · B. Rana
Practical Action, Kathmandu, Nepal

Abstract In this chapter, we explore the challenges of achieving a level of awareness of disaster risk, by each person or organisation receiving a warning, which
allows them to take actions to reduce potential impacts while being consistent with
the warning producer's capabilities and cost-effectiveness considerations. Firstly
we show how people respond to warnings and how the nature and delivery of the
warning affects their response. We look at the aims of the person providing the
warning, the constraints within which they must act and the judgement process
behind the issue of a warning. Then we address the delivery of the warning, noting
that warning messages need to be tailored to different groups of receivers, and see
how a partnership between warner and warned can produce a more effective result.
We include illustrative examples of co-design of warning systems in Argentina and
Nepal, experience in communicating uncertainty in Germany and the Weather-
Ready Nation initiative in the USA. We conclude with a summary of aspects of the
warning that need to be considered between warner and decision-maker when
designing or upgrading a warning system.

Keywords Decision-maker · Emergency responder · Response · Media ·
Vulnerability · Confidence · Behaviour

3.1 Introduction

In this chapter, we explore the challenges of achieving a level of awareness of disaster risk by each person or organisation receiving a warning, which allows them to
take actions to reduce potential impacts while being consistent with the warning
producer's capabilities and cost-effectiveness considerations. We show that:

- A successful warning provides the receiver with useful information in a usable
  form and is used effectively.
- Warnings are issued in a complex and challenging environment, in which needs
  are constantly changing.
- Success depends on the warner providing 'fit for purpose' information and on the
  response of the receiver as well as on the accuracy of the information and the
  technology to deliver it.
- Information sources, social and environmental cues, channel access and the
  receiver's characteristics influence behavioural response.
- Warnings are issued on uncertain information, and the level of confidence should
  be reflected in the warning.
- The 'warning to decision' process is not only about exchanging information but
  also about establishing relationships.
- A successful relationship requires assessment of the receiver's needs, beliefs,
  values, behaviours and decision-making processes.
- Warning messages should be tailored to different groups of receivers. The
  Common Alerting Protocol (CAP) can facilitate this while minimising the overheads of using multiple channels.

- Critical evaluation is the foundation for improvement. It allows receivers to feedback on warning effectiveness and reinforces communication.

3.2   Needs of the Receiver

3.2.1   Who Receives Warnings?

Warnings are produced for use by a variety of receivers in different situations. The job of a professional emergency manager is to be aware of the risks they are responsible for and to be prepared to respond to a warning and hazardous event. They may be highly knowledgeable about relevant hazards and familiar with technical language, though perhaps less familiar with the specific forecasting methods used by the warner. There is a much larger group of responders who are given an emergency response role as part of a wider job, but who nevertheless will generally have had some training in the risks and in the responses that are needed. Both of these groups will typically have some understanding of the relationship between the hazard and the impacts that they are concerned with, so will look to obtain supplementary information on the hazard. Whereas these groups have been given responsibility to respond primarily on behalf of others, the vast majority of receivers are responsible for their own safety and that of their families, friends, dependents and businesses (Lazo et al. 2020). They have a wide mix of understanding of hazard and risk, but in general are less interested in the hazard and are more interested in the specific impact on them and in understanding what they should do. Members of the public are increasingly able to prepare their personalised response and preparedness plans (e.g. multiple mobile phone applications now provide these decision support tools). However, levels of preparedness and guidance on the responses people should be undertaking can vary considerably.

Multi-hazard early warning systems are encouraged to be 'people-centred' (Basher 2006; UNISDR 2015), empowering 'individuals and communities threatened by hazards to act in sufficient time and in an appropriate manner to reduce the possibility of personal injury and illness, loss of life and damage to property, assets and the environment' (WMO 2018, p. 3). In people-centred warning systems, the public is a central element and resource: 'communities' (...) inputs to warning system design, and their ability to respond ultimately determines the extent of risk associated with natural hazards' (UNISDR 2013). For example, some individuals might act as local champions or peer educators to fellow residents to increase their level of preparedness and capacities to respond to warnings. A stronger involvement of the public and communities can also lead to stronger political and budgetary support for warning systems and reduction of conflicts related, for example, to decisions about warning system implementation (Kuhlicke et al. 2011). One of the main drivers for change towards people-centred systems is also the changing requirements of users, in response to increased information availability and growing threats (World Bank 2019). Some key characteristics of people-centred warning systems include a stronger focus on stakeholder engagement and responsibility sharing, enhanced communication supported by technological innovations and institutional capacity building, including stronger inter-agency collaboration (Scolobig et al. 2015).

In order to become people-centred, those implementing a warning system must know who their audience is and conduct meaningful engagement to understand their information requirements for an optimal response (Zhang et al. 2019). Thus, a critical question is: who are the receivers of weather warnings? Warnings are received by a wide range of sectors, including various publics (including tourists, business owners/operators, vulnerable communities, event organisers and attendees, the education sector, the horticulture and agriculture sectors, community groups, outdoor enthusiasts and motorists), agencies with emergency response and mitigation roles (including emergency services, emergency management from local to national levels and operations), the health sector, lifelines and infrastructure network agencies (including water, electricity, gas, telecommunications, road, rail), media, insurance, marine, aviation, science and monitoring agencies for cascading hazards, private weather services, global re-users (e.g. Google, IBM/The Weather Company) and non-governmental organisations.

This wide range of sectors demonstrates the immense challenge of issuing a severe weather warning that meets the needs of every receiver. Another challenge is to provide warnings that are co-designed between those implementing and those receiving them. An example of the co-design of a new warning product based on a receiver's needs assessment in Argentina is reported below.

Box 3.1 Users' Needs Assessment for the New Early Warning System in Argentina
Julia Chasco

In 2014, the National Meteorological Service of Argentina (NMS-AR) presented, as a result of cooperation with academic institutions, the Alert.Ar project. One objective was to better understand users' meteorological information needs for the management of severe weather events. Users were mainly emergency managers and operators. The needs assessment was carried out by triangulating different methods and tools that were conceptualised by a multidisciplinary team of social scientists: sociologists, anthropologists and geographers. Over approximately 2 years, multiple workshops, surveys and in-depth interviews were carried out, revealing to the NMS-AR that the information included in the meteorological warnings did not fit the needs of emergency managers.

Figure 3.1 shows the results of a simple exercise NMS-AR conducted with decision-makers who were all given the same weather warning in text format with the instruction, without interacting with each other, to draw their interpretation on a map of the country. As we can see, each decision-maker made a very different interpretation of the same warning.

(continued)

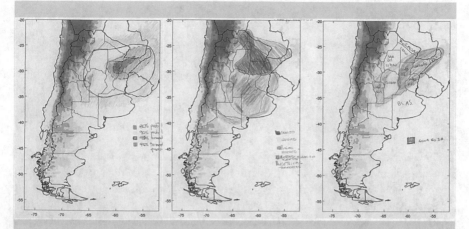

Fig. 3.1 Different interpretations of the same warning

This type of exercise allowed NMS-AR to identify weaknesses in the warnings: difficulties in understanding coverage areas and phenomena, little effectiveness in communicating spatial variations in severity and overly technical language in the message text.

Following this, the NMS-AR decided that this group of social scientists would become a permanent Department of Meteorology and Society with the mission of interacting with users to identify their needs and to use this information to improve services.

Between 2018 and 2020, the NMS-AR worked intensively on the implementation of a new early warning system oriented towards users' needs and decision-making. Some of the main changes in the new early warning system include:

1. Interoperable messages in traffic light formats and map displays.
2. Categorisations of phenomena with simple unified nomenclatures.
3. Publication of warning thresholds.

The Department of Meteorology and Society tested the nomenclatures, the iconography and definition of colours and the warning thresholds with different users. To meet the needs of emergency agencies, services were developed to facilitate monitoring over large areas, e.g. the NMS-AR has developed a graphic product, only for emergency agencies, which is sent by email twice a day at 6 am and 6 pm with alerts of the day for each province. Most emergency organisations communicate in large groups on WhatsApp, so the product was designed to be shared easily on that platform. NMS-AR developed systems for requesting special forecasts for specific operations or events as well as a special access platform in case there are problems with the website, so they can independently access alerts, nowcasting and radar and satellite images.

(continued)

The system also provides alert messages for persistent phenomena not covered by weather warnings, such as reduced visibility due to smoke, dust, ash and high or low temperatures. This information is in addition to the already established heat wave warning system, co-designed with the Ministry of Health and based on mortality data.

As part of any implementation made by NMS-AR since the creation of the social scientists' work unit, any system change is accompanied by training options for decision-makers, which implies an investment of several months. The Department of Meteorology and Society concluded that for improved warning understanding, co-production processes should be developed and tested at multiple levels of implementation, although they may be time-consuming.

3.2.2   The Nature of the Decision

The need to take protective action can arise at any time and in any place, not just when an emergency manager is at their desk, or a member of the public is at home. When a warning is received, the actions will be different for someone at home, at the office, working outdoors, engaged in sport or at a public gathering such as a festival. Those at risk need to be reached on a back-country hike or a sailing trip, woken in the middle of the night, interrupted in a business meeting, with information relevant to their situation. Once alerted, they need to confirm that a response is needed, perhaps requiring access to additional information, regardless of when, where and how they received the warning. A warning informs situational awareness and decision-making in a variety of ways, depending on factors such as the role and responsibilities of the receiver, their familiarity with the location and the hazard, costs and benefits of the decision, emotions and confidence in the information. While confidence is low, it may still be worth taking low cost, 'no regrets', protective actions that will either reduce the need for later action or prepare for more disruptive and costly actions when confidence becomes higher.

Responding agencies use warning information to plan their response and allocation of resources and may generate further communications to the public and other responding organisations such as infrastructure companies (Kox et al. 2015). Their tasks and duties are strongly influenced by national legislation and vary between countries. Responders have pressures inherent to their decision-making, including acceptable risk thresholds, time pressure and constantly changing conditions (Doyle & Johnston 2011). Weighing up the costs and benefits of the decision, while also taking into account the level of uncertainty associated with the warning, is a difficult challenge for receivers. For responding agencies, the decision to act on the warning can have very high stakes, such as the lives of exposed populations. However, the cost of reacting to a warning, such as by ordering an evacuation, is also high. Responding agencies must determine acceptable risk thresholds, which incorporate likelihood and potential consequences. Members of the public use weather warnings as one source of information to determine their behavioural response. Lindell and Perry (2012) describe how the decision-making process follows a series of ideal stages:

- Risk identification (assessing perceptions of level of threat).
- Risk assessment (expected level of personal impact).
- Protective action search (retrieving information about what to do to reduce the impact).
- Protective action assessment (determining which of the options is most suitable).
- Protective action implementation (when does the determined action need to occur).

Often the decision will be to look for further information to increase understanding of any of these factors (Wood et al. 2017). More generally, a number of factors related to the hazard, the situation and other respondents' characteristics (see Sect. 3.2.4) determine whether respondents are able to assess all of the decision-making steps described above.

Responding agencies may conduct risk assessments as part of their decision. Risk assessments can be quantitative (e.g. dollar losses or fatalities), or qualitative (e.g. low, medium, high), and use various tools. The result may fall into the categories of 'acceptable risks', where the risk is seen to be minimal and no mitigation measures are required; 'tolerable risks', where it is determined that the benefits of living with the risk outweigh the potential cost, and some mitigation may be required; and 'intolerable risks', where the risk is seen as being high and mitigation actions are required (e.g. Standards New Zealand 2009).

3.2.3   Which Information Sources Do Receivers Use?

Identifying the information sources receivers use is important for understanding the relative value of each source, and the contributions of those sources to the provided information and messages (Lazo et al. 2009). Information sources vary between users, according to their needs. They may include official sources, private weather services, environmental cues (such as seeing storm clouds), social cues and/or culturally indigenous cues. The information often travels through multiple channels, including the traditional and social media (such as tabloid newspapers), social connections (such as from friends, family, neighbours, colleagues and education facilities) and/or response agencies (including emergency services, local government, infrastructure companies and emergency management). Each of these channels, and the way the information travels through secondary sources, can have its own challenges.

Official sources of weather information tend to be National Meteorological and Hydrological Services (NMHS). Depending on national legislation, many NMHS are entitled alerting authorities, listed in the WMO Register of Alerting Authorities (https://alertingauthority.wmo.int/). Some receivers of information (e.g. airline companies) pay for a subscription service to receive more detailed or specific data to meet their needs.

Official information can be disseminated and often altered through other channels (such as TV stations/media or emergency managers) to provide additional information. In fact, most people access weather forecasts from the media and via the private sector, rather than directly from official information sources (Lazo et al. 2009, Hayes et al. 2014). As a result, most people use multiple sources of information when making response decisions, such as evacuating from a hurricane (e.g. Dow & Cutter 1998), to meet their individual needs. The majority of private sector weather forecasts are directly or indirectly based on official weather services (Lazo et al. 2009), including using public-sector global observations and models, and atmospheric research (Thorpe & Rogers 2018). The private sector is often highly interested in incorporating the 'authoritative voice', such as issued warnings, into their products and services (e.g. Google Crisis Response/Public Alerts – https://crisisresponse.google/products/public-alerts/) (Kaltenberger et al. 2020), for consistency, and to add further information or features to make their products more user-oriented. An increasing interest in impact-based forecasts and warnings by the private sector is expected due to the focus on user needs (WMO 2015; Thorpe & Rogers 2018).

Environmental and social cues are key sources of information that can influence the receiver's perceptions and actions (Lindell & Perry 2012). Environmental cues, such as a gathering storm, a funnel cloud, a roaring sound or heavy rain, are often considered when determining a response, such as during the Joplin, Missouri and US tornado warnings (Kuligowski et al. 2014). However, environmental cues can be hidden by low light (at night) and muffled by other noise (e.g. flash flooding muffled by heavy rain). Some cues are subtle or ambiguous, whereas others are much more obvious. Perceptions of environmental cues can also be altered depending on the time of day and other activities that are prioritised by the recipient (Ruin et al. 2014). Social cues stem from observing what other people are doing (e.g. packing a car to evacuate) and receiving information from community networks, such as family/friends and neighbours. This can be via a range of channels, including social media, phone calls, text messages and face-to-face. The availability of assistance, including the provision of transport and shelters, also provides cues (Lindell & Perry 2012).

In addition to the above sources, responding agencies sometimes have their own information sources. These can include trained storm spotters, staff members collecting impact or hazard data, crowdsourced reports, video cameras in public areas (e.g. traffic cameras) and their own monitoring equipment.

Finally, personal experience and cultural knowledge including, for many people, indigenous knowledge are used as information sources or a lens through which to interpret cues, prepare and respond. In many areas of the world, such knowledge is drawn upon to interpret environmental cues, such as the behaviour of wildlife, to help forecast future weather and seasonal events (e.g. Pareek & Trivedi 2011). Traditional knowledge is also a source of information to inform mitigation plans, and preparedness and response activities, such as terracing hillsides to prevent run-off and erosion and building sturdy houses that can withstand strong wind. The relevance of such knowledge is often lost when people migrate, especially into urban environments, but can be replaced if communities make efforts to build a new shared experience, e.g. by marking the heights of historic floods or the limit of inundation by tsunamis, by acknowledging anniversaries of disasters and by ensuring that schools teach their children about the hazards and responses appropriate to where they live and go to school.

3.2.4   Behavioural Influences on Decision-Making

Why do some people not respond in the way that they were directed to by a weather warning? There are many things that influence people's behavioural response to a warning. Several theoretical models describe these influences, such as the Protective Action Decision Model (PADM; Lindell & Perry 2012). Other theories and models are described in the best practice guidelines for risk communication and behaviour by the NOAA Social Science Committee (2016).

Lindell and Perry (2012) describe the factors influencing the 'pre-decisional processes' of the receiver (Fig. 3.2). They include the receiver's characteristics, such as their age, gender, primary language, mental models (general understandings and misconceptions), economic resources, social resources and physical abilities. As discussed in Sect. 3.2.3, behavioural responses to warnings may also be influenced

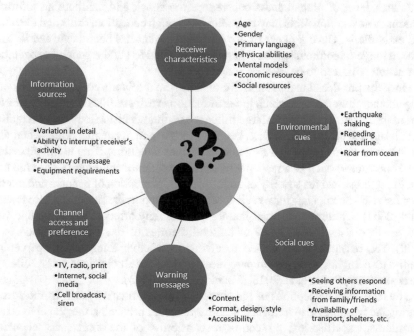

Fig. 3.2 Factors influencing pre-decisional processes towards protective actions. (From Potter (2018), based on the Protective Action Decision Model (Lindell & Perry 2012))

by indigenous knowledge and cultural norms, environmental and social cues and information sources (Shaw 2009). Message characteristics (particularly content and format, e.g. Wood et al. 2015; Potter 2018) and channel access and preference (see Sect. 3.3.3) are further influences, with relevant channel characteristics including the level of detailed information, precision of the message, the frequency that the message is sent out, the equipment requirements and how much it interrupts the receiver's activities.

The pre-decisional processes inform the decision that is being made according to the level of exposure the receiver had to the cues or warnings, the attention they paid (including intrusiveness of the alert) and their level of understanding of the message – including weather literacy (Fleischhut et al. 2020) and understanding of forecast uncertainty (Joslyn & Savelli 2010; Morss et al. 2010).

This demonstrates how using simple language, avoiding jargon and using a range of languages, when possible, influence the response. The influences mentioned here can take place very quickly and without much conscious thought. Theory from cognitive science suggests that in situations of high stress, people may make decisions using a faster decision pathway that is rather emotion-driven, while in less stressful situations, they are more likely to base their decisions on information (Weyrich et al. 2020b). In such cases, more information is not necessarily better. Moreover, which decision-making pathway people utilise may depend on the context. At the same time, cognitive theory has been hard to test in the field, because of the ethical challenges of submitting people to actually dangerous conditions.

A high level of preparedness can reduce the shock and facilitate an informed response to a warning. This may be achieved through education, drills and exercises and early alerts. Once a likely future hazardous event has been identified in forecasts, a range of communication formats can be used to build a general expectation that action will need to be taken.

Once the pre-decisional process is complete, the receiver's core perceptions provide another filter through which the information is assessed. The receiver's perception of the threat (or risk) has a large influence on their decision to respond (e.g. Kox & Thieken 2017). Risk perception refers to the judgments people make when they are asked to characterize and evaluate hazardous situations. This is influenced by numerous factors such as a respondent's prior experience, familiarity with the hazard, expected/perceived severity of consequences, perception of benefits and control over the risk source, etc. (for a review of risk perception literature, see Slovic et al. 2004, 2007). Cultural and social factors also strongly affect how people live with and perceive risks (for a review of cultural theories of risk, see Thompson et al. 1990). Yet, research shows mixed results as to whether a higher risk perception relates to a higher level of preparedness or response behaviour (e.g. Mileti & Sorensen 1990, Potter et al. 2018, Wachinger et al. 2013).

The receiver's perceptions of the source of the information (such as the agency issuing the warning) also influence their response. In particular, the perceived trustworthiness, expertise and protection responsibility of the source are taken into account (Arlikatti et al. 2007). On the one hand, institutional trust, for example, in NMHS issuing weather warnings, and credibility of the source, increases the likelihood for protective action (Potter et al. 2018, Ripberger et al. 2015). On the other hand, self-efficacy, including the belief in their own ability to inform and protect themselves against severe weather, is also an important driver for protective action (Kox & Thieken 2017).

A receiver's perception of protective actions can influence their decision to respond (e.g. Demuth et al. 2016; Johnston et al. 2005). An action perceived as effective in reducing the risk has a higher likelihood of being acted on, as do actions that are considered to be affordable and achievable and take little time and effort and that don't require coordination with other people. However, much of the research that has been conducted requires more testing in a warning context, rather than a general preparedness context, and in a 'real event' context, as opposed to a hypothetical context.

Following the pre-decisional processes and the triggering of the core perceptions, the decision to respond is undertaken (Lindell & Perry 2012). Once the risk has been identified, people look for information about the risk, involving thinking about personal impacts and length of time until the impacts are likely to occur. Options to increase their safety include drawing on their memory and past personal experiences as well as those they have seen in the media or read about, and education initiatives (Demuth et al. 2016; Sutton & Woods 2016). The receiver may seek further information, observe social cues and/or use guidance messages from warnings. Alternative actions are compared with continued normal activities. If it is determined that multiple actions need to be taken, the order of these needs to be assessed. A response plan may be determined only if there is time and a good enough knowledge to do so. The plan may indeed range from vague to detailed, potentially including factors such as evacuation route, destination and means of travel. These decision-making processes can cause delays to taking action as further information is sought, which is referred to as 'milling' (Wood et al. 2017). This highlights the importance of including 'what to do' guidance messaging in the warning message, and hyperlinks to detailed information to help people determine their actions. Once an action plan has been developed, the situation may hinder its implementation. For example, roads could be blocked, children and pets need to be found, or there may be a lack of access to transport, prompting alternative actions to be assessed, decided on and implemented.

3.3 Capabilities of the Warner

3.3.1 Who Issues Warnings?

While many countries have legal limitations on who can issue a public warning, in reality, both professional responders and the public have access to a variety of sources of warnings. Government agencies dominate as sources of information, while police and/or fire and rescue often have the final authority to order people out of an area of immediate danger. For infrastructure operators, however, while the advice may come from a government agency, it will often be an in-house emergency manager who will issue the emergency response order to the operations department. More generally it is a variety of public, private and non-governmental organisations that issue warnings to the public, in a wide range of formats and through a variety of media. Some of these organisations may simply repeat the message issued by the originator, but more often than not, there will be some degree of translation and reformatting. Indeed, a compelling presenter, even if they have minimal knowledge of how the warning was created, can have a huge influence on the level of response – and may become, in themselves, the basis for the trust of the audience.

Underlying observations and model predictions are generally provided by NMHS. This information is then used by themselves and private sector meteorologists to produce forecasts and warnings (Thorpe 2016; Pettifer 2015). These different – public and private – sources do not always provide consistent warning information: through the use of different colour-coding and warning thresholds, by interpreting model outputs differently, or by conveying different overall messages, they create perceived or real inconsistencies in the warning information received by the public (Weyrich 2019). National weather services and emergency managers have suspected that these conflicting cues exist and negatively influence public behaviour; however, more research is needed. Existing empirical evidence clearly shows that contradictory visual and textual information have negative effects on public behaviours (Lindell & Brooks 2013; Williams & Eosco 2021).

The challenge of consistent warning information grows with the number of agencies involved in the warning decision process. Inter-agency coordination is needed to ensure the delivery of consistent messages by multiple agencies within impact-based forecasting and warning systems (Potter et al. 2021). On a regional scale, cross-border high-impact weather situations have a greater potential for conflicting warning information from multiple-channel sourcing information from different NMHSs. This is not just because of differently designed warning systems but also because of a lack of standard operating procedures (SOPs) in exchanging information among neighbouring NMHSs (Kaltenberger et al. 2020). Regional programmes, such as Meteoalarm in Europe (https://meteoalarm.eu) and WMO GMAS-Asia (https://gmas.asia), aim to foster regional cooperation and information exchange that improves cross-border warning information consistency. Such consistency is of fundamental importance both to global re-users and to global responders such as the UN and international NGOs.

3.3.2   Types of Warnings

Warnings may be classified in many ways, including how they are produced, their message structure or the mode of delivery. For speed of response, automated warning is optimal, e.g. a fire alarm based on a smoke sensor or a flood siren that sounds when an upstream river gauge registers above a critical threshold. Such systems depend on recipients being familiar with what the alert means and what they should do to respond. In the past, warnings were often produced using bespoke templates and with highly compressed language to minimise the number of characters or words to be transmitted. Warnings of this type remain in use, such as in maritime and aviation safety. They typically use fixed hazard thresholds related to impacts that are meaningful to a specific audience. Free-form warnings give more options for the warner to include information specific to an event or for multiple audiences, but also increase the risk of misunderstanding or lack of clarity.

In order for everyone to have a common understanding of the severity of the hazards that are forecast, are observable or have occurred in the recent or distant past, they are often given a label. This understanding can have many uses, one of which is to help raise awareness in the public about an impending event so as to prompt preparedness actions. Some systems are numerical and continuous (e.g. earthquake magnitudes), some are divided into categories (such as the Saffir-Simpson hurricane wind scale), others have a small number of levels (such as mete-orological 'outlook, watch, warning' systems), and some are binary (e.g. fire alarms). When designing a warning or category system for a hazard, one of the many decisions that needs to be made is what it will be based on (the 'foundation'; Potter et al. 2014). This determines the trigger for issuing a warning. Options range from hazard through to guidance on response, i.e. what people should do. Each has its own benefits and challenges.

At one end of the spectrum is basing the foundation on the hazard. This means that warnings are triggered by the severity, extent, duration or magnitude of the peril, regardless of whether people will be exposed to it. Examples include expected wind speed, rainfall intensity and earthquake magnitudes (such as the Richter scale, moment magnitude and local magnitude), where the scales have fixed thresholds based only on their intensity. Volcanic alert-level systems primarily use the magnitude or severity of volcanic activity as a foundation for warnings (e.g. New Zealand's, Potter et al. 2014; and the US system, Gardner & Guffanti 2006). The benefits of these systems are that scientists can issue information quickly, based on their understanding and data; little coordination is needed with agencies who hold information on impacts, vulnerability, exposure and mitigation procedures. The main challenge is that phenomena-based systems may not initiate the most appropriate or timely responses by members of the public. Including impact information and guidance on response can improve effectiveness (e.g. WMO 2015).

Further along the spectrum, the severity thresholds chosen for each warning or category level may be fixed according to the severity of damage it would cause to people or property if they were exposed to the hazard. These systems assume someone or something could always be exposed; i.e. the thresholds do not vary over space or time. Examples of this type of system include the enhanced Fujita scale, Saffir-Simpson hurricane wind scale and the modified Mercalli intensity (earthquake shaking intensity) scale. Once the thresholds vary over space and/or time (e.g. climatology-based thresholds), the system is heading towards being an impact-based warning system. This can take account of the dynamic situation, including antecedent ground conditions (e.g. prior rainfall causing wet catchments and therefore accelerated flooding), variable populations (such as rush hour) and specific impacts such as airport closures (WMO 2015). These systems require collaboration between agencies who hold the information about meteorology, society and impacts, which has the potential to cause delays in issuing warnings (Potter et al. 2021). Impact-based forecasts and warnings are thought to increase the level of understanding about the situation by the public, and raise risk perceptions, but there are mixed results as to their effectiveness in prompting a behavioural response (e.g. Weyrich et al. 2018; Potter et al. 2018). Determining which impacts or consequences to base the warning on becomes important, whether it is for safety of life, injuries and well-being, damage, disruption, or economic or environmental impacts. Risk modelling can help with mitigation and hazard management (Crawford et al. 2018) and may become increasingly utilised in real-time situations.

Personalised warning messages that include information about local disruptions and impacts may help to prompt effective responses and would generally be issued by partner agencies who hold the roles and responsibilities to issue them (WMO 2015). These 'impact-oriented' warnings require substantial improvements in impact data collection and storage (Kaltenberger et al. 2020). Finally, some warning systems are based on the action required by the population at risk. Higher levels may include evacuations of large areas, and lower levels may promote increased awareness. Examples include New Zealand's COVID-19 pandemic alert-level system, Japan's volcanic alert-level system, fire alarms or tsunami sirens requiring an evacuation and the seatbelt sign in aircraft. These systems tend to promote compliance and require receivers to understand the actions relating to the levels or alerts. Further investigation into these types of systems would be beneficial to identify how underlying observational data support the decision-making to trigger a warning.

3.3.3 Communication Channels

Getting the warning to the receiver is essential if it is to have any value. The diverse types of dissemination channels – print, mechanical, electronic and face-to-face – have very different characteristics, for instance with respect to the dissemination rate or precision (Lindell & Perry 2012). A wide range of channels should be used according to the needs of receivers, including local or national TV (including cable), radio (including specific weather radio channels), newspapers, friends/family, co-workers, neighbours and smart/cell phones. Increasingly, web pages and mobile applications are important sources of weather information, including cell broadcast alerts and social media (Hayes et al. 2014). A siren can make people instantly aware of a threat but has a limited reach and provides no further information. Mobile phone applications can provide instant messaging with greater information content, but the recipient needs to have a (functioning) mobile phone and to consult it to receive the message. Newspapers are slow, but can provide detailed information and context, and are ideal for early alerts. Television can strongly engage the viewer, but with limited airtime, and may miss out important information. A trusted neighbour or official is a compelling source in an emergency but is very resource intensive and time-consuming.

Different channels support a different range of warning formats. For instance, some channels are restricted to text (SMS on cell phone), audio (radio) or video (regular TV), while others can contain a variety (websites, smartphone apps, weather TV). A mixture of formats can reinforce the message if consistent and if they are quickly and easily accessible. For instance, the Met Office weather app provides colour-coded warnings identifying the type of hazard on a map of the UK. Each warning can be expanded to a higher-resolution map, identifying affected towns and cities, supported by a brief explanation of the hazard and its source, what the likely impacts will be and the status of the warning (including when it was last updated and why). Links are provided to guidance on how to respond, and to additional information including a more detailed context and a list of the administrative authorities covered. Supporting material is also provided in the form of the TV weather forecast video, which includes the same information communicated verbally and visually by a presenter. Care must be taken when using visual formats as many people have difficulty reading maps, colours should be distinguishable by those who are colour blind, and colour scales and icons are not universally recognised (but see Guemil (2021) for an initiative towards an internationally accepted set of emergency icons). Consideration needs to be given to how many and which languages will be used, and to reaching the visually impaired and those with severely limited language skills.

Channels differ in the extent that they reproduce the authoritative warning or add or provide their own or other independent warning information, but it is essential to acknowledge that only in an ideal world should they all tell the same story, but in reality inconsistencies persist. Thus it is essential for agencies to recognise and to live with inconsistencies. Where the information source is known and trusted, the dissemination channel should provide source attribution and branding, as these will speed up recognition by the user, and help to distinguish the message from other potentially conflicting or misleading messages. Indeed, as the information landscape becomes more crowded, an increasingly important role of the information provider must be to monitor these multiple sources and try -to the extent possible- to issue corrective statements to counteract false information before it spreads.

New technologies, based on smartphones, are not only a channel to disseminate information. They can also be used to test the effectiveness of different communication strategies, such as by disseminating different types of messages (e.g. impact and non-impact based) to the receivers and by enabling verification – including via surveys – of which types of messages lead to adaptive behaviours (Weyrich et al. 2020a). They can also open a window for two-way communication through social media and for collecting data through crowdsourcing, for example.

Sights and sounds that indicate hazard onset are important inputs for people's emergency response (Lindell & Perry, 2012), but may come too late. For example, the presence or absence of thunder and lightning during a storm may influence responses by providing evidence supporting the imminent threat. Where the threat is not yet sufficiently close, CCTV, webcams and social media videos can provide equivalent cues at a distance. Using two-way communication such as this can be a valuable component of a warning service. Not only can the local information provided through social media be used to reinforce the warning in places not yet reached by the storm, but it can also help the forecaster to keep abreast of an evolving situation and to fine-tune their forecast for use in updated warnings.

Many challenges currently hinder the incorporation of social media and crowd-sourced data in warning practices. For example, agencies are afraid that social media will produce harmful and inaccurate information and that it can be difficult to evaluate the credibility and validity of user-generated content (Weyrich et al. 2020a,b, Goolsby 2009, Kaplan & Haenlein 2010). Also, the enormous amount of information can be overwhelming, and ethical concerns can further discourage agencies from fully exploring their potential. These issues include the potential for breaches of privacy, even from anonymised datasets, the lack of consent involved and the possibility of misuse by commercial entities interested in surveillance (Maxmen 2019). Other problems arise from unequal access to social media. For example, in 2016–2017 nearly 1.3 million households had no Internet connection in Australia, and lower digital inclusion was observed in already vulnerable groups, including the unemployed, migrants and the elderly (Howarth 2018). As a result, gaps between privileged and marginalised people may grow wider. It must also be acknowledged that the private weather sector may view these social dilemmas and challenges differently, because of their interest in the commercial aspects of warning dissemination.

Opportunities for warning agencies to use these newer channels include a better understanding of public debates about warning-related issues, monitoring dangerous situations and interacting with receivers, promoting crowdsourcing and other collaborations as well as extending the reach of organisational information and improving transparency, visibility and reputation. In the future, artificial intelligence may increasingly be used to monitor these channels and to sift out the critical pieces of information that indicate a change to a warning is needed.

This plethora of media and formats can seem daunting when designing a warning system to a restricted budget. The international standard Common Alerting Protocol (CAP) provides an increasingly valuable tool for minimising the overheads of using multiple channels (FEMA 2020). Provided a warning is produced in the XML-based CAP format (OASIS-Open 2010), standard software is available to convert it for delivery through a wide and increasingly varied set of channels, including a degree of automatic tailoring to the needs of specific user communities. As an internationally supported standard, training, support and implementation guidelines are available to facilitate the use of CAP in new and existing warning systems (e.g. WMO 2013).

The following points highlight where CAP-enabled alerting can provide benefit:

- Enables effective dissemination through new media, using smartphones and the Internet of Things, as well as existing mass media such as radio and television.
- People with special needs or a language barrier can be served.
- Alerting areas can be precise, reducing receipt of alerts outside the area at risk.
- Simplifies issuing of alerts to a single message.
- Facilitates sharing of situational awareness between emergency managers and across boundaries to create a 'common operating picture'.
- Enables links to immediate response alarms and automated controls.

3.3.4   Influences on the Warner

People who issue warnings are influenced by many factors, both individually as a person and corporately as a member of an organisation. These range from the quality and quantity of available information to their role and responsibility in providing science advice (legislative), to managing perceived risks from over-warning and warning fatigue by the public, through to personal and group psychological biases. Warners should be familiar with the nature of their audiences, including the way in which warnings will reach them. While helpful in general, this can easily lead to biased warnings and ultimately to a degraded response. Many procedures (e.g. tick boxes, thresholds and templates) are in place to help overcome biases and other influencing factors, and to retain consistency over time and between different warners and agencies (Fig. 3.3).

The legal, institutional and political context can also influence the warner and his/her behaviours and attitudes (see Chap. 2). For instance, in some countries there may be an increase in legal conflicts related to the dissemination, use and interpretation of risk, forecasts and warning information (Altamura et al. 2011) and, more generally, in disaster risk management (Lauta 2014). This can lead to defensive, self-protective behaviours of warners to avoid personal blame and liability. For example, following the trial and prosecution of scientists and officials in Italy following the communication of risk information and a subsequent impactful earthquake in the town of L'Aquila, the number of false alarms was observed to increase in Italy (Altamura et al. 2011), with consequent impacts on weather-related decision-making. On the other hand, a US National Weather Service policy decision to reduce false alarms for tornadoes by switching from county-based to storm-based warning areas resulted in a dramatic reduction in the size of warning areas (Sutter & Erickson 2010). The institutional framework can also weaken the warning process, for

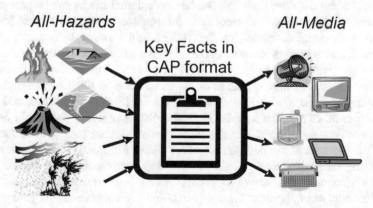

Fig. 3.3 Reaching a plethora of receivers with different needs from a single information source through multiple channels using the Common Alerting Protocol. (Source: Eliot Christian, based on ITU 2019)

instance by failing to define stakeholders' distinct roles and responsibilities (Thorpe 2016). This can impede collaboration and cooperation between the actors involved in the warning chain. Institutional priorities may also determine whether warnings are primarily aimed at reducing impacts on individuals or on society as a whole (Potter et al. 2021).

Technical and financial constraints mean that a warner is almost certainly working under time pressure and with access to information limited by the speed of a desktop computer. They usually have access to only a subset of the observational data used in the forecast, and to a limited range of forecast products, especially if the forecast has come from a global forecasting centre. They may be trying to provide direct information by phone to key individual responders while preparing their warning for the public and a script for a live broadcast. They should also be monitoring the current state of the hazard and information circulating in the media, including social media. Under these pressures, the warner's attitudes and behaviours will be influenced by a lack of resources, inadequate communication protocols and ineffective engagement with the media and the private sector (UNISDR 2013, NOAA 2016, Handmer & Dovers 2007).

Timeliness is of particular importance to the user since response actions take time and must be completed ahead of the arrival of the hazard. When forecast information is produced across several organisations, such as a weather service, a flood agency and an emergency response agency, there may be a substantial time delay before the warner even receives the information. This can be exacerbated with impact-based forecasts and warnings, which may require information from several agencies to be collated (Potter et al. 2021).

Behaviour may also be affected by organisational competitiveness, whether for profit, funding, prestige or power (such as with other ministries, businesses or nations), by inappropriate organisational or individual target setting (e.g. concentrating on volume or accuracy over user value) or by excessive protectionism (including requiring that everything is produced in-house) or outsourcing (so that internal roles are devalued). Misinformation on external media can be particularly challenging; interventions to correct false information carry the risk that both the false information and the correction distort the original message.

Warners, and experts in general, tend to perceive risk differently from each other, and from lay people (Bostrom 1997). While experts tend to use formal definitions of risk and emphasise the magnitude of the hazard and its likelihood, members of the public may include other factors, such as the familiarity of the threat and dread (Fischhoff et al. 1978). The social and institutional contexts of risk are important, such as motives and values. Experts also translate verbal probabilities (such as 'likely') to numerical probabilities (such as 70%) differently from non-experts, highlighting the importance of using probability translation tables (Doyle et al. 2011) as encouraged by the WMO (WMO 2008). Interpretations of probabilities also differ between experts and non-experts, and even within expert groups (Kox et al. 2015). The decision-making that goes into assessing risk inevitably requires a series of judgements to be made, to refine the circumstances or scope of the situation.

Many psychological biases can influence judgements at an individual or group level. In a setting of scientists/warners making decisions, these can include the desire to conform to the group (Asch 1952), the influence of a minority (Crano & Chen 1998), the groupthink phenomenon (Janis 1982), obedience to authority (Milgram 1974) and potentially the presence of an audience affecting performance (Dashiell 1930). The ways in which questions requiring judgement are asked can influence the outcome, with care needing to be taken to not bias the result (Potter 2014). Another influence on warner decision-making is avoidance of potential false alarms. While warners are often very concerned about the 'cry wolf' effect of false alarms, several studies have shown that their effect can be minimised by careful messaging before, during and after events. For example, Kox and Lüder et al. (2018) show that emergency managers state their discomfort about warning fatigue and high false alarm rates but indicate that they would respond to the warning as common practice. In general, policymakers are much more concerned about misses than false alarms. Various alternatives exist to support decision-making, such as a structured approach (e.g. Bayesian), or a more naturalistic approach (Doyle & Johnston 2011).

3.4   The Bridge Between Warner and Receiver

3.4.1   Building a Relationship

Those issuing warnings must understand the information needs of the receivers. This is achieved by building a relationship between the receiver, the warner and any other stakeholders involved in the process. This relationship can take many forms, depending on the laws and culture of the country, the governance structure of the warning organisation and the nature of the receiver – whether an individual, an organisation or a whole community. Relationships take time to build, require active management to thrive, depend on flexibility and compromise on all sides and ultimately need to commit to solutions that are beneficial to all parties. Relationships are equally necessary whether the warner is in a public or private sector organisation, and whether the receiver is a member of the public, a global business or a non-governmental relief organisation. However, the nature of the relationship will be different in each case. Sustaining the relationship requires continued and regular two-way interaction, as personnel change and technical capabilities evolve, including periodic training and exercising.

The level of engagement underpinning these relationships ranges from one-way informing (e.g. a radio announcement about a hazard/event) to 'empowering' the end users to make the decision (as described by the International Association for Public Participation; https://cdn.ymaws.com/www.iap2.org/resource/resmgr/pillars/Spectrum_8.5x11_Print.pdf). There is a vast literature maintaining that effective warning communication is achieved through a two-way exchange of information between parties and that involving stakeholders in the communication, planning and discussion stages can increase their commitment, and ultimately foster the adoption of effective behaviours (for a review, see Macintyre et al. 2019). Indeed, this relationship and, in general, the whole warning communication process are circular and two-way as it also depends very much on the environmental, social and technical context, as well as the degree of partnership among the involved stakeholders. Two-way exchange of information also includes co-design and co-production of information, in which the receiver's role shifts from pure user to collaborator and partner (Kox, Kempf et al. 2018).

There are multiple successful examples of co-design of people-centred early warning systems and citizen science engagement processes (Preuner et al. 2017, Scolobig et al. 2016). These show how it is essential to adopt a warning value chain approach to engage with multiple stakeholders and to co-design warning system options. They also provide evidence that the inclusion of different stakeholder perspectives on the technical, social, economic, legal and institutional characteristics of a warning system helps to address decision stalemates and conflicts, e.g. about funding allocation or priorities.

Yet, despite many authors highlighting that communication should be a two-way exchange of information, there is an operationalisation gap in many countries. The current literature and established practices suggest that most warning communication with the public remains unidirectional, from decision-maker to an uninvolved public, rather than a dialogue. Without doubt, there are multiple barriers to true two-way communication (see also Sect. 3.3.3). For example, given the large number of sectors and groups receiving warnings (as described in Sect. 3.2.1), it is difficult for agencies issuing warnings to build relationships and establish multidirectional engagement with all of them. Where stakeholder engagement is used in the co-design process, it can be difficult to involve some vulnerable groups such as women with young children, housebound elderly or those with chronic illness, and even when present, for these groups to feel equally engaged, especially if they lack good language skills. A combination of group meetings and individual interviews is needed, but community acceptance of such distributed inputs requires a high level of transparency in presenting and interpreting the evidence. It especially requires dedication of time and resources to develop successful processes. Co-production is also not suited for all kinds of warning communication. While it has proven successful for planning and evaluation, e.g. co-design of warning system options and evaluation of warning communication effectiveness, it is more challenging for fast two-way communication. This is currently the preserve of a small but rapidly growing group of receivers, such as storm spotters and organised groups of amateur observers (see, e.g. Elmore et al. 2014). Combining the concept of trained storm spotters or 'trusted spotters' with quality management of the data received was rated best practice by the European Meteorological Society (Krennert et al. 2018). Where funding is available for a tailored service to a specific business or sector, the opportunities for building a relationship are much greater than for a general public service. Yet, they are often missed, with available resources focused on technical capability at the expense of achieving a mutual understanding of the problem and its solution. With the growth of web-based warning services, tailoring has begun to be feasible for a much wider range of receivers. Tailoring by location has been in use in the USA for many years, firstly on NOAA Weather Radio and more recently on cell phones (Wireless Emergency Alerts), targeting messages to individual cities or counties (NOAA 2021). The ability to set locations of interest is also well established, for instance in the Red Cross warnings app (e.g. British Red Cross 2021). However, the potential for much greater use of targeting is shown by its increasingly sophisticated use in online advertising, for instance by Google and Facebook. Such targeting lies not only in selecting the information of interest but also in presenting it according to the selected or revealed preferences of the receiver. Ethical issues surround the use of this technology, but as it gains wide acceptance through online platforms like Google, there will undoubtedly be increasing application in the communication of tailored forecasts and warnings.

An effective long-term partnership provides an environment within which to build up many of the conditions for successful use of warnings, including education and training between events, shared experience and interpretation of historical events, mutual understanding of forecast capability and trust in the reliability of the warning system, all of which influence the way a warning is both produced and received (Parker et al. 2009).

Two examples follow that exemplify current good practices, via the establishment of strategic partnerships aimed at strengthening the relationships between the warner and the receiver.

Box 3.2 The Global Weather Enterprise
Sally Potter

The Global Weather Enterprise (GWE) 'is an enabling environment fostering global engagement between public, private, and academic sectors that share the common goal of providing accurate and reliable weather information and services that save lives, protect infrastructure, and enhance economic output' (World Bank 2019: 3) and is described by Thorpe and Rogers (2018). It focuses on increasing value through the full chain from scientific research and observations to models, forecasts, products and services. It is an initiative aiming to acknowledge and address obstacles in the collaboration of these three sectors (particularly public and private), including funding pressures on the public sector, growth of the private sector capabilities and international financing structures. It encourages a fair market for weather forecasts and services, and clear governance of taxpayer-funded open-access data in relation to commercialisation.

Indeed, business can derive economic value from weather knowledge – estimated at $US13 billion in a report by the US National Weather Service (NOAA/NWS 2017). Increasing the recognition of the value of underpinning publicly funded weather services in leveraging the private sector may help ease the funding pressures on the NMHSs (Thorpe & Rogers 2018).

(continued)

Additionally, it has been suggested that the public sector could make more use of private sector data, with the recognition that open-access data and long-term reliability of data be considered (Hayes et al. 2014). This includes exploring potential business relations with the private sector as part of pilot projects (World Bank 2019). The GWE also calls for clear roles and responsibilities, particularly around data ownership, as the private sector moves towards the provision of data services and away from infrastructure (Thorpe & Rogers 2018). At the same time, there are also opportunities and good practices related to the role played by the private sector that can be exemplary to improve warning communication. For a public audience, the ability to respond online to social media posts and media stories, and provide information through crowdsourcing mobile applications, enables a higher level of engagement than the traditional one-way communication of hazard information.

Box 3.3 NOAA's National Weather Service Strategic Goal: Building a Weather-Ready Nation
Douglas Hilderbrand

In 2011, a series of extreme events across the USA, including tornadoes, floods, hurricanes and wildfires that killed over 1000 people, became the driving force behind a profound movement to improve the country's 'weather services'. The National Oceanic and Atmospheric Administration (NOAA) and its daughter agency, the National Weather Service, started a long-term, strategic goal to improve forecast accuracy, communication and delivery of information to the public. The strategic goal to build a 'Weather-Ready Nation' (WRN) galvanised the operational and research arms of NOAA to improve not only forecasts and warnings but also their value through better societal responses (Hilderbrand, 2014). Internally, WRN became the impetus to measure value not just from a forecast accuracy perspective but also by a societal outcome perspective. This change in mindset has resulted in four focus areas:

- Delivery of Impact-Based Decision Support Services (IDSS).
- More effective communication of preparedness and protective 'call-to-action' statements.
- Integration of physical and social science in products and services.
- Better ways to deliver information in a timely and relevant manner.

Some WRN successes include:

- Creation of storm surge inundation maps that better communicate where and how much storm surge can be expected.

(continued)

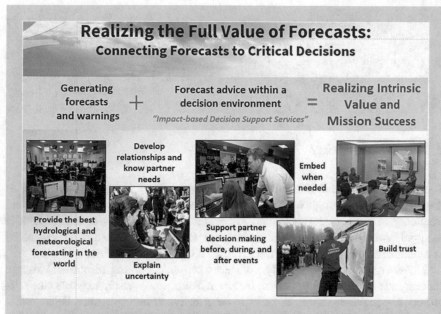

Fig. 3.4 Components of the NWS Weather-Ready Nation initiative © NOAA 2021

- Use of social media such as Facebook and Twitter for two-way interactions (e.g. posting safety messaging and forecast information and receiving storm reports by followers).
- Emergency alerting on cell phones sent via the nearest cellular tower.
- NWS personnel on location at emergency management operations centres to deliver forecast advice within the broader decision environment.

Beyond the changes made internally, mentioned above, WRN also became a commitment to collaborate at the community level (Fig. 3.4). Government could not achieve a Weather-Ready Nation alone, but rather needed to embrace external partnerships at the federal, state and local levels, and across industry, non-profits and other community organisations. To show its commitment to partnership and give others 'partial ownership' of WRN, NOAA launched the Weather-Ready Nation Ambassador programme in 2014. As of August 2020, NOAA has recognised over 11,000 organisations as ambassadors, sharing the goal of making communities ready, responsive and resilient to extreme weather, water and climate events. WRN Ambassadors are as diverse as community needs – from global corporations to small non-profits. With these WRN Ambassadors acting as force multipliers, weather safety and life-saving forecast/warning information can reach many more people in communities across the USA. The wide range of skill sets collectively across these ambassador organisations allow for innovative collaborations with NOAA and even other ambassadors.

(continued)

Looking ahead, WRN continues to build momentum on tough challenges such as quantifying and communicating forecast certainty, folding probabilistic forecasting into the decision-making process and finding new ways to inspire the public to take appropriate actions so that when extreme weather threatens, communities will be ready.

3.4.2   What Works?

The content and format of a message differs depending on varying user needs, as does the channel through which they would like to receive it, and the timing and frequency of updates. As described in Sect. 3.1.1., extensive user engagement is required to understand these needs in advance, to ensure the information is received in a useful and usable way (e.g. Becker et al. 2019, Kox, Kempf et al. 2018).

The warning message directly influences people's warning response. Message content and style are thus important factors in determining whether people take self-protective behaviour or not. In order to be effective at inducing such behaviour, a message should contain the information elements of hazard (nature and magnitude), location (area affected by the hazard), time (occurrence time or time to impact), guidance (action recommendations) and source (Mileti & Sorensen 1990). Previous research also shows that, to be effective, warnings should describe the exact nature of the threat (including potential impacts), provide a source of confirmation and be personally relevant (Weyrich et al. 2018, Lindell & Perry 2012, Mileti 1999). Technology developments are increasingly allowing the inclusion of links to further information, facilitating faster information seeking and response decision-making. The order of these elements and the length of the message can themselves influence responses, with social science research showing that relative to shorter messages, longer messages can reduce people's inclination to search for further information, which then shortens the delay before responding (Wood et al. 2017). In addition, each of the information elements should be addressed by the five stylistic dimensions of a warning message, which are specificity, consistency, accuracy, certainty and clarity (Mileti & Sorensen 1990).

Familiarity and education have a key role to play. When a response is practised frequently, it becomes an almost automatic reaction. Fire alarms are a good example. Practising fire evacuations is mandatory for organisations in many countries, to ensure that employees not only recognise the warning but know what to do without thinking about it. Most hazard warnings are more complex than that but achieving an initial reaction through familiarity is still important. Beyond that, the response will be informed by knowledge, which requires education. The best time to learn is at school, but education about hazards and their impacts, warnings and their capabilities should be continuous and focused especially on those lacking familiarity with the area they live or work in. Education may take many forms, ranging from talks and workshops, through paper exercises and online games, to drills and full-scale exercises (ITU 2019). These build on one another and can be targeted at different stakeholders. Opportunities for knowledge reinforcement ahead of periods of increased risk should be taken, whether provided by changes in the seasons or by long-range weather forecasts. A critical issue is to mainstream warning-related learning units designed for different age groups in different subjects in order to guarantee that risk education is part of the curriculum.

Weather forecasts and warnings are inherently uncertain. Yet, reducing uncertainty is not the same as managing the effects of scientific uncertainty and communicating it (Brashers 2001). Notably, reducing and communicating uncertainty require completely different types of knowledge and expertise, ranging from natural to social sciences. Research shows that people understand that there is uncertainty inherent in weather forecasts, as well as that different scientific factors shape this uncertainty (Joslyn & Savelli 2010). With respect to communication, some researchers argue that communicating scientific uncertainty (e.g. in probabilistic forecasts) leads to better outcomes (Fischhoff & Davis 2014), as it will increase understanding (Stirling 2010). LeClerc and Joslyn (2012) found that providing weather forecasts (of high consequences but low probability) with uncertainty information enhances the chances that users take precautionary action. However, conclusions about how scientific uncertainties should be communicated (quantitative vs. qualitative, graphs/maps vs. text, etc.) are not yet clear and depend on the user, the context and the type of information.

Technological improvements continually reduce the level of scientific uncertainty in weather forecasts and warnings. Over the last decades, messages have been improved as warnings are made with greater accuracy, geographic precision and lead time. However, the information is not always clear, specific or consistent. Information may be more or less available, information from different sources may be inconsistent or contradictory, and information can increase or decrease the perception of uncertainty (Brashers 2001). For example, many people do not understand the standard phenomenon-based warnings and have difficulties translating a 'heavy' rainfall warning (e.g. indicating 100 mm of rain) into effective impacts. Communicating specific impacts, for instance on road and rail transport, and possibilities of delays, could improve warning effectiveness (Weyrich et al. 2018). Moreover, it is important that a warning message is consistent within itself and across different messages (Mileti & Fitzpatrick 1992). This means that the underlying meaning of a message and potentially the colours and terminology used are similar or uniform, including from different providers at a given point in time (Williams & Eosco 2021).

Unlike in some weather forecasts, including probability of rain forecasts, there is almost no room for the communication of uncertainties within most current public warning systems. Warnings are issued by a forecaster when expected weather events reach a subjective level of certainty. The inclusion of probabilistic information could enable the forecaster to better communicate the varying degrees of certainty associated with each warning situation. However, the price of this type of probabilistic information can be the risk of misinterpretation or lack of understanding within the target audience (Kox, Kempf et al. 2018). One means of dealing with these complications that has gained credibility in climate forecasting is to develop storylines based on forecast scenarios (Shepherd 2019). For instance, during December 2011, an extratropical cyclone was forecast to move up the English Channel, producing extensive snow in central England (Mylne 2012). However, within the range of ensemble predictions, there was a low probability that the cyclone would turn north, moving the area affected by snow and allowing damaging winds to affect the extreme south-east of England. Rather than limit his presentation to the most likely forecast, or attempt to present probabilities, the TV forecaster chose to convey the uncertainty by presenting these two scenarios as alternative storylines of likelihood and impact.

Communication of uncertainty is further complicated because it is multifaceted: not just scientific but also social, legal, institutional and political uncertainties affect how a warning is perceived and acted upon. Moreover, perceptions of uncertainty vary between people and social groups: one person may have an amount of information that other people would deem sufficient to make a decision, yet she/he may still feel uncertain on what to do (Brashers 2001).

While the general inclusion of probabilistic information in public warnings remains challenging and contentious, there is no doubt that it is an important component of the communication with many professional responders. In some cases, they have sufficient evidence to calibrate their actions directly on the probability of a specific threshold being crossed. In these cases, a role of the relationship is to identify these thresholds and the nature of the uncertainties, so that the warner can ensure that the information reaching the receiver contains the required level of probabilistic detail, and that it is unbiased.

Box 3.4 Probabilistic Weather Information for Emergency Managers in Germany
Nadine Fleischhut

Although probabilistic forecasts are key to informing decisions under uncertainty, probabilistic weather forecasts are still rarely communicated to lay audiences due to fears that they are difficult to communicate clearly and that lay people may be reluctant to use them. Yet there is growing evidence that people understand probabilistic information if it is presented transparently (Hoffrage et al. 2000), that it can improve decisions (e.g. Joslyn & LeClerc 2013) and that it is preferred by the public (Morss et al. 2008). In contrast, deterministic warnings can hinder informed decisions, since forecasters must issue warnings without knowing the needs of their users, who are left to guess the uncertainty of the forecasts (e.g. Fleischhut et al. 2020, Joslyn & Savelli 2010). Probabilistic forecasts, however, enable everyone to apply the decision thresholds that fit their needs.

(continued)

In Fundel et al. (2019), we described how we evaluated the usefulness of probabilistic forecasts for emergency managers' decisions in Germany. As a large and diverse group, their required lead times and probability thresholds differ considerably (Kox et al. 2015), likely reflecting varying institutional constraints and capacities (Demeritt et al. 2010).

Using a new approach, we designed FeWIS Pro: five uncertainty representations implemented in parallel within the fire brigade weather information system (FeWIS) of Germany's national meteorological service. The representations display forecasts for wind, rain and thunderstorms (48-hour lead time) as probabilities for binary events (exceeding warning thresholds) or as probability thresholds for continuous variables (e.g. for wind or precipitation). All representations were designed based on evidence from risk communication research in a range of fields, including medicine. The approach made it possible to observe and quantify emergency managers' preferences under real operational constraints and over a longer period of time.

We analysed which representations emergency managers preferred for two severe storms (Xavier and Herwart) in Germany in October 2017 (see Fig. 3.5). In general, emergency managers used probabilistic forecasts frequently, indicating that they found the forecasts informative and useful. During both storms, the most frequently consulted representation was a map displaying probabilities for exceeding warning thresholds. The map, which provides a clear overview of the areas most likely to be affected by the storms, may be useful for emergency managers coordinating emergency services, vehicles and personnel.

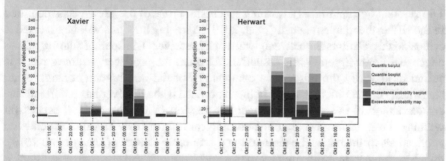

Fig. 3.5 Search for information before, during and after storms Xavier and Herwart. Plots show how often each representation was selected by time, from 48 hours before to 24 hours after the storm. Dotted lines mark when the first weather watch was issued; solid lines mark the first weather warning. Red lines along the x-axis indicate when the storms passed through Germany. For each storm, the analysis includes the behaviour of emergency managers who should expect to be affected first by the storm, defined here as all users for which the 90 percentile of the forecast was ≥110 km/h during the first third of the storm ($N = 93$ collective users for Xavier; $N = 114$ for Herwart). Overall, they selected representations 439 times during Xavier and 722 times during Herwart. For details, see Fundel et al. (2019)

(continued)

The other four representations presented timelines for a selected area. Of these, the representation of probabilities of exceeding warning thresholds was used most frequently during Xavier. Box plots showing the likely range of wind speeds were used most frequently during Herwart; they were only slightly less popular during Xavier. Quantile information (e.g. box plots) was consulted frequently shortly before – and even more so during – a storm.

Emergency managers' focus on warning thresholds may reflect that their main duties are reactive and that warnings trigger a range of decisions, such as the declaration of an emergency. Representations displaying probabilities of exceeding warning thresholds may thus be particularly useful during an early weather watch. In contrast, quantile plots such as box plots make it easier to evaluate how wind speeds might develop; they may therefore help emergency managers maintain the ability to respond during an event and prepare for daily operations afterwards (see also Kox, Lüder et al. 2018).

FeWIS Pro was developed as part of the interdisciplinary research project WEXICOM, funded by the Hans Ertel Centre for Weather Research. WEXICOM aims to improve the communication, understanding and use of uncertainty in weather and impact forecasts. The work and results reported here have previously been published in more detail in Fundel et al. (2019).

3.4.3   Measuring Success

The question of what makes a warning successful is one of the key aspects identified in the HIWeather context (Zhang et al. 2019). Agreeing how this will be assessed is critical to the relationship between warner and receiver. Transparent sharing of evidence and its interpretation is essential. Yet often the producer assumes that it is sufficient to verify the observable components of the information content, using statistically correct methods, and is then surprised if the receiver has a quite different perception of the value of the warning service. While the roots of successful warnings lie in having accurate meteorological information, its value depends on applications of the social, behavioural and economic sciences (Zhang et al. 2019), thus becoming an interdisciplinary matter.

To be successful, the warning must enable its recipients to make the right decisions to protect themselves and their communities (Golding et al. 2019; Taylor et al. 2018). In order to achieve an effect in the sense of risk-reducing behaviour through warning communication, warnings must generate attention (wake-up signal), indicate potential impact and give guidance for adequate response.

Recipients may differ in their needs and requirements for weather information and warnings, subject to their responsibilities and competencies. Thus, what constitutes useful or 'good' information varies according to the areas of activity an end user represents. In an essay on the goodness of weather forecasts, Murphy (1993) mentions three general types of forecast goodness: consistency, quality and value.

In order to be consistent, the forecast (oral or written) should be the best possible estimate or assessment of the weather situation by the forecaster. The forecast may be inconsistent if it contains (more or less) spatial or temporal specificity, or if the uncertainty in the judgement of the forecaster is not accurately reflected in the corresponding forecast, either in words or numbers. Quality refers to the degree of agreement (or similarity) between the forecast and observed events, expressed in terms of distortion, accuracy or skill. Finally, the value refers to an increase in benefit to a forecast user as a result of using the forecast. End users of forecasts and warnings place particular emphasis on the value aspect of a forecast's goodness (Kox, Kempf et al. 2018). Economic perspectives are most commonly used to define the value of weather forecasts and warnings (e.g. Lazo et al. 2009), but weather forecasts do not have an intrinsic value in an economic sense, they rather have a specific value for a user when he/she takes measures, and these measures avoid or reduce damage costs (Murphy 1993, 1994, Mylne 2002). Kox and Thieken (2017) add that a purely economic perspective does not apply to situations where monetary damage costs are difficult to allocate, such as loss of life or social or political prestige. In other situations, people may want to act, but may not be able to do so due to professional constraints or limited resources.

Fire brigades are a good example of such an end-user group operating outside of a simple cost-loss analysis, while road maintenance services responsible for salting roads have clearer cost calculations (Kox, Kempf et al. 2018). Accordingly, the end-user perspective may vary on what is understood as the goodness of a forecast or the success of a warning. A warning message that is of high value to one user may be useless to another. In this context, the ability to tailor warnings to individual needs (Joslyn & Savelli 2010) and to provide access to additional meteorological information is of importance to achieve high value for all end users.

Keeping all that in mind, the measurement of success is a difficult task. It gets further complicated by the possibility of 'grades of success' (Golding et al. 2019), for instance in near-miss situations, when the spatial or temporal extent is only slightly in error (Sharpe 2016), or in situations where no damage occurs due to successful warning. For example, in the case of road icing, observations of road state will not show that the road is covered in ice if it has been treated in advance, though the temperature may still show the road surface to be below freezing. In this instance the meteorological trigger for the hazard (sub-zero temperature) is verifiable, but the hazard itself (slippery road) is not. Here, the absence of such impacts might be an indicator of success especially if other impacts of the hazard are observed.

Continuous evaluation of warnings is essential if the benefits of improvement are to be identified and any degradation is to be arrested quickly. Changes in technology and external conditions can otherwise lead to a warning system rapidly losing its impact. Regular surveys of a range of users, especially following the issue of warnings, should be followed up with one-to-one interviews to identify negative issues before they lead to a loss of trust, and to reinforce positive changes.

To improve weather warnings in all three dimensions of forecast goodness, a broad range of challenges need to be addressed. A strong collaboration between the main users and the national weather services in the form of an ongoing dialogue and discussion of critical needs is important for the success of weather warnings (Kox, Kempf et al. 2018). This dialogue should be particularly active following any unsuccessful or only partially successful warning, to bring together the perceptions of the two sides and to conclude a joint evaluation. In parallel, it is also important to explore new options (e.g. for warning communication) from the perspectives of different stakeholders and to test these options theoretically and sometimes practically, to determine what is critical to improve the process from warning to decision (World Bank 2019).

Relationships are not static, and it is essential that the success of a particular relationship is reviewed frequently. With rapid population growth and urbanisation occurring in much of the world, vulnerabilities are changing, while in a warming climate, the hazards to which audiences are exposed are changing, too. And with new technical capabilities, both the forecast quality and the ability to communicate them are shifting. Among all these external changes, the lifestyles and expectations of warners and receivers, and the structures and personnel of the partner organisations, are also evolving. For a relationship to survive and thrive in this dynamic environment, it requires active management – including an open channel for feedback on any aspect of the warning service. However, it is also important to periodically review the health of the partnership: is the relationship still the right one? Does it have the right membership? Are the outcomes improving? Is the cost affordable? All of these questions should be addressed periodically – perhaps every 5 years – and if the answers are 'no', the relationship needs overhauling. Below is an example of rapid changes in the development of an early warning system.

3.5   Example

Box 3.5 Community-Based Early Warning System in Nepal
Dharam Uprety and Bikram Rana

Based on work undertaken in 2018.

Extreme and regular flooding in Nepal results in significant loss of life, property and livelihoods (Practical Action 2009, 2020; Bhandari et al. 2018). In response, several flood early warning systems (EWSs) have been established, in an attempt to reduce the number of people affected and killed by floods. However, there are still challenges in these systems, especially in communicating flood warning to the most vulnerable and ensuring they have the skills and resources to be able to respond:

> "If we can get information before the floods come, it will save our lives. We may not be able to rescue everything, but our children and families will be saved."

The roll-out of flood EWSs began about 20 years ago and was initially focussed on manual observation towers. These towers provided peace of mind

(continued)

for people in their immediate vicinity but had numerous limitations such as maintaining observers 24 hours each day as well as communicating warnings during heavy rainfall events. Their success created the momentum to work with the national Department of Hydrology and Meteorology (DHM) to explore more comprehensive systems, with expansion into real-time river level monitoring, automation of gauge stations and subsequently the adoption of technologies including forecasting and SMS messaging. The evolution of the system is indicated in Fig. 3.6.

The evolution of the EWS has been marked by parallel improvements in the local area. For example, social capital has been enhanced between the upstream and downstream communities, with a substantial number of households informing their neighbours immediately after receiving flood warning, and this is the primary warning source for many households. The increased warning time gives people more time to respond; human capital has been built as individuals have adapted from 'fright and flight' to learning what to do, including moving vulnerable assets to high ground or stored on upper floors prior to evacuating, with these responses reinforced by annual drills. This additional time has enhanced learning, and community members have constructed simple mitigation measures, such as waterproof storage facilities in their houses for grains and less movable assets, improving their physical

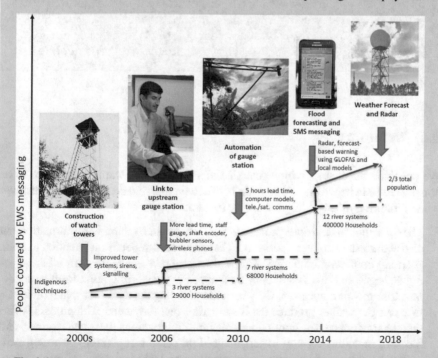

Fig. 3.6 Development and use of technology for flood early warning systems in Nepal

(continued)

capital. As local communities have experienced the positive benefits of the EWS, some people have started to express a willingness to pay for the EWS services, offering a boost in financial capital and perhaps a long-term and sustainable mechanism to cover the EWS operational costs long term. This has been further enhanced with the two major telecommunication companies in Nepal (NTC and NCELL) joining hands with DHM to send free mass SMS warnings to all mobile users living in the flood plains of major river basins during the monsoon:

> "If we can make early warning effective and efficient, the necessity for rescue can be reduced – if we can manage better with what we have I don't think we need 'big' technologies, we don't need more resources and we don't need extra personnel."

However, the EWS is far from comprehensive and many challenges remain. Firstly, climate change is a huge uncertainty in the Himalayas with implications for the appropriateness and robustness of the evolving system. Secondly, while progress has clearly been made in the monitoring and forecasting, integration of sociocultural aspects needs to be strengthened that can make early warning information accessible to the most vulnerable. Thirdly, engagement and ownership with local communities must be maintained to ensure the system is co-designed to deliver information tailored to meet their diverse needs for rapid dissemination and timely protective action. Finally, for flood risk communication to bridge 'the last mile' in terms of reaching the most vulnerable in the community, it must take account of their distinct social, economic and political experiences in both the content and the delivery of the information.

3.6 Summary

We conclude the chapter with a checklist of aspects of the warning that need to be considered to improve the relationship between warner and decision-maker/receiver when designing or upgrading a warning system:

- The 'warning to decision' process is not only about exchanging information but also about establishing relationships. Effectiveness depends on attention to both.
- A strong collaboration between the warner and receiver in the form of an ongoing dialogue and discussion of critical needs is the starting point for the success of warning communication. Only by understanding the decisions that individuals face can the warner produce the information that they need. The process is further characterised by continuous, flexible interactions between warner and receiver, including support in personalised preparedness planning, warning evaluation, co-design of warning system options and co-production of information, e.g. through citizen science. In these cases, the receiver's role shifts from pure user to collaborator and partner.

- To be effective, a warning message should contain the information elements of hazard, impact, location, time, guidance, source and a link to further information. In addition, each of the information elements should be addressed – to the extent possible – by the five stylistic dimensions of specificity, consistency, accuracy, certainty and clarity. Besides these general characteristics, a warning message should be tailored to different audiences. The international standard Common Alerting Protocol (CAP) minimises the overheads of using multiple channels and increases interoperability of systems.
- Not only addressing needs but also personalising the message by including information about local disruptions and impacts through impact-based forecasts and warnings may help to prompt effective responses. These warnings require substantial improvements in impact data collection and storage.
- Information sources, social and environmental cues, channel access and preferences and receiver's characteristics are key factors influencing a behavioural response after a warning is received.
- Reducing uncertainty is not the same as managing the effects of uncertainty and communicating it. There is no 'one fits for all solution' for managing and communicating uncertainty. Not only scientific but also legal, social, institutional and political uncertainties need to be taken into account for effective warning communication, together with the complete range of behavioural and psychological responses to uncertainty.
- Working with trusted sources and the public, testing different message options and evaluating the results of communication and cooperation efforts are critical. New technologies increasingly allow evaluation of communication effectiveness, sometimes even in real time using smartphone applications. The evaluation of warning communication as standard practice is also critical to guarantee that lessons are learnt and needed reforms are implemented. Evaluation also allows the public to provide robust feedback on warning message effectiveness and the warner to establish a permanent communication channel.

References

Arlikatti S., M. K. Lindell and C. S. Prater, 2007. Perceived stakeholder role relationships and adoption of seismic hazard adjustments. *Int. J. Mass Emergencies and Disasters,* **25**(3), 218.

Altamura M, L. Ferraris, D. Miozzo, L. Musso, F. Siccardi, 2011. The legal status of uncertainty. *Nat. Hazards Earth Sys. Sci.,* **11**(3), 797–806. doi:https://doi.org/10.5194/nhess-11-797-2011

Asch, S. E., 1952. *Social psychology.* NJ: Englewood Cliffs. Prentice-Hall. doi:https://doi.org/10.1037/10025-000

Basher R., 2006. Global early warning systems for natural hazards: systematic and people-centred. *Phil. Trans. Roy. S., A: Mathematical, Physical and Engineering Sciences. Roy. S.,* **364**(**1845**), 2167–2182. DOI:https://doi.org/10.1098/rsta.2006.1819

Becker J. S., S. H. Potter, S. K. McBride, A. Wein, E. E. H. Doyle and D. Paton, 2019. When the earth doesn't stop shaking: How experiences over time influenced information needs, communication, and interpretation of aftershock information during the Canterbury Earthquake Sequence, New Zealand. *Int. J. Disaster Risk Reduction.* 34:397–411. doi: https://doi.org/10.1016/j.ijdrr.2018.12.009

Bhandari, D., M. Uprety, G. Ghimire, B. Kumal, L. Pokharel and P. Khadka, 2018. *Nepal flood 2017: wake up call for effective preparedness and response.* Practical Action, Rugby, UK. https://www.preventionweb.net/publications/view/59123 (Accessed 16/6/2021)

Bostrom A., 1997. Risk perceptions: experts vs. lay people. *Duke Environ. Law & Policy Forum,* **8,** 101–113. https://scholarship.law.duke.edu/delpf/vol8/iss1/8/ (Accessed 2/9/2021)

Brashers, D. E., 2001. Communication and uncertainty management. *J. Communication,* **51(3),** 477–497. https://doi.org/10.1111/j.1460-2466.2001.tb02892.x

British Red Cross, 2021. *Free Emergency Apps.* https://www.redcross.org.uk/get-help/prepare-for-emergencies/free-emergency-apps##. Accessed 9-Feb 2021.

Crano, W. D. and X. Chen, 1998. The leniency contract and persistence of majority and minority influence. *J. Personality and Social Psychology,* **74(6),** 1437–1450. DOI:https://doi.org/10.1037/0022-3514.74.6.1437

Crawford M.H., K. Crowley, S. H. Potter, W. S. A. Saunders and D. Johnston, 2018. Risk modelling as a tool to support natural hazard risk management in New Zealand local government. *Int. J. Disaster Risk Reduction,* **28,** 610–619. doi: https://doi.org/https://doi.org/10.1016/j.ijdrr.2018.01.011

Dashiell, J. F., 1930. An experimental analysis of some group effects. *J. Abnormal and Social Psychology,* **25(2),** 190–199. https://doi.org/10.1037/h0075144

Demeritt, D., S. Nobert, H. Cloke and F. Pappenberger, 2010. Challenges in communicating and using ensembles in operational flood forecasting. *Meteorol. Appl.,* **17,** 209–222. DOI:https://doi.org/10.1002/met.194

Demuth J. L., R. E. Morss, J. K. Lazo and C. Trumbo, 2016. The effects of past hurricane experiences on evacuation intentions through risk perception and efficacy beliefs: A mediation analysis. *Wea. Climate Soc.,* **8(4),** 327–344. doi:https://doi.org/10.1175/wcas-d-15-0074.1.

Dow K. and S. L. Cutter, 1998. Crying wolf: Repeat responses to hurricane evacuation orders. *Coastal Management,* **26(4),** 237–251. DOI:https://doi.org/10.1080/08920759809362356

Doyle E. E. and D. M. Johnston, 2011. Science advice for critical decision making. In: Paton D, Violanti JM, editors. *Working in high risk environments: developing sustained resilience.* Springfield, Ill.: Charles C. Thomas Publisher. pp. 69–92. ISBN 9780398086923

Doyle, E. E. H., D. M. Johnston, J. McClure and D. Paton, 2011. The Communication of Uncertain Scientific Advice During Natural Hazard Events. *New Zealand J. Psychology,* **40(4),** 39–50. https://www.psychology.org.nz/journal-archive/NZJP-Vol.40-No.4-Distributionfinalpp39-50.pdf (Accessed 2/9/2021)

Elmore, K. L., Z. L. Flamig, V. Lakshmanan, B. T. Kaney, V. Farmer, H. D. Reeves and L. P. Rothfusz, 2014. MPING: Crowd-sourcing weather reports for research. *Bull. Amer. Meteorol. Soc.,* **95,** 1335–1342. https://doi.org/https://doi.org/10.1175/BAMS-D-13-00014.1.

FEMA, 2020. Common Alerting Protocol, https://www.fema.gov/emergency-managers/practitioners/integrated-public-alert-warning-system/technology-developers/common-alerting-protocol (Accessed 2/9/2021)

Fischhoff, B., P. Slovic, S. Lichtenstein, S. Read and B. Combs, 1978. How safe is safe enough? A psychometric study of attitudes towards technological risks and benefits. *Policy sciences,* **9(2),** 127–152. DOI:https://doi.org/10.1007/BF00143739

Fischoff, B. and A. L. Davis, 2014 Communicating Scientific Uncertainty. *Proc. Nat. Acad. Sci.,* **111,** 13664–13671. https://doi.org/10.1073/pnas.1317504111

Fleischhut, N., S. M. Herzog and R. Hertwig, 2020. Weather Literacy in Times of Climate Change. *Wea. Climate Soc.,* **12,** 435–452. DOI:https://doi.org/10.1175/wcas-d-19-0043.1

Fundel, V. J., N. Fleischhut, S. M. Herzog, M. Göber and R. Hagedorn, 2019. Promoting the use of probabilistic weather forecasts through a dialogue between scientists, developers and end-users. *Quart. J. Roy. Meteorol. S.,* **145(S1),** 210–231. doi:https://doi.org/10.1002/qj.3482

Gardner C. A. and M. C. Guffanti, 2006. *U.S. Geological Survey's alert notification system for volcanic activity.* 2006–3139. https://pubs.usgs.gov/fs/2006/3139/ (Accessed 2/9/2021)

WMO, 2008. *Guidelines on communicating forecast uncertainty.* Geneva, Switzerland: World Meteorological Organisation. https://library.wmo.int/doc_num.php?explnum_id=4687 (Accessed 2/9/2021)

Golding, B., E. Ebert, M. Mittermaier, A. Scolobig, S. Panchuk, C. Ross and D. Johnston. 2019. A value chain approach to optimising early warning systems. Contributing paper to *Global Assessment Report on Disaster Risk Reduction 2019*. UNDRR. 30pp. https://www.preventionweb.net/publications/view/65828 (Accessed 2/9/2021)

Goolsby, R., 2009. Lifting elephants: Twitter and blogging in global perspective, in *Soc. Comput. Behav. Model.*, Springer: pp1–6. https://doi.org/https://doi.org/10.1007/978-1-4419-0056-2_2

Geumil, 2021, Icons for Emergencies, http://guemil.info. (Accessed 7/7/2021).

Handmer, J. and S. Dovers, 2007. The Handbook of Disaster and Emergency Policies and Institutions. Routledge. ISBN-10: 1844073599. ISBN-13: 978–1844073597

Haworth, B. T., 2018. Implications of Volunteered Geographic Information for Disaster Management and GIScience: A More Complex World of Volunteered Geography, *Ann. Am. Assoc. Geogr.* **108**. 226–240. https://doi.org/10.1080/24694452.2017.1321979

Hayes, J., J. Abraham and H. Ahluwalia, 2014. Report of the three special panels on creating a "Global weather and climate ready society." *World Weather Open Science Conf*, Montreal, QC, Canada, WMO/ICSU/EC/NRC, 25 pp., www.cmos.ca/document/2679 (Accessed 2/9/2021)

Hilderbrand, D., 2014. Building a Weather-Ready Nation. WMO Bulletin, 63(2), 26–27. https://library.wmo.int/doc_num.php?explnum_id=6982 (Accessed 8/9/2021)

Hoffrage, U., S. Lindsey, R. Hertwig and G. Gigerenzer, 2000. Communicating statistical information. *Science*, **290**, 2261–2262. DOI:https://doi.org/10.1126/science.290.5500.2261

ITU, 2019. *Guidelines for National Emergency Telecommunication Plans*, https://www.itu.int/en/ITU-D/Emergency-Telecommunications/Documents/2019/NETP_Global_guideline.pdf. (Accessed 9/2/2021)

Janis, I. L., 1982. *Groupthink: Psychological studies of policy decisions and fiascoes* (2nd ed.). Boston: Houghton Mifflin Company.

Johnston D., D. Paton, R. K. Crawford, B. Houghton and P. Burgelt, 2005. Measuring tsunami preparedness in coastal Washington, United States. *Nat. Hazards*, **35**. doi:https://doi.org/10.1007/s11069-004-2419-8.

Joslyn, S. and J. LeClerc, 2013. Decisions with uncertainty: The glass half full. *Current Directions in Psychological Science*, **22**, 308–315. DOI:https://doi.org/10.1177/0963721413481473

Joslyn, S. and S. Savelli, 2010. Communicating forecast uncertainty: public perception of weather forecast uncertainty. *Meteorol. Appl.*, **17(2)**, 180–195. DOI:https://doi.org/10.1002/met.190

Kaltenberger, R., A. Schaffhauser and M. Staudinger, 2020. "What the weather will do"–results of a survey on impact-oriented and impact-based warnings in European NMHSs. *Adv. Sci. Res.*, **17**, 29–38. https://doi.org/10.5194/asr-17-29-2020

Kaplan, A. M. and M. Haenlein, 2010. Users of the world, unite! The challenges and opportunities of Social Media. *Business Horizons*, **53**, 59–68. https://doi.org/10.1016/j.bushor.2009.09.003

Kox, T., H. Kempf, C. Lüder, R. Hagedorn and L. Gerhold, 2018. Towards user-orientated weather warnings. *Int. J. Disaster Risk Reduction*, **30, Part A**, 74–80. https://doi.org/10.1016/j.ijdrr.2018.02.033

Kox, T., C. Lüder and L. Gerhold, 2018. Anticipation and Response. Emergency Services in Severe Weather Situations in Germany. *Int. J. Disaster Risk Science*, **9(1)**, 116–128. https://doi.org/10.1007/s13753-018-0163-z

Kox, T. and A. H. Thieken, 2017. To act or not to act? Factors influencing the general public's decision about whether to take protective action against severe weather. *Wea. Climate Soc*, **9(2)**, 299–315. DOI:https://doi.org/10.1175/WCAS-D-15-0078.1

Kox, T., L. Gerhold and U. Ulbrich, 2015. Perception and use of uncertainty in severe weather warnings by emergency services in Germany. *Atmos. Res.*, **158–159**, 292–301. DOI:https://doi.org/10.1016/J.ATMOSRES.2014.02.024

Krennert, T., R. Kaltenberger, G. Pistotnik, A. M. Holzer, F. Zeiler and M. Stampfl, 2018. Trusted Spotter Network Austria – a new standard to utilize crowdsourced weather and impact observations, *Adv. Sci. Res.*, **15**, 77–80, https://doi.org/10.5194/asr-15-77-2018.

Kuhlicke, C., A. Scolobig, S. Tapsell, A. Steinführer and B. De Marchi, 2011. Contextualizing social vulnerability: Findings from case studies across Europe. *Nat. Hazards.*, **58**. 789–810. DOI: https://doi.org/10.1007/s11069-011-9751-6.

Kuligowski, E. D., F. T. Lombardo, L. T. Phan, M. L. Levitan and D. P. Jorgensen, 2014. *Technical investigation of the May 22, 2011, tornado in Joplin, Missouri.* Final report, national institute of standards and technology. https://doi.org/10.6028/NIST.NCSTAR.3

Lazo J. K., R. E. Morss and J. L. Demuth, 2009. 300 billion served: Sources, perceptions, uses, and values of weather forecasts. *Bull. Amer. Meteorol. S.,* **90(6)**, 785–798. doi:https://doi.org/1 0.1175/2008bams2604.1.

Lazo, J. K., H. R. Hosterman, J. M. Sprague-Hilderbrand and J. E. Adkins. 2020. Impact-Based Decision Support Services and the Socioeconomic Impacts of Winter Storms. *Bull. Amer. Meteorol. S.,* **101(5)**, E626–E639. DOI: https://doi.org/10.1175/BAMS-D-18-0153.1

Lauta, K. C., 2014. *Disaster law.* Routledge, London. ISBN 9781138212336

LeClerc, J. S. and Joslyn, 2012. Odds ratio forecasts increase precautionary action for extreme weather events. *Wea. Clim. Soc.,* **4**, 263–270. DOI:https://doi.org/10.1175/ WCAS-D-12-00013.1

Lindell M. K. and H. Brooks, 2013. An Integrated Agenda for Research on Severe Storms. *Int. J. Mass Emergencies Disasters,* **31(3)**, 429–454. http://ijmed.org/articles/639/download/

Lindell M. K. and R. W. Perry, 2012. The protective action decision model: theoretical modifications and additional evidence. *Risk Analysis,* **32(4)**, 616–632. DOI:https://doi. org/10.1111/j.1539-6924.2011.01647.x

Macintyre E., S. Khanna, A. Darychuk, R. Copes and B. Schwartz, 2019 Evaluating risk communication during extreme weather and climate change: a scoping review. *Health Promotion Chronic Disease Prevention Canada.* **39**(4), 142–156. https://doi.org/10.24095/hpcdp.39.4.06

Maxmen, A., 2019. Surveillance Science. Nature, **569**, 614–617. https://media.nature.com/original/magazine-assets/d41586-019-01679-5/d41586-019-01679-5.pdf

Mileti, D. S., 1999. Disasters by Design: A Reassessment of Natural Hazards in the United States. National Academies Press, 371 pp. https://doi.org/10.17226/5782.

Mileti, D.S. and C. Fitzpatrick, 1992. The causal sequence of risk communication in the Parkfield earthquake prediction experiment. *Risk Analysis,* **12**, 393– 400. https://doi.org/10.1111/j.1539-6924.1992.tb00691.x

Mileti D. S. and J. H. Sorensen, 1990. *Communication of emergency public warnings - a social science perspective and state-of-the-art assessment.* Oak Ridge National Laboratory. 166. https://doi.org/10.2172/6137387

Milgram, S., 1974. *Obedience to authority: An experimental view*: Taylor & Francis.

Morss, R. E., J. K. Lazo and J. L. Demuth, 2010. Examining the use of weather forecasts in decision scenarios: results from a US survey with implications for uncertainty communication. *Meteorol. Appl.,* **17**, 149–162. https://doi.org/10.1002/met.196

Morss, R.E., J. L. Demuth and J. K. Lazo, 2008. Communicating uncertainty in weather forecasts: A survey of the U.S. public. *Wea. Forecast.,* **23**, 974–991. https://doi.org/10.1175/2008 WAF2007088.1

Murphy, A.H., 1993. What is a good forecast? An essay on the nature of goodness in weather forecasting, *Wea. Forecast.* **8**, 281–293. https://doi.org/https://doi.org/10.1175/1520-0434(199 3)008<0281:WIAGFA>2.0.CO;2

Murphy, A. H., 1994. Assessing the economic value of weather forecasts: an overview of methods, results and issues. *Meteorol. Appl.* **1**: 69–74

Mylne, K. R., 2002. Decision-making from probability forecasts based on forecast value. *Meteorol. Appl.,* **9**, 307–315. DOI:https://doi.org/10.1017/S1350482702003043

Mylne, K., 2012, Personal Communication: presentation to Royal Society meeting on Handling Uncertainty in Weather & Climate.

NOAA Social Science Committee, 2016. *Risk communication and behavior: Best practices and research findings.* 63. http://www.performance.noaa.gov/wp-content/uploads/ Risk-Communication-and-Behavior-Best-Practices-and-Research-Findings-July-2016.pdf. (Accessed 9/2/2021)

NOAA, 2016. Risk Communication and Behavior: Best Practices and Research Findings. NOAA Social Science Committee, Washington D.C., http://www.performance.noaa.gov/wp-content/uploads/Risk-Communication-and-Behavior-Best-Practices-and-Research-Findings-July-2016.pdf (Accessed 2/9/2021)

NOAA/NWS, 2017. *National Weather Service enterprise analysis report: Findings on changes in the private weather industry.* Tech. Rep., 24 pp., www.weather.gov/media/about/Final_NWS%20Enterprise%20Analysis%20Report_June%202017.pdf (Accessed 2/9/2021)

NOAA/NWS, 2021. *NOAA Weather Radio.* https://www.weather.gov/nwr/. (Accessed 9/2/2021).

OASIS-Open, 2010. *Common Alerting Protocol version 1.2,* OASIS Standard. http://docs.oasis-open.org/emergency/cap/v1.2/CAP-v1.2-os.html (Accessed 2/9/2021)

Pareek, A. and P. C. Trivedi, 2011. Cultural values and indigenous knowledge of climate change and disaster prediction in Rajasthan, India. *Indian J. Traditional Knowledge,* **10 (1),** 183–189.

Parker, D.J., S. J. Priest and S. M. Tapsell, 2009. Understanding and enhancing the public's behavioural response to flood warning information. *Meteorol. Appl.,* **16,** 103–114. doi:https://doi.org/10.1002/met.119

Pettifer, R.E.W., 2015. The development of the commercial weather services market in Europe: 1970–2012. *Meteorol. Appl.,* **22,** 419–424. DOI:https://doi.org/10.1002/met.1470

Potter, S. H., 2014. *Communicating the status of volcanic activity in New Zealand, with specific application to caldera unrest.* (Ph.D. thesis). Massey University, Wellington, New Zealand. http://mro.massey.ac.nz/handle/10179/5654 (Accessed 2/9/2021)

Potter S. H., 2018. Recommendations for New Zealand agencies in writing effective short warning messages. 28. *GNS Science Report* 2018/02. https://doi.org/10.21420/G20H08

Potter, S., S. Harrison and P. Kreft, 2021. The benefits and challenges of implementing impact-based severe weather warning systems: Perspectives of weather, flood, and emergency management personnel. *Wea. Clim. Soc.,* **13(2),** 303–314. doi:https://doi.org/10.1175/wcas-d-20-0110.1

Potter S. H., G. E. Jolly, V. E. Neall, D. M. Johnston and B. J. Scott, 2014. Communicating the status of volcanic activity: Revising New Zealand's volcanic alert level system. *J. Appl. Volcanology.,* **3(13).** DOI:https://doi.org/10.1186/s13617-014-0013-7

Potter S. H, H P. Kreft, P. Milojev, C. Noble, B. Montz, A. Dhellemmes, R. J. Woods and S. Gaudenling, 2018. The influence of impact-based severe weather warnings on risk perceptions and intended protective actions. *Int. J. Disaster Risk Reduction,* **30** (Special Issue on Weather and Communication), 34–43. doi:10.1016/j.ijdrr.2018.03.031.

Practical Action-Nepal, 2009. *Early warning saving lives,* https://cbdrmplatform.org/resources/early-warning-saving-lives-establishing-community-based-early-warning-systems-nepal (Accessed 3/9/2021)

Practical Action, 2020. *Practical Action and Early Warning Systems* http://hdl.handle.net/11283/622753 (Accessed 3/9/2021)

Preuner P., A. Scolobig, J. Linnerooth-Bayer, D. S. Hoyer and B. Jochum, 2017. A Participatory Process to Develop a Landslide Warning System: Paradoxes of Responsibility Sharing in a Case Study in Upper Austria. *Resources,* **6,** 54. doi:https://doi.org/10.3390/resources6040054

Ripberger, J. T., C. L. Silva, H. C. Jenkins-Smith, D. E. Carlson, M. James and K. G. Herron, 2015. False alarms and missed events: The impact and origins of perceived inaccuracy in tornado warning systems. *Risk Anal.,* **35,** 44–56. https://doi.org/10.1111/risa.12262.

Scolobig A., T. Prior, D. Schröter, J. Jörin and A. Patt, 2015. Towards people-centred approaches for effective disaster risk management: balancing rhetoric with reality, *Int. J. Disaster Risk Reduction,* **12,** 202-212. DOI: https://doi.org/10.1016/j.ijdrr.2015.01.006

Scolobig A., L. Pellizzoni and C. Bianchizza, 2016. Public Participation and Trade-Offs in Flood Risk Mitigation: Evidence from Two Case Studies in the Alps, *Nature Culture,* **11.** DOI: https://doi.org/10.3167/nc.2016.110105

Sharpe, M. A., 2016 A flexible approach to the objective verification of warnings. *Meteorol. Appl.* **23,** 65–75. DOI: https://doi.org/10.1002/met.1530

Shaw, R., 2009. *Indigenous knowledge for disaster risk reduction.* United Nations International Strategy for Disaster Reduction, Policy Note. https://www.preventionweb.net/files/8853_IKPolicyNote.pdf (Accessed 3/9/2021)

Shepherd, T. G., 2019. *Storyline approach to the construction of regional climate change information. Proc. Roy. Soc.* A, **475 (2225)**, 20190013. doi: https://doi.org/10.1098/rspa.2019.0013

Slovic, P., M. L. Finucane, E. Peters and D. G. MacGregor, 2004. Risk as Analysis and Risk as Feelings: Some Thoughts about Affect, Reason, Risk, and Rationality. *Risk Analysis*, **24**, 311–322. https://doi.org/10.1111/j.0272-4332.2004.00433.x

Slovic, P., M. L. Finucane, E. Peters and D. G. MacGregor, 2007. The affect heuristic. *Eur. J. Oper. Res.*, **177**, 1333–1352, https://doi.org/10.1016/j.ejor.2005.04.006.

Standards New Zealand, 2009. *Risk Management-Principles and guidelines.* AS/NZS ISO 31000:2009. https://webstore.ansi.org/Standards/SAI/NZSISO310002009#:~:text=AS%2FNZS%20ISO%2031000%3A2009%20Risk%20management%20-%20Principles%20and,any%20public%2C%20private%20or%20community%20enterprise%2C%20or%20group. (Accessed 3/9/2021)

Stirling, A., 2010. Keep it complex. Nature, **468**, 1029–1031. DOI: https://doi.org/10.1038/4681029a

Sutter, D. and S. Erickson, 2010. The time cost of tornado warnings and the savings with storm-based warnings. *Wea. Climate Soc.,* **2**, 103–112. DOI: https://doi.org/10.1175/2009WCAS1011.1

Sutton J. and C. Woods, 2016. Tsunami Warning Message Interpretation and Sense Making: Focus Group Insights. *Wea. Climate Soc.*, **8(4)**, 389–398. doi:https://doi.org/10.1175/wcas-d-15-0067.1.

Taylor, A., T. Kox and D. Johnston, 2018. Communicating High Impact Weather: Improving warnings and decision making processes. *Int. J. Disaster Risk Reduction*, **30A**, 1–4. doi: https://doi.org/10.1016/j.ijdrr.2018.04.002

Thompson, M., R. Ellis and A. Wildavsky, A., 1990. *Cultural Theory.* Boulder Colo.: Westview Press: Westport, Conn.: Praeger. ISBN 0813378648

Thorpe, A., 2016. *The Weather Enterprise: A Global Public–Private Partnership.* Geneva: World Meteorological Organization, Bulletin, **65(2)**, 16–21. https://elib.dlr.de/135049/1/Carlson-Eyring-WMObull-2016.pdf (Accessed 3/9/2021)

Thorpe, A. and D. Rogers, 2018. The Future of the Global Weather Enterprise: Opportunities and Risks. *Bull. Amer. Meteorol. S.,* **99(10)**, 2003–2008. doi:https://doi.org/10.1175/bams-d-17-0194.1.

UNISDR, 2013. *Implementing the Hyogo Framework for Action in Europe: Regional Synthesis Report 2011–2013*, United Nations International Strategy for Disaster Reduction. https://climate-adapt.eea.europa.eu/metadata/publications/implementing-the-hyogo-framework-for-action-in-europe-regional-synthesis-report-2011-2013/11269598 (Accessed 3/9/2021)

UNDRR, 2015. Sendai Framework for Disaster Risk Reduction 2015-2030. Geneva, Switzerland: United Nations Office for Disaster Risk Reduction, 37. https://www.preventionweb.net/files/43291_sendaiframeworkfordrren.pdf. (Accessed 2/9/2021)

Wachinger, G., Renn, O., Begg, C. and C. Kuhlicke, 2013. The Risk Perception Paradox—Implications for Governance and Communication of Natural Hazards. *Risk Analysis,* **33(6)**, 1049–1065. DOI:https://doi.org/10.1111/j.1539-6924.2012.01942.x

Weyrich, P., A. Scolobig, D. N. Bresch and A. Patt, 2018. Effects of impact-based warnings and behavioral recommendations for extreme weather events. *Wea. Clim. Soc.,* **10**, 781–796. DOI:https://doi.org/10.1175/WCAS-D-18-0038.1

Weyrich, P., A. Scolobig and A. Patt, 2019. Dealing with inconsistent weather warnings: Effects on warning quality and intended actions. *Meteorol. Appl.*, **26**, 569–583. DOI: https://doi.org/10.1002/met.1785

Weyrich P., A. Scolobig, F. Walther and A. Patt, 2020a. Do intentions indicate actual behaviours? A comparison between scenario-based experiments and real-time observations of warning response. *J. Contingencies Crisis Management*, **28**, 240–250. https://doi.org/10.1111/1468-5973.12318

Weyrich P., Scolobig A., Walther F., Patt A., 2020b, Responses to severe weather warnings and affective decision-making. *Nat. Hazards Earth Syst. Sci.*, **20**, 2811–2821. https://doi.org/10.5194/nhess-20-2811-2020

Williams, C. A. and G. M. Eosco, 2021. Is a Consistent Message Achievable?: Defining "Message Consistency" for Weather Enterprise Researchers and Practitioners. *Bull. Amer. Meteorol. S.*, **102**, E279-E295. https://doi.org/10.1175/BAMS-D-18-0250.1

WMO, 2013. *Guidelines for Implementation of Common Alerting Protocol (CAP)-Enabled Emergency Alerting*. WMO-No. 1109, available at: https://library.wmo.int/?lvl=notice_display&id=14699 (Accessed 18/4/2021).

WMO, 2015. *Guidelines on Multi-Hazard Impact-Based Forecast and Warning Services*, 2–15. WMO-No. 1150, available at: https://library.wmo.int/doc_num.php?explnum_id=7901 (Accessed 18/4/2021).

WMO, 2018. Multi-hazard Early Warning Systems: A Checklist, Outcome of the first Multi-hazard Early Warning Conference, 22-23 May 2017, Cancun, Mexico. Geneva, Switzerland: World Meteorological Organization, 20. https://etrp.wmo.int/pluginfile.php/21553/mod_page/content/18/MultihazardChecklist.pdf (Accessed 2/9/2021)

Wood M., H. Bean, B. F. Liu and M. Boyd, 2015. *Comprehensive testing of imminent threat public messages for mobile devices: updated findings*. MD, USA. https://www.dhs.gov/sites/default/files/publications/WEA%20-%20Comprehensive%20Testing%20of%20Imminent%20Threat%20Public%20Messages%20for%20Mobile%20Devices%20Updated%20Findings.pdf (Accessed 3/9/2021)

Wood M. M., D. S. Mileti, H. Bean, B. F. Liu, J. Sutton and S. Madden, 2017. Milling and Public Warnings. *Environment and Behavior*. **50(5)**, 535–566. doi:https://doi.org/10.1177/0013916517709561.

World Bank, 2019. *Weathering the Change: How to Improve Hydromet Services in Developing Countries?* Washington, DC: World Bank. https://www.gfdrr.org/en/publication/weathering-change-how-improve-hydromet-services-developing-countries (Accessed 3/9/2021)

Zhang, Q., L. Li, E. Ebert, B. Golding, D. Johnston, B. Mills, S. Panchuk, S. Potter, M. Riemer, J. Sun, A. Taylor, S. Jones, P. Ruti and J. Keller, 2019. Increasing the Value of Weather-Related Warnings". *Science Bulletin*, **64**, 647–649, ISSN 2095-9273. https://doi.org/10.1016/j.scib.2019.04.003.

Chapter 4
Connecting Forecast and Warning: A Partnership Between Communicators and Scientists

Cheryl L. Anderson, Jane Rovins, David M. Johnston, Will Lang,
Brian Golding, Brian Mills, Rainer Kaltenberger, Julia Chasco,
Thomas C. Pagano, Ross Middleham, and John Nairn

Abstract In this chapter, we examine the ways that warning providers connect and collaborate with knowledge sources to produce effective warnings. We first look at the range of actors who produce warnings in the public and private sectors, the sources of information they draw on to comprehend the nature of the hazard, its impacts and the implications for those exposed and the process of drawing that information together to produce a warning. We consider the wide range of experts

C. L. Anderson (✉) · J. Rovins
Massey University, Palmerston North, New Zealand

D. M. Johnston
Massey University, Palmerston North, New Zealand

WMO/WWRP HIWeather project, Geneva, Switzerland

W. Lang · R. Middleham
Met Office, Exeter, UK

B. Golding
Met Office, Exeter, UK

WMO/WWRP HIWeather project, Geneva, Switzerland

B. Mills
University of Waterloo, Environment and Climate Change Canada, Waterloo, Canada

WMO/WWRP HIWeather project, Geneva, Switzerland

R. Kaltenberger
Zentralanstalt für Meteorologie und Geodynamik, Vienna, Austria

WMO/WWRP HIWeather project, Geneva, Switzerland

J. Chasco
National Meteorological Service of Argentina, Buenos Aires, Argentina

WMO/WWRP HIWeather project, Geneva, Switzerland

T. C. Pagano · J. Nairn
Bureau of Meteorology, Melbourne, Australia

© The Author(s) 2022
B. Golding (ed.), *Towards the "Perfect" Weather Warning*,
https://doi.org/10.1007/978-3-030-98989-7_4

who connect hazard data with impact data to create tools for assessing the impacts of predicted hazards on people, buildings, infrastructure and business. Then we look at the diverse ways in which these tools need to take account of the way their outputs will feed into warnings and of the nature of partnerships that can facilitate this. The chapter includes examples of impact prediction in sport, health impacts of wildfires in Australia, a framework for impact prediction in New Zealand, and communication of impacts through social media in the UK.

Keywords Warning producer · Impact · Communication · Social media · Trust · Information broker · Tailored warning · Evaluation

4.1   Introduction

This chapter examines the ways that warning providers connect and collaborate with impact experts to improve and communicate warnings. We first look at the role of the warner, then the impact forecaster and finally the linkages between them. In doing so, we shall see that:

- A successful warning is used to take action. It is as much a compelling narrative as information.
- A skilful impact forecast identifies who or what will be impacted, where, when and by how much.
- Impact data are often confidential, requiring partnership with the data owner and a clear mutual understanding of the objectives of any impact prediction tool.
- Partnerships between information producers and warning communicators can be facilitated by intermediaries.

4.1.1   Warnings

Warnings are produced by a wide range of actors in the public and private sectors, based on information from weather and hazard forecasts, on science related to weather or hazards and on estimates of the anticipated impact of the hazard, produced using a variety of tools and technologies. The intent is to provide enough lead time to reduce the risk from the hazard and thus prevent a disaster. Expert risk information must be presented in a form that enables the creation of a convincing warning narrative that ultimately supports decision-makers and encourages action. Research in the last decade has demonstrated that strengthening warnings with impact information significantly reduces the disaster (WMO 2015, Casteel 2016, Anderson-Berry et al. 2018, Potter et al. 2018).

In producing a warning, the warner is as much an artist as a scientist, crafting a persuasive story out of a selection of uncertain facts, using their experience of context and precedent and fitting the result into a variety of formats to be delivered

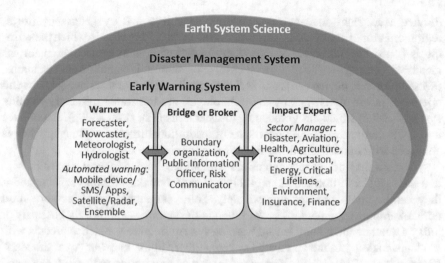

Fig. 4.1 Interactions from the Warner to Impact Expert embedded in complex systems. (Source: Adapted from Ruti et al. 2020; Beaven et al. 2016; Pielke Jr. 2007)

through different media, all while under considerable time pressure. Warning information must reach decision-makers in broad, varied disciplines and fields of service and expertise. The available information is constrained by forecasting capabilities, which vary with lead time and location and which may be more or less relevant to the end user. Temporal scales of the potential threat and the time taken to issue warnings add further complexity. Partnerships that help the warner to have confidence in the sources of their information are critical.

The "bridges" connecting the information provider to user may be complex because of their situation within complicated, embedded systems and should therefore be designed prior to hazard occurrence (see Fig. 4.1). Research shows that intermediaries, which may be organisations or individuals within an organisation, can aid in connecting people throughout the warning chain.

4.2   The Warner and Warning Information

4.2.1   The Warner

The warner of each potential threat will vary with the type of threat and the roles that define positions in organisations and governments, including authority to provide warnings and the systems in which the threat is evaluated. Warnings may be categorised by the type of threat (hazard type and complexity, science and technology that provides analyses), by role (forecaster/nowcaster, modeller, public information officer/risk communicator, emergency manager), by discipline (hydrology, meteorology, physical science, social science), by authority (Meteorological

Service, emergency management, government entity) and by geography (local, regional, national or transboundary scale of the threat). The ways in which the warning is issued depend on all of these factors with additional consideration for the means of warning – official press release, television/radio announcements, sirens, SMS/DM mobile device alerts and social media. These researchers, technicians and operators, information systems and technologies are integrated into early warning systems (EWS) which aid in reducing disaster risks (Tan et al. 2020a). Experience in the UK demonstrates the importance of institutional trust in warnings, which is enhanced with impact-based warnings (Taylor et al. 2019).

Warnings may be entirely automated without any human input if a warning system has sufficient information and analytical skill to make warnings reliable and if they can be communicated appropriately. Some situations – such as very short notice warnings – are perhaps better suited to the automated approach, but many of today's weather and natural hazard warning systems require a mixed approach, with the human adding to the automated system, either "in the loop" or "over the loop" (Pagano et al. 2016). Issuing warnings of this type requires expertise in science and the art of communication in equal measures. It is not a purely *mechanistic* process, which can be easily automated, and there will always remain an element of subjectivity, but it should at least be a *methodical* process. The methods adopted will vary depending on circumstances, but all should look to ensure a balance between the scientific, the practical and the useful and should ensure a level of consistency by limiting differences of opinion, biases and risk appetites.

Invariably, no one person has all of the expertise, information or experience across these fields, which is why warning creation needs to be a collaborative and multi-disciplinary process. The resulting diversity of perspectives, experiences and insights is both a strength and a challenge of this collaborative approach. The warner operates within guidelines of EWS design, which may be an automated system managing big datasets and information sources, or a human forecaster analysing hydrometeorological conditions, and/or may further contain interpretation of the nature of the impact. The means of communication and the target audience must also be considered, together with wider aspects of decision-making from the individual to broader governance systems (Tan et al. 2020b).

4.2.2 Warning Content

The warning must focus on what the warner is trying to achieve with the warning. Information is needed both as warning content and for decisions on the importance and timing of the warning. The warner must trust their information sources (official and unofficial), be able to select key aspects of information received and determine ways to interact with information to produce the warning in different situations and contexts.

"An effective warning…specifies the exact nature of the threat" (Casteel 2016). The content should be clear and understandable. The more that the warning includes information about the hazard impact, the more actionable it is and the more effective the warnings become (Lazo 2020; Potter 2018). "The increased specificity provided by the "hazard" portion…should therefore enhance personal relevance and potentially increase the likelihood that the recipient takes protective action" (Casteel 2016).

Even though technological advances have improved the information available, it must be interpreted for sector-specific use. Disaster managers and emergency responders need to know the nature, severity and geographical extent of the disaster, the timing (start and end) and how soon evacuations must occur. The agriculture sector needs to know the likelihood of a threat such as flooding, its geographical extent (how many crops will be affected) and scale (one district or several, cross-jurisdictional boundaries) to effectively enact readiness measures that protect crops, livestock and agricultural livelihoods. The energy sector requires knowledge of the timing and type of event to ensure that energy can be supplied to users, including emergency responders and at-risk populations.

The information required by the warner evolves as the threat approaches. At the early warning stage, there are likely to be few sources, though they may reach the warner through multiple routes. At this stage there is also more time available to refer back to experts for clarification and to carefully craft a convincing narrative for the receiver. As deadlines for specific actions approach, the message needs to be oriented to the impacts that are relevant to those actions, with emphasis on the level of confidence and on alternative outcomes. Once the threat is imminent, there is no longer time for careful analysis, but details of changes in track, intensity, timing and associated impacts may be critical to responders' actions and safety. At this stage the warner will look for multiple data sources, up and down and outside the warning chain, to maintain situational awareness: of the hazard, of responses to warnings and, once it arrives, of the actual impacts. These will all inform warning updates.

4.2.3 Warning Creation

In preparing a warning, the warner aims to create a compelling narrative that will convince the receiver to take notice and then to take action appropriate to their situation. Selecting the warning level is a critical part of the process. It requires careful interpretation of the available information, especially the predicted impact and the level of confidence, as well as the context of the warning, for instance, if those affected are already dealing with the impacts of another hazard. The content of the warning is selected and presented with all this in mind. Central will be the expected impact on the receiver. In support will be sufficient information on the causal hazard including both the prediction and its confidence, supported, when available, by evidence (e.g. links to CCTV at upwind/upstream locations). Where specific vulnerabilities are relevant, these need also to be included, together with the level of confidence. Where there are significant uncertainties in the impacts, the range of outcomes may be usefully represented by two or more scenarios or storylines, while stressing the need for preparation ahead of the situation becoming clearer.

Taking a typical impact-based weather warning system as an example, it is useful to subdivide the warning process into distinct components.

Weather   assessment → *Risk   assessment* → *Change   assessment* → *Utility assessment*

- Weather assessment: What the weather will be.
- Impact/risk assessment: What the weather will do.
- Change assessment: Does this change my perception of the "story"?
- Utility assessment: Who needs to know, and how?

Most studies have focused on the weather and risk assessment aspects, but equal emphasis should be given to consideration of the latter components which are related to decision-making and communication.

The range of sources of information for the weather assessment can be very large, typically greater than can realistically be absorbed and processed by even the most experienced meteorologist. For this reason, the information must be filtered, either by limiting the sources used (procedurally or technically) or through intermediary systems which can sift and extract signals from, and summarise, the information.

A range of inputs may be used for the impact/risk assessment:

- Modelled.
- Empirical/heuristic based on individual or collective experience.
- Specific, current knowledge modifiers relating to exposure, vulnerability or the prediction of the hazard itself (e.g. based on assessment of current model performance).

One of the most difficult challenges for the warner is to acknowledge that their assessment has changed sufficiently to change the warning. Once a hazard "story" has been defined, it possesses inertia; it can be difficult to accept that it no longer reflects the best interpretation, even in the face of new, conflicting data. This is the psychological phenomenon known as "anchoring", which, among other things, is why forecasts and warnings can be most prone to change following handovers between shifts.

Even when the warner feels that other criteria for issuing a warning may have been met, they must consider whether the warning information will be useful to end users. This consideration acts as another filter, with the warner playing an editorial role to determine what user(s) need to know. For example, matrix impact-based warning systems should arguably result in far more long-period, low-probability warnings than they do. This is because warners judge that issuing too many warnings is counterproductive.

Warnings are not issued in isolation. They exist in relation to other warnings, to additional communications and of course to the previous and future versions of warnings for the same event. While a warning to take protective action now for high weather impacts in the next 6 hours may be usable by some recipients, it gives little time for preparation. In this sense, no warning should come as a surprise! So the warner should use a succession of communications: advisories, watches, warnings, press releases, blogs, tweets and other advice to manage uncertainties and expectations well before the event, such that the final "take action now" message is expected when it comes. Best practice is to think strategically over the whole period from initial indications of severe weather up to the event itself and plan the issue of warnings and other communications to best inform users and allow them to prepare while avoiding overcommitting resources should the worst conditions *not* occur.

The key to preparing high impact weather warnings is the development of expertise: in the needs of the warning's users, in the reliability of the various sources of guidance (both meteorological and socio-economic) and in the behaviour of the weather system causing the hazard. Expert warning forecasters assimilate the incoming data, generating a conceptual model of the situation and enabling a level of situational awareness sufficient to anticipate events (Klein 1989). Anticipation enables the expert warner to filter the voluminous information and focus on the most relevant aspects for use in decision-making. Decisions are always subject to judgement in the face of uncertainty. Given the uncertainty and the impact of warning decisions, there is never enough information.

Looking forward, research is focusing on using machine learning systems to undertake the process of filtering, so as to identify the key areas of risk uncertainty requiring human judgement. One way of facilitating this is for hazard and impact predictions to be formulated into a first-guess warning, combining the probabilistic and severity elements. In some situations, this might provide the route towards fully automated warnings, but more generally it should be accompanied by tools for the warner to probe the individual components, assess the sensitivity of the outputs and amend the resulting warning.

4.2.4   Tailored Warnings

"Forecasts will occasionally take into account some societal factors (e.g., extending a warning's timing to cover when schools are releasing students), but often do not directly account for human factors related to decision-making prior to, and during life-threatening extreme events" (Uccellini & Ten Hoeve 2019). Engaging with social scientists to work with specific groups of users in the design of tailored warnings can lead to better warning responses. Research shows that tailoring warnings to the needs of recipients increases their effectiveness. However, this benefit, in better warning response, has to be set against the cost, complexity and potential for inconsistency of doing so. The benefit is not restricted to economics but may include a variety of non-economic benefits and ethical issues of human rights as well. We consider several aspects of tailoring here. The first few options relate to selection of data, while the latter ones relate to presentation.

Selecting the Best Forecast for the User   When working with a decision-maker to improve their access to predictions of hazards, the basis for selecting the source of that information will include several factors, of which accuracy or skill is likely to be an important one. On the other hand, a provider of information, trying to optimise the value to users of the information they provide, has to choose which improvements to their prediction system to implement. In both cases it may appear to be beneficial to focus the evaluation on the conditions of interest to the user, i.e.

the hazard itself or perhaps the extreme weather conditions that give rise to the hazard – both things that occur rarely.

If the exercise is being undertaken following a disaster, it may even be felt that the most important consideration should be that the disaster would have been predicted with the new system or data source. This approach can be very dangerous, typically leading to over-prediction and loss of confidence in the warning system. While there are unbiased methods of assessing extreme values of weather variables and of assessing the skill in predicting occurrence of an event, in both cases they are not immediately intuitive and so are not the most widely used evaluation approaches. In addition, the small number of data points available for extreme events means that the error bars in the score are likely to be very broad.

Even if different information sources are reliably identified as giving the best guidance for different phenomena, the risk of inconsistent predictions is considerable. For instance, if one information source gives better hurricane track predictions and another gives better hurricane intensities, it would be foolish to rely entirely on the track from one and the intensity from the other. For the warner this means that decisions on the scenario for the warning should be based on as much information as possible about the current situation, about the evolution predicted by each source and about the performance of that source in similar situations. It is essential that this information includes full probability distributions and that biases have been removed.

Selecting the Information to Communicate Having obtained a skilful forecast source, there are a myriad of products that could be extracted. Information for inclusion in warnings is unlikely to be the same as that used for routine forecasting. For instance, extracting the probability of key thresholds being passed enables the user to focus immediately on their specific concerns. Having said that, it is advisable to standardise if possible, so that users receiving warnings that contain different thresholds do not perceive an inconsistency. Other means of tailoring the information include recalibration, bias correction, and derivation of user-relevant variables (both physical and socio-economic). Thus a flood forecast may be presented as flow in the river, as water level above the river bed, as a map of flood depth, or more specifically in terms of the depth on roads to the Highways Department, the extent of residential property flooding to the public, and the probability of reaching a critical depth at an electricity substation to a power company. In the pressured environment of an emergency response team, the more precise and actionable statements are likely to produce more effective responses with less room for error. It is recommended that warnings of the highest identified risks should use tailored prediction products incorporating the probability of specific hazard thresholds being crossed and information about exposure and vulnerability of communities in the threatened areas.

Geographical Tailoring When considering tailoring of information, making allowance for geography is perhaps the most obvious and most necessary. For instance, when forecasting wintry weather in complex terrain, low-lying valleys may have rain, while higher up the slopes, snow is accumulating. The meteorologist may use the height of the snow line, but to communicate this may require particular locations to be identified. Proximity to rivers is a key driver of risk from flooding, but few people know the distance to their nearest river and city dwellers may not realise that a river exists, let alone that they could be flooded by it. Warnings that refer to settlements at risk or escape routes that are safe or threatened may use names that are only current in the immediate vicinity, so that visitors are unaware of their meaning. For instance, reference to numbered highways or to numbered junctions on a highway will be incomprehensible to a portion of the travelling public. Maps can help in avoiding these difficulties, provided they are clearly located relative to major towns, highways and other widely known features.

Tailoring the Communication This area offers the greatest opportunity for tailoring with the minimum risk of confusion. For instance, warnings should be disseminated in multiple languages, according to the make-up of the population, and through different media (newspaper, TV, radio, Internet, mobile app, social media, etc.) according to accessibility by the population. It is crucial to use geographical names that are generally understood. Maps can be powerful tools for communicating the proximity of a warning area to dispersed communities – but only to those that are able to read them. Colours can provide powerful support to communication - the green-amber-red sequence of traffic lights is understood in many cultures, but not all, and care must be taken to cater for those with colour blindness. The use of cartoon characters to communicate has been very successful in some cultures, but not all. Since users will often seek confirmation from friends and family before responding, it is important that different means of communicating the information are consistent.

Other Areas for Tailoring Many aspects of warning design affect how particular groups receive, interpret and respond to information. Cultural cues can be important, e.g. the colour red has particular and conflicting cultural meanings. Similarly, the idioms used in the language can be as important as the words. Phrases such as "snowing handkerchiefs" or "raining cats and dogs" are meaningful to some and not to others. Gender is of particular importance in most countries. However, when considering tailoring for women, it is necessary to consider the route by which the warning will reach them. A direct route, e.g. by social media, will require different tailoring from an indirect one, e.g. through a village chief or street warden. Another potential area of tailoring comes from study of the psychological response to challenging situations. Some people typically respond positively, seeking to turn it to their advantage, while others are followers of the crowd, and yet others will fight against change. In the West, marketing companies have learnt to target these groups differently, and it is likely that similarly targeted warning messages may be effective, though research has yet to demonstrate this.

Tailoring for Specialists Where there is an emergency manager for a large infrastructure facility that will affect thousands of people, the case for tailoring very specifically to that role's needs is very strong. It is essential that the response is pre-planned and that it is carried out quickly and effectively when the warning threshold is reached. This may involve simplifying the warning down to a simple code word, which is learnt and practised by each provider and user. The same holds for organisers of large public events such as pop festivals. Tailoring for major public facilities such as schools or hospitals is more complex, but equally important. Candidate specialist users for tailored warnings include power generators and suppliers, water suppliers, dam operators, telecommunications operators, road transport, rail transport, air transport, marine transport, food retailers, education, emergency responders, health services, waste disposal, public event organisers, major employers and businesses. Such users should not be using generic public warnings to take decisions. They should have carried out a risk assessment for their business, which identifies the hazards they are exposed to and the level of risk for each. They should also have a risk response plan including trigger points at which action must be taken, together with the information needed, both to identify the trigger and to inform the action. The receipt of a tailored warning should be the primary trigger for preparatory actions ahead of weather-related hazards. Activation of response plans should be tied to the receipt of a warning.

Co-Design in Generic Warnings Currently a high degree of tailoring cannot be justified for public warnings. The alternative is to co-design a compromise generic warning system that meets most needs and to use education to embed its characteristics in the users' cultures. Such co-design activities must be very carefully planned to ensure an adequate voice for all sections of the community. Evidence also suggests that a feedback loop is required in which community representatives first identify what they feel are the problems with current capability and then criticise successive sets of upgrade options in an iterative fashion. Not only does this help to produce effective warnings, but it also builds a sense of ownership in the community that helps with the adoption and use of the final product. Looking to the future, social profiling in combination with machine learning techniques, e.g. as used for selecting online advertisements, has the potential to enable individual tailoring of weather information based on individual risk profiles. However, the implications of getting it wrong mean that warnings are likely to adopt such techniques more slowly than other environmental forecasting services.

4.2.5 Evaluating the Warning

While evaluation is important throughout the entire warning value chain, it is particularly critical where risk information is translated into actionable messages for those at risk and those responsible for mitigating and managing hazard-related impacts (e.g. emergency managers). NMHSs have a long history of using statistical methods to verify weather forecasts (Ebert et al. 2015), but relatively less experience in evaluating the use and efficacy of their products and services.

While it would be desirable to demonstrate benefit by observing a decreasing trend in metrics of death, distress and damage as warnings improve, it is rarely possible to do this. Since the objective of warning is to help the recipient make better decisions, surveys of people's actual receipt and reaction to warnings are probably the best available tool. A baseline is required, so surveys should be designed and established before the introduction or upgrade of a warning system and continued after it is complete, using the same format throughout. If surveys have been part of a co-design process, it may be appropriate to continue these, bearing in mind the difference between anticipated response and actual response.

By definition, evaluations are comparative: over time; between places, jurisdictions and populations; or between other contrasting features or situations. For example, the introduction of a warning service or modification to add impact information should be evaluated using assessments before and after service implementation. Traditionally, one-off evaluations often take place long "after-the-fact" making it difficult to collect and interpret the information. Ideally, evaluation should be undertaken continuously throughout the life cycle of any significant change, permitting course corrections as the service is developed and introduced.

A useful approach is to treat weather warnings as a form of programme intervention, not unlike a campaign to increase vaccination uptake or use of masks in disease prevention. Whether explicitly or implicitly defined, weather warnings are provided to influence awareness, risk perception, behavioural intent, decisions, behaviours and, ultimately, outcomes—all of which can potentially be measured. The theory of change is a methodology for planning and evaluating social change programmes (see, e.g. Taplin & Clark 2012) that is now widely used in international development and is very relevant to the challenge of evaluating warnings. It approaches a social intervention of any kind by first determining the desired outcomes and then associating measurable success indicators with each. It involves documenting the actors and processes through which a service is expected to affect outcomes, together with any intermediary factors (e.g. awareness and comprehension of warning information, trust, beliefs, etc.). The analysis may draw on personal experience, expert opinion or evidence and models obtained from social science research (e.g. Theory of Planned Behaviour, Ajzen 1991; Risk Information Seeking and Processing, Griffin et al. 1999).

The process of confirming a theory of change naturally leads to working hypotheses that may be examined and tested using qualitative and quantitative research. Each approach has strengths and weaknesses, so it is beneficial to adopt multiple lines of inquiry and triangulation over the course of an evaluation. For instance, observational field research, focus group sessions and mental modelling interviews (Morgan et al. 2002) are often helpful in documenting change theories and underlying constructs among those who are developing, providing and utilising warning services. Surveys, however, may be better suited to assess the representativeness of such findings across groups of actors (e.g. emergency managers, Hoss et al. 2018) and the effect of intermediary factors (e.g. trust, perceived threat) on behavioural intent and recalled responses and outcomes, particularly following memorable severe events (e.g. Winter Storm Doris, Taylor et al. 2019).

Experimental research using hypothetical or simulated scenarios allows for selective control of variables that might influence protective decisions and so is particularly useful in comparing multiple formats and content options prior to implementation (e.g. Casteel 2018, Potter et al. 2018). The disadvantage of such flexibility is that hypothetical situations may not adequately capture the context and responses that only fully materialise during actual threat events (Weyrich et al. 2020a). More generally there is a question as to how well stated intent and recalled responses correspond to actual behaviour (Weyrich et al. 2020b). Both limitations can be partially alleviated through the application of experience-based sampling techniques (Hektner et al. 2007) that attempt to measure warning-related variables in near real time. Finally, it is also possible to understand warning efficacy through analyses of behavioural outcomes (e.g. injuries, damage) using cohort or case-control observational study designs (e.g. fall-related and motor vehicle collision injuries, Mills et al. 2020).

4.3 Capabilities of the Impact Forecast

The impact forecast enhances the underlying hazard forecast, incorporating information on vulnerability and exposure to estimate the impact of the hazard. Impact experts provide critical information to "core partners responsible for public safety". Impact-based decision support services help to better understand and utilise forecasts and warnings when dealing with extreme events (Lazo et al. 2020).

Currently impact information is often generated by the warner based on their experience of previous events and is thus limited by the experience of each warner or their understanding of their relevance to the current situation. Sometimes these analogues may be documented and semi-quantitative (e.g. US water supply forecasts put the current forecast on a scatterplot relative to past years).

"A growing number of experts are suggesting that standard warning information should be augmented with additional information about these factors" (Weyrich et al. 2018). Their expertise is often applied offline to develop tools that either enable the forecaster to convert a hazard forecast into an impact forecast or enable the decision-maker or warner to convert the decision threshold into a hazard threshold. For instance, people are increasingly making real-time forecasts of hurricane damages, particularly in the USA (e.g. this hurricane is expected to cause $750 million in damages if it follows the expected track).

Since they are often not involved in the real-time issue of warnings, the relationship of an impact expert may be more academic and detached than that of some of the other actors in the chain. On the other hand, their studies likely include analysis of events that caused major social and economic loss. By developing warnings within specific hazard early warning systems, the warnings for a single event may link to consequences of actions and decisions and will be better able to deal with potential impacts of cascading events where multiple responses from different sectors will be necessary.

4.3.1 Sources of Impact Information

A fundamental limitation to our ability to estimate impacts comes from the lack of routine collection of data on weather-related socio-economic impacts. Chapter 5 will cover this in more detail, but most available data are highly aggregated – national scale census, production, health, infrastructure performance, etc. More local data are typically not available except to specifically accredited researchers. For health data, this is because of patient confidentiality, while for infrastructure and business performance, it is to preserve commercial confidentiality. The result is that models generated using these data sources cannot be replicated or inter-compared, while those from open sources are mostly too coarse to be useful.

Attempts have been made to overcome this barrier using media reports to catalogue impacts. This approach is used in the International Disasters Database (EM-DAT) (https://www.emdat.be) coordinated by the Centre for Research on the Epidemiology of Disasters (CRED) within the Université Catholique de Louvain in Brussels and the United Nations to categorise and identify disasters globally. While originally dependent on manual interpretation, recent research has demonstrated the use of automated methods for classifying reports. According to a recent survey, 19% of NMHSs in Europe are collecting media reports to an in-house impact database, 9% are storing impact observations from storm spotter organisations and 13% are collecting other types of human impact observations (Kaltenberger et al. 2020). Among other sources, media reports of impacts of severe weather are systematically monitored, quality checked and fed into the European Severe Weather Database (ESWD, e.g. Dotzek et al. 2009). Some NMHSs and DRM organisations in least developed countries are also using media reports to gather impact information for use in establishing impact-based warning services.

Another approach that is likely to grow in the future uses automatic data collection from the Internet of Things. For instance, autonomous vehicles carry sensors for the weather, but also record information about speed, traffic density, etc. Taken together such data could provide a major step forward, both for training impact models and for evaluating forecasts and warnings of highway conditions.

4.3.2   Capabilities of Different Impact Estimation Methods

We can identify some key aspects of impact estimation tools that affect performance. The strongest evidence comes from repeatable laboratory testing and is often used as the basis of impact estimation for engineered structures. Certainly, it is important to know the failure modes and thresholds of the materials of concern. However, reproducing conditions in the real world is very demanding, e.g. ageing of materials, combinations of wind and rain, the complex motions of the sea against a barrier, etc. In designing a structure, the remaining unknowns are often dealt with by adding a safety factor. An appropriate way of dealing with this needs to be included in any failure prediction tool. Examples of this approach are wind impacts on concrete bridges, flood impacts on retaining walls and wind impacts on moving vehicles.

Where there is a clearly identifiable set of processes leading to failure, it may be possible to model these and to calibrate the model parameters using experimental evidence. For instance, the ways in which flood water damage a building are well established for particular construction methods, so a relation between flood depth and cost of recovery can be developed (Penning-Rowsell et al. 2013). In a similar way, the response of networks can be modelled. So, for instance, if a road is closed by a fallen tree or accident, or if a telecommunications link is broken, the resulting impact on road or communication traffic can be modelled as a function of the expected loadings for the time, day and season. This approach can also be used for the spread of vector-borne diseases if the behaviour of the vector (which is typically weather-dependent) can be modelled. These approaches have several limitations. Models are unlikely to be complete, so missing processes may, on occasion, be significant. The models and their parameters are typically validated for a limited range of conditions – which may not include extremes. To minimise the risk of misleading information, models should be run with hazard inputs sampled from across the uncertainty range – preferably from an ensemble prediction system – and using a range of parameter settings consistent with the validation data. The resulting probability distribution should then be interpreted for use in the warning, e.g. by extracting the most likely, the probable worst case or the probability of exceeding a particular damage threshold.

In most cases, however, the processes are hidden or too complex for modelling. In that case, prediction tools must rely on the statistical analysis of historical data to extract relevant relationships. This approach is most developed in the field of epidemiology (Armitage, Berry & Matthews 2002), but similar tools apply in many other impact areas including in the atmospheric sciences (Wilks 2006). Traditional approaches have been based on fitting an appropriate statistical distribution to data by selecting the parameters of the distribution that optimise the fit. Increasingly, these approaches are being replaced by machine learning techniques such as neural networks. In order to extract a useful relationship, data should be pre-processed to remove the influence of extraneous factors, such as time of day, day of the week, holiday periods, policy changes, etc., and to remove trends. It is also essential that all factors that may be expected to influence the impact data are represented. For instance, if a stormy period is being compared with a non-stormy period to study the relationship between weather and accidents, the different mix of people travelling – perhaps less elderly or less women – could bias the results unless allowed for in the analysis. Like the process models, the resulting statistical models should not be used in parameter ranges that are rare or missing in the training data. Standard statistical techniques can be used to assess the uncertainty in the association, and this information should always be incorporated in any predictive model so as to avoid overconfidence.

The normal statistical approach is to look for a repeatable association between the hazard and its impact. We might call this the forward model. However, where there is a unique decision to be made at a specific threshold, it may be more appropriate to predict the probability that this threshold will be exceeded. This involves less complex statistical analysis and provides the probabilistic information directly. However, the influence of confounding factors, trends, etc. can still produce misleading results.

All statistical models must be evaluated using a dataset that is uncorrelated with that used for training the model. It must also be large enough to provide statistical significance in the parameter ranges of importance for hazard impacts.

4.3.3 Sector-Specific Impact Tools

Risks vary across sectors and policy areas, such as health, environment, water/power supply, transport, technology, security, insurance and finance, so this is the first aspect that must be considered before developing warnings (Eiser et al. 2012). In some sectors, advances in data gathering, modelling and computing have increased the ability to provide critical data in their decision-making timeframes (Ruti et al. 2020, Yu et al. 2018). The health sector will be concerned with impacts of death and injury, need and capacity for hospital admissions and services and use impact assessment tools such as epidemiology, transmission, and exposure. The energy sector may be concerned with circuit failures and loss of service to critical users and will rely on detailed engineering modelling. The water sector will have numerous types of threats from lack of supply for drinking and for critical infra-structure support, threats from drainage overflow and contamination and additional health threats; therefore, the water sector must be engaged in detailed modelling of infrastructure. Emergency management is concerned with threats to all critical infrastructure, lifelines and services, such that problems with transport, power, water, energy, agriculture, environment and financial protection must be factored into the types of threats, but also impacts that may result in cascading events and multiple types of emergencies.

4.4 Structures that Facilitate Warning Information

The concept of a bridge between warner and impact expert, across which informa-tion flows back-and-forth, reflects a much more complex reality of multiple connec-tions between different types of warners and numerous experts using multiple communication tools for interaction. It is important to have "an integrated warning system that is built on social science research and ensures full communication between all actors throughout the entire emergency management process" (Lazo et al. 2020). In recent research, the development of mitigation actions emerges from inputs of forecasts and warnings through impact-based decision support services, which feed into reducing asset damage, service interruptions and human health.

4.4.1   Relationship Between Information Provider and Warner

The relationship between information provider and warner is critical, and it is important to understand the research structures and methods of working that facilitate the applicability and application of research. Frameworks that link members through early warning systems, disaster management systems and earth system science have established structures and relationships that move from the development of science, forecasts and models to effective communication with emergency managers and decision-makers, including the general public (Beaven et al. 2016). Information needs to be shared across group boundaries, specifically by knowledge brokers (Ali et al. 2019). Boundary organisations and individuals that link research with communication and knowledge application are key facilitators of these relationships (Pielke, Jr. 2007).

Coordinated structures, such as the Natural Hazards Research Platform in New Zealand (Beaven et al. 2016) and the Natural Hazards Partnership in the UK (Hemingway & Gunawan 2018), provide mechanisms to improve tools and models and to evaluate warning systems and improve capabilities. Such structures enable discussion of caveats and uncertainties that may prevent the warner from using the information incautiously or out of context.

Within relationships, tensions between the policy and science domain create a hybrid zone in the "bridge". Science becomes "applied science" as information turns into action. "Development of…impact-focused information and advice is supported by coordinated access to cutting-edge science and natural hazard impact research" (Hemingway & Gunawan 2018). Policy relevance requires interdisciplinarity and will likely be time-sensitive, driving a move to shorter-term actions. The general public and non-experts require simplified information, but this should not compromise the understanding of uncertain, complex information (Beaven et al. 2016).

4.4.2   Communicating Impacts and their Uncertainty

The objective of the interaction between the warner and the impact specialist is to provide the warner with the means to incorporate relevant impact information in the warning. Typically, this is achieved by providing a model or tool. There are several dangers that must be recognised by those involved if the exercise is to be successful in making the warning more effective. Great care must be taken to identify the impacts that matter to the decision-maker and to avoid simply predicting the impacts for which there are good data or simple models. It is also important to avoid generating complex sets of output that overwhelm the warner with data. Since the warner is aiming to produce a narrative that will help the receiver to act, it may be helpful to consider producing storylines (Shepherd, 2019) of hazard impacts that describe one or more impacts together with their uncertainty. Such an approach could be especially suitable when the impact specialist provides real-time interpretation as part of the warning chain.

One of the challenges faced by intermediaries between warning and impact information is the discussion of uncertainty. An important aspect of building trusted relationships is that there should be discussions of how accurate the forecasts are at different time and space scales. When these conversations are combined with hazard impact and sector impact models, knowledge of the certainty of each of these models and the ways in which the impacts interact will be critical. "The non-communication of these is problematic as interdependencies between them, especially for multi-model approaches and cascading hazards, can result in much larger deep uncertainties" (Doyle et al. 2018). It is important for uncertainty to be communicated effectively to best inform decision-makers and to ensure action is taken that best protects the community.

The full range of uncertainties throughout the warning process must be allowed for (from defining the problem, computational issues, initial conditions, verification and beyond). Scientists must set realistic expectations concerning uncertainties, recognise cultural differences between disciplines, and ensure that engagement develops mutual understandings of the issue and supports decision-makers. "When visualizing uncertainty, the focus must be on the data and uncertainty relevant to the decision" (Doyle et al. 2018).

Communication of uncertainty increases levels of trust (Joslyn & LeClerc 2015). The message should be precise about the sources of uncertainty involved and how to effectively present disagreements between experts in a way that does not minimise the message or credibility. It is also important that the impacts are well-understood by communicators and that they include in their warnings "decision-relevant time frames, including information on *when the uncertainty may be reduced*" (Doyle et al. 2018). Developing partnerships and communicating uncertainty prior to the need to issue the warning increase trust and confidence.

4.4.3 Exchanging Information About Tools

A general principle across the whole warning chain is that users have greater confidence in warnings if they understand how they were produced. The greatest challenge to achieving that lies in impact prediction, which is often hidden in statistical "black boxes". It is therefore an essential part of any partnership between impact modeller and warner to convey the basis of the model, the predictors used and the uncertainty bounds in the predictions. The warner should have access to routine verification and be able to query unexpected results. These requirements place demands on the information produced during tool development and handover and on the availability of ongoing expert support. They also require that users, impact scientists, IT developers and warners are all involved throughout the development process. When a new or upgraded tool is handed over, users should be inducted into its use through presentations and supported hands-on practice. Detailed instructions should be provided, describing its operation, including operating limits and how to deal with any failure modes. It should also be accompanied by a comprehensive test report, together with datasets and any ancillary software required to reproduce the test results. The test report should clearly state the ranges of input data that have been validated and any caveats about tool outputs, including situations where performance will be below the norm, and should identify the sources and magnitudes of uncertainty in the results. In order to guard against overconfident messaging, uncertainty ranges should be provided as standard outputs from the tool. Metrics used in the evaluation should be clearly described, together with the reasons for using them and their limitations.

These technical aspects of handover are important to ensure that the warner does not inadvertently produce misleading information. However, they also contribute to helping the warner to have enough confidence to use and accept the information that is generated. To fully achieve this acceptance, warners should be involved throughout the development process, to ensure that the tool is designed to produce the information that they feel is required, that the developers test the tool in circumstances identified as important by the warners and that performance can be challenged by those who will use it. Ideally the relationship between developer and warner should be personal, but if not, regular contact throughout the development process will help ensure that the tool contributes to better warnings once it is handed over.

4.4.4   Challenges of Evaluating Tools

In order to ensure that warning information is used and useful, it is important to conduct evaluations. The data and models need to provide actionable results, and the results of the models should be validated and verified. Evaluations of each aspect of the system can be complicated, as the warning may be based on integrated, ensemble models with impact scenarios and simulations that are then communicated using various infrastructure and tools throughout the early warning system. Studies have been conducted to determine decisions that are made from warnings, using the determinant that protective action occurred as a measure of success (Gutter et al. 2018). Each of these stages will need evaluation, but finally the decision-making processes must be considered and whether or not action was taken.

4.5 Examples

Box 4.1 A Structure for Warner and Impact Expert Interaction in New Zealand
Cheryl L. Anderson

Research has found that the framework or structure for interactions of the communities issuing and receiving the warnings is critical for ensuring that lives are protected. One example of this type of framework is the New Zealand Natural Hazards Research Platform (NHRP) that was organised to ensure that hazard research and science informed disaster policy. NHRP served as a boundary organisation to facilitate collaboration on disaster risk reduction, with one of the key areas being early warning systems. The interactions of scientists and policy advisors in the boundary organisation aided in developing trusted relationships (Beaven et al. 2016).

The Sendai Framework for Disaster Risk Reduction commits signatory countries to establish coordinating governance arrangements to increase the integration of stakeholders across domains, sectors and levels and to "foster cooperation among scientific and technical communities, other relevant stakeholders and policymakers in order to facilitate a science-policy interface for effective decision-making in disaster risk management (UNDRR 2015, 13). The NHRP facilitated cross-sector collaboration, including the activities of advisory bodies, international climate change and biodiversity initiatives and collaborative approaches to the management of shared resources.

Box 4.2 Research Demonstration Projects at the Olympic Games
Cheryl L. Anderson

The Olympic Games have been used to advance an understanding of the complexities of forecasting and nowcasting since 2000. The WMO World Weather Research Programme (WWRP) organised Forecast Demonstration Projects and Research Development Projects that advanced the development of warning infrastructure, training and use of warning systems and methods for distributing information quickly (WMO 2017). The process involves building relationships with the Olympic committees to understand the end-user needs for the event and developing methods to deliver these needs. The Sydney 2000 Olympics was the first demonstration project, and international teams used the opportunity to install a radar system and learn to provide rapid nowcast warnings, primarily for wind and rain (Wilson et al. 2004, WMO 2017). Knowledge from the Sydney games fed into the Beijing Summer Olympic Games, where improvements in Numerical Weather Prediction (NWP) models, capacity-building in communicating the nowcasts through web interfaces

(continued)

and visualisations and direct weather briefings with Olympic officials established long-term working relationships across the international forecasting community. The Winter Games have provided more challenges. Events such as downhill skiing require wind, precipitation and visibility forecasts at multiple elevations to ensure that events are fair and that competition can proceed. Olympics nowcasting in the 2010 Vancouver Winter Olympics left infrastructure that has benefitted aviation and transportation and advances in forecasting precipitation by improving timing of storms and visibility. It also deepened relationships among forecasters, the Olympics Committee, and events coordinators and managers. Conversations about event needs for information on visibility, snowfall, and wind speeds led to the development of thresholds for making decisions about postponement and delays for each event (Isaac et al. 2014; Joe et al. 2010; Joe et al. 2014; Joe et al. 2004; WMO 2017).

Box 4.3 Linking Fire and Health Impacts to Action in Australia's Summer of 2019/2020
John Nairn

Australia's 2019/2020 summer of cascading multi-hazards ceased with flooding rains. Bushfire smoke produced the highest documented human health impact with 417 excess deaths (Borchers-Arriagada et al. 2020) compared to 33 bushfire deaths (Commonwealth of Australia 2020). An extremely active fire season produced unprecedented bushfire intensity, area burnt, significant mortality and property and animals destroyed. Seasonal forecasts set expectations for an extremely intense bushfire season. Fire and emergency services agencies performed rigorous pre-fire season briefings incorporating antecedent climate and seasonal outlook intelligence as the basis for scenario plans, resource allocation and testing of community message systems. Health impact information from the season's dust, heat waves, fires, persistent smoke and flash floods could benefit from co-design of impact forecast products tailored to public health needs. Public health's response to the persistent smoke hazard indicated a lack of coordination, with disparate community advice undermining community confidence. An increased focus on pre-season scenario planning would allow the public health sector to achieve the same level of preparedness as is evident with Australia's fire authorities but extended across multiple hazards.

(continued)

Box 4.4 Creative Collaborations in Social Media Communication at the UK met Office
Ross Middleham

Creative collaborations and partnerships can help us learn from others, share best practice and accomplish mutual goals. Before entering into a partnership, it's important to understand your own organisational goals and what activities align with your purpose. At the Met Office, our purpose is to keep people safe and able to thrive, so every decision we make must support this.

Every opportunity starts with a conversation. We actively seek collaborations that can help support our messages, reach new audiences, position us as experts and the authoritative source or bring insight and learning to the organisation. These partnerships can be formal or ad hoc, paid or organic depending on the benefit and impact that will be delivered (Fig. 4.2).

We actively share and support messaging with partners who align with our brand. The key here is that we have a common aim, so it's natural for us to share and support each other's messages. For example, we work with the Royal Automobile Club and Royal National Lifeboat Institute to amplify safety messages.

We actively seek partners who can help position us as a trusted source of information. For example, we worked with Facebook on their Climate Science Information Centre to become an international partner which sees our climate science content being pulled into their hub. We actively seek creative collaborations that can help our content reach new audiences. We identify people and organisations who share a similar purpose but have their own engaged audience who follow and trust what they say. For example, we approached the

Fig. 4.2 Joint Met Office RAC travel safety video on YouTube. (© Crown Copyright 2021, Met Office)

(continued)

Jamie Oliver Group after he mentioned weather and climate on one of his TV programmes and then worked with him to co-create climate and food security content for his 8.4 million followers on Instagram.

Working with others has many advantages, but it's not an easy thing to do. It takes time: time to identify opportunities, time to build your network and time to develop an idea and make it happen.

The power of LinkedIn and Twitter to approach the organisations you want to speak to should not be ignored. You may need to consider ways to grab their attention, even just to have an initial conversation. That might be doing a mock-up of your idea or sending a demo video. Be prepared for your initial chat by researching the organisation and understanding their objectives. Then act quickly when responding to follow-up emails and idea sharing to maintain momentum.

Partnerships aren't just about making your own messages go further. For example, we actively seek creative collaborations that inspire and bring insight into the team. Over the years we've run lots of workshops in university design studios around the country. We share what we've learnt with young people, and in return they give us a different perspective on our problems and give us insight into their worlds, offering us a way to creatively test our ideas. We also actively seek creative collaborations to inspire and drive innovation. For example, we've worked with One Minute Briefs to crowd source ideas through mass design participation on Twitter.

In summary, we actively seek creative collaborations to keep us evolving. But why is that important? Because we know that the way we do things now will not stay the same. The digital landscape is becoming noisier and noisier and we continue to fight for attention. Ever-changing algorithms change the way our content is served up on channels, and the way people consume information is continually changing. For example, we'll soon need to think of ways to reach a whole new user group. The ones growing up gaming, being home-schooled, communicating virtually and who rely on YouTube. We need to work with others to help us understand that audience.

We'll continue to keep our eyes on the horizon and actively seek opportunities to ensure our information is trusted, listened to and acted upon, helping to keep people safe and able to thrive (Fig. 4.3).

(continued)

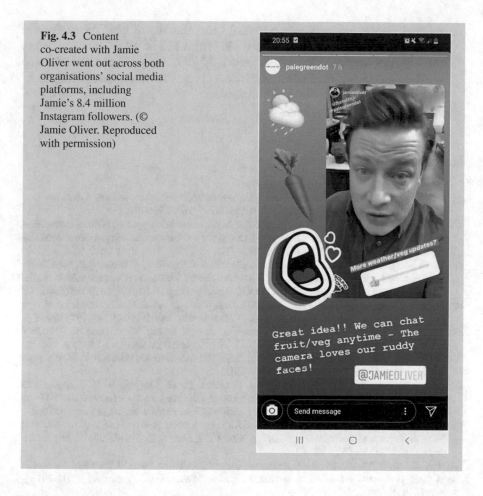

Fig. 4.3 Content co-created with Jamie Oliver went out across both organisations' social media platforms, including Jamie's 8.4 million Instagram followers. (© Jamie Oliver. Reproduced with permission)

4.6 Summary

- The success of a warning is that people listen, understand the message and use it to take action that protects lives, livelihoods, the environment, property and infrastructure. Impact information is one ingredient in helping this to happen.
- Expertise in weather-related hazard impacts is widely distributed. Impacts data are often difficult to access and analysis methods can be very specialist. It is therefore important to identify which impacts matter, who has access to relevant data and who has the requisite analysis skills.
- In order to circumvent issues with proprietary and confidential data, hazard and weather forecasters must be prepared to make their data available to the impact specialist in a form that enables the impact specialist to match it with impact data and develop a model or tool. As the impact data cannot generally be shared, it is

essential that the partners have a mutual understanding of what the analysis is trying to achieve, why a tool is needed and how it will be used.

• Relationships between the "warner" and the "impact expert" can be facilitated within boundary organisations where "honest brokers" serve as intermediaries to effectively translate and convey information.

References

Ajzen, I., 1991. The theory of planned behavior. *Organizational behavior human decision processes*, **50(2)**, 179–211. https://doi.org/10.1016/0749-5978(91)90020-T

Ali, F., M.-N. S. Sharipudin and K.-S. Fam, 2019. Information Sharing across Group Boundaries by Knowledge Brokers during a Disaster - Lessons for the Tourism Industry. *Asian J. Business Research,* **9(2)**, 76–94. DOI: 10.14707/ajbr.190061.

Anderson-Berry, L., T. Achilles, S. Panchuk, B. Mackie, S. Canterford, A. Leck, D. K. Bird, 2018. Sending a message: How significant events have influenced the warnings landscape in Australia. *Int. J. Disaster Risk Reduction*, **30**, 5–17. DOI: https://doi.org/10.1016/j.ijdrr.2018.03.005.

Armitage, P., G. Berry and J.N.S. Matthews, 2002. *Statistical Methods in Medical Research*, Fourth Edition. Wiley Print ISBN:9780632052578. DOI:https://doi.org/10.1002/9780470773666

Beaven, S., T. Wilson, L. Johnston, D. Johnston, and R. Smith, 2016. Role of Boundary Organization after a Disaster: New Zealand's Natural Hazards Research Platform and the 2010-2011 Canterbury Earthquake Sequence. *Nat. Hazards Rev.*, **05016003**, 1–4. DOI: https://doi.org/10.1061/(ASCE)NH.1527-6996.0000202.

Borchers-Arriagada, N., A. J. Palmer, D. M. Bowman, G. G. Morgan, B. B. Jalaludin and F. H. Johnston, 2020. Unprecedented smoke-related health burden associated with the 2019–20 bushfires in eastern Australia. *Med. J. Aust.*, **6**, 282–283. https://doi.org/10.5694/mja2.50545

Casteel, M. A., 2016. Communicating Increased Risk: An Empirical Investigation of the National Weather Service's Impact-Based Warnings. *Wea. Clim. Soc.,* **8(3)**, 219–232. DOI: https://doi.org/10.1175/WCAS-D-15-0044.1.

Casteel, M. A., 2018. An empirical assessment of impact-based tornado warnings on shelter in place decisions, *Int. J. Disaster Risk Reduction*, **30A**, 25–33. https://doi.org/10.1016/j.ijdrr.2018.01.036

Commonwealth of Australia, 2020. Royal Commission into National Natural Disaster Arrangements. https://naturaldisaster.royalcommission.gov.au/publications/html-report/introduction. (Accessed 19/6/2021).

Dotzek, N., P. Groenemeijer, B. Feuerstein and A. M. Holzer, 2009. Overview of ESSL's severe convective storms research using the European Severe Weather Database ESWD. *Atmospheric research*, **93(1-3)**, 575-586. DOI: https://doi.org/10.1016/j.atmosres.2008.10.020

Doyle, E. E. H., D. M. Johnston, R. Smith and D. Paton, 2018. Communicating model uncertainty for natural hazards: A qualitative systematic thematic review. Int. J. *Disaster Risk Reduction,* **33**, 449–476. https://doi.org/10.1016/j.ijdrr.2018.10.023.

Ebert, E., B. Brown, J. Chen, C. Coelho, M. Dorninger, M. Goeber, T. Haiden, M. Mittermaier, P. Nurmi, L. Wilson and Y. Zhu, 2015. Numerical prediction of the Earth system: Crosscutting research on verification techniques, Ch 21 in Brunet, G., S. Jones and P.M. Ruti (eds), *Seamless Prediction of the Earth System: from Minutes to Months*. WMO-No. 1156, Geneva, Switzerland. pp403–418.

Eiser, J. R., A. Bostrom, I. Burton, D. M. Johnston, J. McClure, D. Paton, J. van der Pligt and M. P. White, 2012. Risk interpretation and action: A conceptual framework for responses to natural hazards. *Int. J. Disaster Risk Reduction,* **1**, 5–16. https://doi.org/10.1016/j.ijdrr.2012.05.002.

Griffin, R. J., S. Dunwoody and K. Neuwirth, 1999. Proposed model of the relationship of risk information seeking and processing to the development of preventive behaviors, *Environmental Research*, **A80**, S230–245. https://doi.org/10.1006/enrs.1998.3940

Gutter, B. F., K. Sherman-Morris and M. E. Brown, 2018. Severe Weather Watches and Risk Perception in a Hypothetical Decision Experiment. *Wea. Climate Soc*, **10**(4), 613–623. DOI: https://doi.org/10.1175/WCAS-D-18-0001.1.

Hektner, J.M., J.A. Schmidt and M. Csikszentmihalyi, 2007. *Experience Sampling Method: Measuring the Quality of Everyday Life*. Sage, Thousand Oaks, U.S.A. ISBN 9781412925570, 9781412984201

Hemingway, R. and O. Gunawan, 2018. The Natural Hazards Partnership: A public sector collaboration across the UK for natural hazard disaster risk reduction. *Int. J. Disaster Risk Reduction*, **27**, 499–511. https://doi.org/10.1016/j.ijdrr.2017.11.014.

Hoss, F. and P. Fischbeck, 2018. Use of observational weather data and forecasts in emergency management: An application of the Theory of Planned Behavior, *Wea. Clim. Soc*, **10**(2), 275–290. DOI:https://doi.org/10.1175/WCAS-D-16-0088.1

Isaac, G. A., P. I. Joe, J. Mailhot, M. Bailey, S. Bélair, F. S. Boudala, M. Brugman, E. Campos, R. L. Carpenter Jr., R. W. Crawford, S. G. Cober, B. Denis, C. Doyle, H. D. Reeves, I. Gultepe, T. Haiden, I. Heckman, L. X. Huang, J. A. Milbrandt, R. Mo, R. M. Rasmussen, T. Smith, R. E. Stewart, D. Wang and L. J. Wilson, 2014. Science of nowcasting Olympic Weather for Vancouver 2010 (SNOW-V10): A World Weather Research Programme Project. *Pure Appl. Geophys.*, **171**, 1–24, https://doi.org/10.1007/s00024-012-0579-0.

Joe, P., C. Doyle, A. Wallace, S. G. Cober, B. Scott, G. A. Isaac, T. Smith, J. Mailhot, B. Snyder, S. Belair, Q. Jansen and B. Denis, 2010. Weather Services, Science Advances, and the Vancouver 2010 Olympic and Paralympic Winter Games. *Bull. Amer. Meteorol. S.*, **91**(1), 31–36. https://doi.org/10.1175/2009BAMS2998.1

Joe, P., B. Scott, C. Doyle, G. Isaac, I. Gultepe, D. Forsyth, S. Cober, E. Campos, I. Heckman, N. Donaldson, D. Hudak, R. Rasmussen, P. Kucera, R. Stewart, J. M. Thériault, T. Fisico, K. L. Rasmussen, H. Carmichael, A. Laplante, M. Bailey and F. Boudala, 2014. The Monitoring Network of the Vancouver 2010 Olympics. *Pure Appl. Geophys.*, **171**(1-2), 25–58. DOI https://doi.org/10.1007/s00024-012-0588-z

Joe, P., D. Burgess, R. Potts, T. Keenan, G. Stumf and A. Treloar, 2004. The S2K Severe Weather Detection Algorithms and Their Performance. *Wea. Forecast.*, **19**, 1, 43–63. https://doi.org/10.1175/1520-0434(2004)019<0043:TSSWDA>2.0.CO;2

Joslyn, S.L. and J.E. LeClerc, 2015. Climate projections and uncertainty communication. *Topics in Cognitive Science*, **8**, 222–241. DOI: https://doi.org/10.1111/tops.12177.

Kaltenberger, R., A. Schaffhauser and M. Staudinger, 2020. "What the weather will do"– results of a survey on impact-oriented and impact-based warnings in European NMHSs. *Advances in Science and Research*, **17**, 29–38. DOI: https://doi.org/10.5194/asr-17-29-2020

Klein, G. A., 1989. Recognition-primed decisions. W. B. Rouse (Eds.). *Advances in man-machine systems research*, 5, 47–92. Greenwich, Conn: JAI Press, Inc. ISBN 155938011X

Lazo, J. K., H. R. Hosterman, J. M. Sprague-Hilderbrand and J. E. Adkins, 2020. Impact-Based Decision Support Services and the Socioeconomic Impacts of Winter Storms. *Bull. Amer. Meteorol. S.*, **101**(5), E626–E639. DOI: https://doi.org/10.1175/BAMS-D-18-0153.1

Mills, B., J. Andrey, S. Doherty, B. Doberstein and J. Yessis, 2020. Winter storms and fall-related injuries: Is it safer to walk than to drive?, *Wea. Clim. Soc.*, **12**(3), 421–434. https://doi.org/10.1175/WCAS-D-19-0099.1

Morgan, M.G., B. Fischhoff, A. Bostrom and C.J. Atman, 2002. *Risk Communication: A Mental Models Approach*. Cambridge University Press, New York. https://doi.org/10.1017/CBO9780511814679

Pagano, T. C., F. Pappenberger, A. W. Wood, M. H. Ramos, A. Persson and B. Anderson, 2016. Automation and human expertise in operational river forecasting. *Wiley Interdisciplinary Reviews: Water*, **3**(5), 692–705. https://doi.org/10.1002/wat2.1163

Penning-Rowsell, E., S. Priest, D. Parker, J. Morris, S. Tunstall, C. Viavattene, J. Chatterton and D. Owen, 2013. *Flood and Coastal Erosion Risk Management: A Manual for Economic Appraisal.* Routledge. 448pp. https://doi.org/10.4324/9780203066393

Pielke, Jr., R. A., 2007. *The Honest Broker: Making Sense of Science in Policy and Politics.* Cambridge and New York: Cambridge University Press. ISBN 9780521694810

Potter, S. H., P. V. Kreft, P. Milojev, C. Noble, B. Montz, A. Dhellemmes, R. J. Woods and S. Gauden-Ing, 2018. The influence of impact-based severe weather warnings on risk perceptions and intended protective actions. *Int. J. Disaster Risk Reduction,* **30**, 34–43. https://doi.org/10.1016/j.ijdrr.2018.03.031.

Ruti, P. M., O. Tarasova, J. H. Keller, G. Carmichael, Ø. Hov, S. C. Jones, D. Terblanche, C. Anderson-Lefale, A. P. Barros, P. Bauer, V. Bouchet, G. Brasseur, G. Brunet, P. DeCola, V. Dike, M. D. Kane, C. Gan, K. R. Gurney, S. Hamburg, W. Hazeleger, M. Jean, D. Johnston, A. Lewis, P. Li, X. Liang, V. Lucarini, A. Lynch, E. Manaenkova, N. Jae-Cheol, S. Ohtake, N. Pinardi, J. Polcher, E. Ritchie, A. E. Sakya, C. Saulo, A. Singhee, A. Sopaheluwakan, A. Steiner, A. Thorpe and M. Yamaji, 2020. Advancing Research for Seamless Earth System Prediction. *Bull. Amer. Meteorol. S.,* **101**, E23–35. https://doi.org/10.1175/BAMS-D-17-0302.1.

Shepherd, T. G., 2019. *Proc. Roy. Soc.* A: 475 (2225). 20190013. ISSN 1471-2946 https://doi.org/10.1098/rspa.2019.0013

Tan, M. L., S. Harrison, J. S. Becker, E. E.H. Doyle and Raj Prasanna, 2020a. Research Themes on Warnings in Information Systems Crisis Management Literature. *17th ISCRAM Conference Proceedings.* Blackburg, Virginia. https://www.researchgate.net/publication/342611823_Research_Themes_on_Warnings_in_Information_Systems_Crisis_Management_Literature (Accessed 3/9/2021)

Tan, M. L., R. Prasanna, K. Stock, E. E.H. Doyle, G. Leonard and D. Johnston, 2020b. Understanding end-users' perspectives: Towards developing usability guidelines for disaster apps. *Progress in Disaster Science,* **7**, 100118. DOI: https://doi.org/10.1016/j.pdisas.2020.100118.

Taplin, D. H. and H. Clark, 2012. *Theory of Change Basics: A Primer on Theory of Change.* New York: Actknowledge. https://www.theoryofchange.org/wp-content/uploads/toco_library/pdf/ToCBasics.pdf (Accessed 3/9/2021)

Taylor, A. L., A. Kause, B. Summers and Melanie Harrowsmith, 2019. Preparing for Doris: Exploring public responses to impact-based weather warnings in the UK. *Wea. Clim. Soc.,* **11(4)**, 713–729. https://doi.org/10.1175/WCAS-D-18-0132.1.

Uccellini, L. W. and J. E. Ten Hoeve, 2019. Evolving the National Weather Service to Build a Weather-Ready Nation: Connecting Observations, Forecasts, and Warnings to Decision-Makers through Impact-Based Decision Support Services**.** *Bull. Amer. Meteorol. S.,* **100(10)**, 1923–1942. https://doi.org/10.1175/BAMS-D-18-0159.1.

UNDRR, 2015. Sendai Framework for Disaster Risk Reduction 2015-2030. Geneva, Switzerland: United Nations Office for Disaster Risk Reduction, 37. https://www.preventionweb.net/files/43291_sendaiframeworkfordrren.pdf. (Accessed 2/9/2021)

Weyrich, P., A. Scolobig, D. N. Bresch and A. Patt, 2018. Effects of Impact-Based Warnings and Behavioral Recommendations for Extreme Weather Events. *Wea. Clim. Soc.,* **10(4)**, 781–796. https://doi.org/10.1175/WCAS-D-18-0038.1.

Weyrich, P., A. Scolobig, F. Walther and A. Patt, 2020a. Responses to severe weather warnings and affective decision-making, *Nat. Hazards Earth Syst. Sci.,* **20,** 2811-2821. https://doi.org/10.5194/nhess-20-2811-2020

Weyrich P., A. Scolobig, F. Walther and A. Patt, 2020b. Do intentions indicate actual behaviours? A comparison between scenario-based experiments and real-time observations of warning response. *J. Contingencies Crisis Management,* **28**, 240–250. https://doi.org/10.1111/1468-5973.12318

Wilks, D.S., 2006. *Statistical Methods in the Atmospheric Sciences* Second Edition. Department of Earth and Atmospheric Sciences Cornell University Academic Press ISBN 13: 978-0-12-751966-1 ISBN 10: 0-12-751966-1

Wilson, J. W., E. E. Ebert, T. R. Saxen, R. D. Roberts, C. K. Mueller, M. Sleigh, C. E. Pierce and
   A. Seed, 2004. Sydney 2000 Forecast Demonstration Project: Convective Storm Nowcasting.
   Wea. Forecast., **19(1)**, 131–150. DOI:https://doi.org/10.1175/1520-0434(2004)019<0131:SFD
   PCS>2.0.CO;2
WMO, 2015. WMO Guidelines on Multi-Hazard Impact-Based Forecast and Warning Services,
   WMO No.1150, https://library.wmo.int/doc_num.php?explnum_id=7901. (Accessed 3/9/2021)
WMO, 2017. *Guidelines for Nowcasting Techniques.* WMO-No. 1198. Geneva: WMO. https://
   library.wmo.int/doc_num.php?explnum_id=3795 (Accessed 3/9/2021)
Yu, M., C. Yang and Y. Li., 2018. Big Data in Natural Disaster Management: A Review. *Geosciences*,
   8(165), 1–26. Basel, Switzerland: MDPI. DOI: https://doi.org/10.3390/geosciences8050165.

Chapter 5
Connecting Hazard and Impact: A Partnership between Physical and Human Science

Joanne Robbins, Isabelle Ruin, Brian Golding, Rutger Dankers, John Nairn, and Sarah Millington

Abstract The bridge from a hazard to its impact is at the heart of current efforts to improve the effectiveness of warnings by incorporating impact information into the warning process. At the same time, it presents some of the most difficult and demanding challenges in contrasting methodology and language. Here we explore the needs of the impact scientist first, remembering that the relevant impacts are those needed to be communicated to the decision maker. We identify the challenge of obtaining historical information on relevant impacts, especially where data are confidential, and then of matching suitable hazard data to them. We then consider the constraints on the hazard forecaster, who may have access to large volumes of model predictions, but cannot easily relate these to the times and locations of those being impacted, and has limited knowledge of model accuracy in hazardous situations. Bridging these two requires an open and pragmatic approach from both sides. Relationships need to be built up over time and through joint working, so that the different ways of thinking can be absorbed. This chapter includes examples of partnership working in the Australian tsunami warning system, on health impact tools

J. Robbins (✉) · B. Golding
Met Office, Exeter, UK

WMO/WWRP HIWeather project, Geneva, Switzerland
e-mail: joanne.robbins@metoffice.gov.uk

I. Ruin
Inst. for Geosciences & Environmental Research, Centre National de la Recherche
Scientifique, Paris, France

R. Dankers
Wageningen University & Research, Wageningen, Netherlands

WMO/WWRP HIWeather project, Geneva, Switzerland

J. Nairn
Bureau of Meteorology, Melbourne, Australia

S. Millington
Met Office, Exeter, UK

for dispersion of toxic materials in the UK and on the health impacts of heatwaves in Australia. We conclude with a summary of the characteristics that contribute to effective impact models as components of warning systems, together with some pitfalls to avoid.

Keywords Economist · Epidemiologist · Engineer · Hazard · Impact · Exposure · Vulnerability · Ethics · Training data · Evaluation

5.1   Introduction

There is a growing recognition that users and decision makers make better informed decisions when warnings incorporate information about potential socio-economic impacts. In this chapter we show that:

- The impact of a hazard depends on the vulnerability of individuals and communities that are exposed to it.
- Weather-related impacts may be human (e.g. death or injury), financial (e.g. property damage or business loss) or service related (e.g. loss of power or transport links).
- Direct impacts can create a cascade of multiple indirect impacts.
- Impact predictions may be produced using process models or statistical models and should be probabilistic.
- Observations of impacts are fundamental for understanding, for monitoring and for verification but are often only accessible through partners.
- When linking impact and hazard models, care must be taken that variables represent the same things in each model, that space and timescales match and that biases are removed.
- Impact forecasters and hazard forecasters often have a very different understanding of the end user's problem. These differences must be shared and reconciled.

5.1.1   Impact and Risk

Impacts of natural hazards can be described in terms of their spatial extent, duration and severity, either focussed on an individual asset (e.g. road transport network) or aggregated to describe impacts across a region. They can be classified as direct or indirect, and tangible or intangible. Direct losses represent the damage, loss of life or economic loss that results directly from the event and tend to map closely to the spatial footprint of the hazard. Indirect losses include reductions in output or revenue, disruptions to markets and distribution networks and impacts to personal well-being and community cohesion. These impacts frequently have broader geographical reach and may have long-term destabilising effects. Tangible impacts are those which can be quantified and are typically well recorded, while intangible impacts are not easily measurable (e.g. place attachment to ancestral land and changes to mental health). Most impact models focus on the tangible and direct losses associated with a hazard, but it is recognised that for effective and sustained recovery, there needs to be a better understanding of indirect and intangible impacts.

Modern risk analysis builds on research conducted under the 'behavioural paradigm' and the 'development paradigm' (Smith & Petley 2009), which together reorientated the focus away from predominantly engineering solutions aimed at containing the hazard, to a better understanding of the social and behavioural drivers of impact variability. Some of the first computer-based risk models were developed within the insurance and reinsurance sector (catastrophe models), where risk simulations are used to quantify the impacts of potential future hazards based on the exposure and vulnerability of the assets in an insurer's portfolio (Grossi & Kunreuther 2005). Such models typically focus on the physical vulnerability of exposed elements, omitting other aspects of vulnerability (e.g. economic, social and attitudinal) that are more challenging to quantify.

Although risk is widely recognised to be a function of hazard, vulnerability and exposure, the ways in which they are expressed can vary significantly across disciplines. For the insurance sector, understanding the financial implications of future hazardous events is critical so that they can deploy capital, and price insurance coverage appropriately. As a result, catastrophe models focus on quantifying the physical damage (number and type of assets damaged or destroyed) and translating that into monetary loss. By contrast, the National Meteorological and Hydrological Services (NMHS) are increasingly adopting impact-based warnings to inform the public and emergency managers of potentially impactful weather in the near future. Impact-based warnings communicate the level of risk, supported by general statements of potential impacts, using predefined impact category descriptors, and the spatial and temporal likelihood of the hazardous event.

For impact and risk models to be effective in early warning, decisions on how to develop these models should be led by the user's needs and determined by the availability of appropriate data. In cases where a general, broad assessment of future potential risk is effective, it may be possible to use general information (e.g. a previous high-impact weather event and its impacts) as an indicator of what the future risk from a similar high-impact weather event might be. However, where a user needs to prioritise emergency decisions, more detail on the vulnerable people and assets within the hazard footprint is essential, as is understanding the potential catastrophic situation that may emerge from unprecedented compound or cascading socio-natural events. Addressing these different styles of risk forecast requires different underpinning data and different approaches to aggregation of the data within the model.

5.2   Impact Forecasting

Approaches to estimating the impact of weather-related hazards vary widely according to the user application and the type of hazard. Nevertheless, there are some common factors that it is worth considering first. Historically, experience and precedent were the main sources of information on severe impacts. The weather forecaster or emergency manager who had experienced a previous event would know what to expect when similar conditions recurred. A warning service would often be instituted based on a review of a particular event, with hazard warning thresholds set based on impact evidence from that event. A recurring practice has been the use of climatological thresholds to describe the variability in severe weather and therefore the anticipated impact a weather event may have. This assumes that areas exposed to regular severe weather will have built resilience to these events, compared to areas whose definition of severe weather is a lower magnitude weather event. However, with the development of more sophisticated methods, formal statistical techniques for identifying the relationship between hazard and impact need to be used. An area of increasing concern is the identification and prediction of indirect and cascading impacts – where one impact leads to other, potentially more serious, impacts. An important part of any hazard-impact assessment is therefore to identify the variety of pathways by which a hazard may have an impact. This is especially evident for urban populations, where remote hazards may interrupt critical infrastructure supplying large numbers of people, but is also relevant for rural populations dependent on neighbouring cities for markets.

5.2.1   Impact Data: Sources and Ethics

For all risk and impact models, the first step is to identify the hazard to impact pathway. This involves understanding what makes an event hazardous and impactful and describing this with available data. The strength of any hazard-impact relationship is dependent on the data used for analysis. A key issue for impact modellers is the availability of impact observations that can be examined in the context of environmental hazards. Impact data (e.g. mortality data, road traffic accident data, insurance claims and financial loss) are collected in many countries, but the drive to collect data is often not for the purpose of risk modelling. In the UK, police who attend traffic accidents are required to record the accident, vehicles involved and causality information in a standard national format. The form includes a section on incidental weather, and therefore one might anticipate that such information would easily support the identification of relationships between different weather conditions and the potential for road traffic accidents. However, the data may be misleading, as the identified incidental weather may not have caused the observed impact. The decision to record incidental weather is also biased by recorder perception. For example, an officer may register that rainfall played a role in the accident because the road surface is wet rather than because of rainfall at the time of the accident. Attributing impacts to a natural hazard type can be very challenging, especially when underlying vulnerabilities (i.e. driver tiredness, ability and responsiveness) are equally likely to have played a role in the observed impact. It should be recognised and, where possible captured, that impact records have their own biases due to the purpose and method of collection, and this needs to be accounted for when relationships are identified.

The quality of historical impact data determines the level of granularity that can be reached when understanding the drivers of observed impacts. Mortality data are collected in most countries, but the level of detail as to the cause and circumstances of death is very variable. The WHO promulgates a standard approach to classifying diseases and related health problems, but frequently the environmental hazards which can trigger these health outcomes are not recorded. Similarly, in developed countries direct economic impacts are most easily obtained from insurance or reinsurance data. However, insurance payouts may not equate to the cost of the damage, either because property was underinsured or because the replacement was of better quality than the original (betterment). Data are often aggregated prior to being made available to the impact modeller, so local variability due to variations in vulnerability and exposure is lost. For all risk and impact assessment, a distinction needs to be made between a 'micro-level' impact recording and a 'macro-level' impact recording. In the first case, the impact is assessed at the individual level and then may be aggregated to community or larger scale. In the second case, the national impact is assessed directly. Impact data obtained from social media or citizen science can be considered 'micro-level' data. One example of this type of data is illustrated through the University of Tasmania's 'AirRater' (https://airrater.org/) smartphone app which both disseminates information on current atmospheric conditions (temperature, smoke, pollution and pollen) and collects clinical symptom reports from registered users (Robbins et al. 2017, Campbell et al. 2020). This allows epidemiological analyses of impacts associated with different atmospheric conditions, and, because the app is targeted towards vulnerable populations, the detailed driving forces of impact variability can be captured. Such data, where available in its raw form, can provide a wealth of detail. However, it should also be used with caution. People's individual perceptions of the magnitude of an impact or the cause of the impact can be biased by their values, beliefs and social demography. Similarly, even the act of asking an individual to self-assess or monitor their activities, behaviour or health can result in a biased picture of reality and result in behaviour modification by the individual before any analysis has taken place. This is where different styles of surveying can help. Survey data are able to capture the incidence of a wider set of symptoms, either as a one-off sample of a representative population (a cross-sectional sample) or through repeated surveying of a cohort, to look at how events affect the same people through time (a longitudinal sample). Such approaches can draw out different types of biases and allow researchers to better understand the relationship between hazard and impact.

'Macro-level' impact data can remove individual perception bias and typically enable an improved view of the overall impact of an event. This is because both positive and negative impacts of an event are captured particularly, for example, where macro-level financial impacts are recorded. The aggregation of data removes small-scale forcings and provides a more holistic interpretation of impacts, which is useful for large-scale comparative studies and for analysing trends in impact behaviour over time and across broad spatial scales. Impact data sources (based on micro- or macro-level reporting) and collection practices influence the way that such data need to be processed prior to analysis. This includes considering what quality control measures may be needed (e.g. there is evidence that the volume of data from social media can be used to self-verify its ability to distinguish events from non-events; see Cavalière et al. 2020), the latency of the available data for analysis and the value of drawing together multiple sources to obtain a better representation of observed impacts. Recording bias, impact data quality and impact attribution are typically managed by impacted sector data custodians, although, as highlighted earlier, not always with risk and impact modelling in mind. To utilise the available datasets robustly and effectively therefore requires sustained collaborative effort.

Almost all impact data are affected by accessibility issues. For example, sources of health impacts include ambulance taskings, hospital admission and general practitioner consultation data, none of which is available for general use due to patient confidentiality requirements. In some countries, anonymised or aggregated data are relatively simple to access, while in others an accredited research licence is required. Similarly, impacts on engineering structures such as utility and road infrastructure can, in principle, be obtained from failure logs, accident and maintenance records, but formal records or reports of engineering failures are rarely available publicly.

Accessibility goes hand in hand with confidentiality. All impact studies must be undertaken within a legal and ethical framework that ensures confidentiality of any data that could be associated with an individual, a business or a specific location. These frameworks have implications on the types of analysis that can be completed and the potential detail of assessment that can be achieved. This is well demonstrated in the health-hazard analysis space. Statistical analysis of health outcomes is constrained by ethical standards which protect the identities of individuals who have experienced a health event. Name, age, sex, current morbidities, family residence, incident location and the nature of the health event are collected and securely stored. How data are collected can mask the contribution of a hazard to a health event, particularly as it is rare for medical systems or practitioners to encode for the presence of a hazard. The strict protocols protecting personal data typically include thresholds for spatial domains (location of event), minimum numbers of people assigned to an impact classification (typically no less than 10) and/or data being grouped by span of time. As most weather-related hazards are on a daily timescale, this will usually result in release of human health impact data across broader domains in order to reach the threshold of ten affected individuals. Human health impacts are classified by either death or type of morbidity. Daily morbidity records are an order of magnitude larger than death records, allowing statistical analysis to be conducted at higher spatial resolution than excess deaths, which are studied across broad regions down to the scale of a large city. The higher case numbers of people living with medical conditions enable ethical standards and statistical significance to be achieved at higher resolution, potentially at suburb scale.

Health-related studies might be assumed to be a special case for ethical consideration. However, even where impact data are openly available (e.g. from social and news media), they should be subject to ethical considerations. Impact data provide insights into vulnerabilities and sensitivities of people, systems and places which can be emphasised when processing, aggregation and visualisation techniques are applied. This can mean that data are no longer neutral or impartial and this can have significant political and ethical connotations. Impact data can highlight 'underperformance' or lack of adherence to globally or nationally recognised guidelines (e.g. humanitarian and UN guidelines or industry standards) or highlight positive and negative adoption of policies or working practices (e.g. within the energy industry where it could be possible to identify variability in the ways companies maintain their assets based on the occurrence of impacts on their networks). Beyond performance aspects, impact data can highlight particularly vulnerable groups or assets when aggregated or visualised in certain ways, and therefore handling of such data needs to be carefully managed to prevent negative targeting of such groups, which can further enhance their vulnerability. This is particularly relevant where natural hazard impacts intersect with fragile and conflict-affected situations.

5.2.2 Impact Relationships: Identifying Pathways from Hazards to People, Service and Financial Impacts

A critical requirement for impact modelling is the identification of the right set of predictors. It may seem obvious that the predictor for building damage from a storm is wind speed and that is reflected in the existence of impact-related wind scales: e.g. the Beaufort scale and the Fujita scale. However, wind direction is critical for many structures, such as walls and roofs, while antecedent or coincident rainfall affects some building materials. Much building damage and transport disruption are caused by falling trees, which are more likely when they are in leaf (if deciduous varieties) and when the ground is wet. For less obvious connections such as health impacts, establishing which environmental predictors a disease is sensitive to, if any, may be a challenge in itself (Fleming et al. 2014). With multiple predictors, care must be taken when sourcing data. For instance, if wind is recorded from the nearby airport, but particulate concentration (PM_{10}, say) is recorded from a city centre monitoring location, the inconsistency may bias any relationship that is found.

In the case of flooding, the hazard comes from critical combinations of precipitation intensity and duration, catchment morphology and land use as well as soil moisture. Those parameters will not only influence the hydrological responses (slow flood versus flash flood) but will also strongly influence the type and severity of impacts. Slow river floods are rarely associated with fatalities but often with large economic losses related to their greater extension and duration. By contrast, flash floods, even if localised, can be much more deadly as they often surprise people during their daily activities (Jonckman 2005; Ruin et al. 2008, 2009, 2014; Diakakis et al. 2020). Studies looking at the circumstances and profiles of flash flood victims allow researchers to make connections between the victims' ages and genders, the type of place and time of their accidents and the flood dynamic (Ruin et al. 2008; Terti et al. 2017). Such detailed analysis of the combination of impactful social and physical circumstances is necessary to comprehend the diversity of predictors that need to be examined and tested for impact prediction (Terti et al. 2015).

Advanced approaches to identifying statistical relationships, using machine learning, can find very sophisticated and indirect relationships that have the potential to greatly strengthen impact-prediction capabilities. However, when less direct impact associations are identified in data, they should be challenged as to cause and effect before being used in a predictive sense. Identifying and removing potential confounding relationships is a key part of this process and requires careful experimental design – either to remove them through sampling or to include their effects as part of the statistical model. Trends should also be removed from data prior to modelling. A linear trend, such as one might find arising from the general improvement in healthcare, is relatively easy to identify and remove, but data jumps may also be present due to changes in the law (e.g. building codes, maximum lorry/truck sizes) or in industry (e.g. new materials) that could easily be misrepresented as linear trends – or ignored altogether. The significance of any relationship should always be scrutinised, and even when a high level of significance is present, the full probability distribution should always be used, rather than just the mean relationship, to ensure that spurious precision does not mislead.

Where statistical relationships are difficult to determine based on available observations, other methods can be adopted. In principle direct impacts on assets such as buildings, dams and road and rail infrastructure can be modelled using detailed, process-based approaches. For example, the probability of a lorry/truck overturning due to strong winds may be assessed based on wind speed and the direction of the wind relative to the vehicle, the height and weight characteristics of the vehicle, its velocity and the underlying road characteristics. A mechanical model can be used to describe the process of vehicle overturning under different hazardous conditions, and this can be used to identify the relationship between hazard and impact, and develop impact-orientated thresholds which can be utilised in forward modelling (Hemingway & Robbins 2020). Likewise, engineered structures such as bridges, dams and overhead power lines are usually designed to withstand relevant hazards such as strong winds or high-water levels up to a specific threshold, beyond which the structure can be expected to be damaged or to fail. Engineers often represent the failure of a structure by a fragility curve that relates the probability of failure to the imposed load. Fragility curves are generally considered confidential and in some cases are national secrets. Engineers can also undertake experiments under controlled conditions to obtain direct evidence of how hazards interact with infrastructure and result in impact. Rolls-Royce undertook testing of its aircraft engine performance in the presence of volcanic ash to produce volcanic ash flight envelopes (Davison & Rutke 2014), for example, while several other studies have used

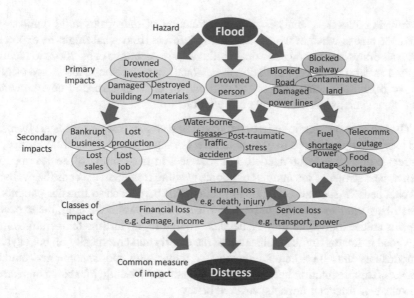

Fig. 5.1 Potential impacts from a flood, classified into financial, human and service losses. (© Crown Copyright 2021, Met Office)

shaking table tests to obtain primary data on the performance of different building types under different seismic scenarios (Fiorino et al. 2019; Xie et al. 2019). Obtaining primary data in this way ensures that the collected information is directly relevant to the researcher (and ultimately the user's) needs and answers the key relationship questions that the researcher would like to model. This reduces the need to filter through secondary data sources which have their own biases. Primary data collection is, however, only as good as the designed experiment, test or survey, and therefore, it is important that care is taken in setting up these activities. It should also be noted that not all impact modellers have the capacity to undertake their own primary data collection and so secondary data sources may be the only viable option for identifying the hazard to impact pathway.

When considering the impacts of weather-related hazards, it is possible to divide them into three main classes: impacts on people, recorded as deaths, injuries and displacement; impacts on property and business, recorded as a financial loss; and impacts on infrastructure, recorded as loss of service (Fig. 5.1). These are not independent: damage to the home can lead to mental health impacts, while personal injuries incur treatment costs and can reduce productivity, and service loss has potential health and cost implications. In all cases, the associated distress is an underlying impact.

Examples of Methods: Impacts on People Studies of the impact of natural hazards on people are part of the science of epidemiology (see, e.g. US Department of Health & Human Services 2012), which is concerned with the frequency and pattern of health events in a population and their causes. Epidemiology relies on the systematic collection, analysis and interpretation of data from valid comparison groups to assess whether what was observed differs from what might be expected. Analysis draws heavily on statistical methods (e.g. Armitage et al. 2002) to identify patterns in time or space, in relation to personal characteristics such as age or gender, or behaviours such as sport or occupation, or to exposure to environmental conditions such as severe weather or pollution.

The use of an epidemiological analysis in impact forecasting must start from the end user and the decisions they need to take. For a government or public service, an aggregate impact may be needed. For instance, in the context of health impacts, such as heat stroke, if the number of people needing treatment exceeds the available hospital beds or specialist equipment, action will be needed to transfer patients or redistribute vital resources. On the other hand, for a public warning, the expected impacts on the individual may be needed. Therefore, night-time temperatures may be a good indicator for hospitalisation of the elderly and chronically ill, but daytime temperatures may be a better indicator for impacts on sportspeople and outdoor workers. This distinction between the aggregate and the individual also appears in the sphere of financial impacts, covered below.

Analytic studies in epidemiology aim to identify and quantify the relationship between an exposure and a health outcome. The hallmark of such studies is the presence of at least two groups, one of which serves as a comparison group. To do this, assumptions must be made about exposure to the hazard in the area covered by the health record unless individual addresses are available. Even in the latter case, assumptions may have to be made about whether the person was at home and whether they were indoors or outdoors. This is especially difficult when assessing slow-acting impacts, e.g. from heat, cold or pollution. Unless there are reasons to study a particular sector of the population, perhaps with specific vulnerabilities, care must be taken that both the exposed and the unexposed populations selected for study are representative of the total population.

In observational case-control studies, subjects are enrolled according to whether they have the disease or not, then are questioned or tested to determine their prior exposure. Differences in exposure prevalence between the case and control groups allow investigators to conclude whether the exposure is associated with the disease. In observational cohort studies (e.g. as reviewed by Raaschou-Nielsen et al. 2013, for air pollution impacts), sample populations are identified and then studied to see how their health responds either to a prescribed exposure or to exposures that occur naturally. In the case of natural hazards, it is typical to select populations after an event has occurred, choosing exposed and non-exposed groups that have the same composition in terms of demography, for example. This is less reliable, since the groups may not be matched in some unknown characteristic of importance, and a large population is required in order to ensure sufficient members are exposed. An alternative methodology for transient impacts is the case-crossover analysis, which uses cases as their own controls (Lombardi 2010). For specific events, a cross-section approach is used, where two groups of people are selected, one exposed and one not exposed, from the same population and for the same time, with careful selection within each to match the overall population characteristics. These are better suited to descriptive epidemiology than for establishing causation.

There are occasions where it may not be possible to clearly delineate exposed and non-exposed groups. For some types of heat event, excess deaths (better described as excess 'all-cause' deaths) are regarded as a reliable indicator of impact. Excess death is calculated against deaths expected for time of year, controlled for the presence of hazards and long-term mortality trend. By linking excess mortality with time-series analysis, it is possible to assess the relationship between known hazardous periods and hazard types, and impact variability (Armstrong et al. 2019). However, some studies will exclude deaths that can be attributed to an external cause, where the hazard is not thought to be a contributor. This can pose an interesting problem for heatwaves, as a population may progressively fatigue and potentially experience more 'external cause' impacts. In other instances, analysis may focus on only the exposed group to understand the variability of risk within the group. This is particularly relevant where hazard exposure varies significantly within a broad-scale footprint, as is the case with multi-hazard events (e.g. tropical cyclones and volcanic eruptions). Interestingly, in Brown et al. (2017), a key challenge was identified relating the recorded deaths associated with different volcanic eruption events with medical (e.g. laryngeal and pulmonary oedema; asphyxiation and blast trauma) and hazard (e.g. pyroclastic density currents) causes, so that spatial and temporal distributions could be assessed and risks to different populations determined.

A relatively new area of study is in mental health impacts following disasters. Munro et al. (2017) used a cross-sectional survey of those displaced by floods and identified significant increases in depression, and post-traumatic stress disorder (PTSD), while Garske et al. (2021) used social media data to track negative emotions during and after superstorm Sandy. The extent of mental ill health can depend not just on the hazard or whether people were warned but also on the challenges associated with recovery (Mulchandani et al. 2019, Schwartz et al. 2017), which can be directly linked to livelihood and poverty status. New technology is also permitting new approaches to overcome some of the challenges of connecting exposure and response to environmental stresses. For instance, 'wearables' can track a person's exposures and physiological response while undertaking sport or other activities and has the potential to dramatically improve clinical research. As a field reliant on expertise from across science, engineering, analytics, healthcare, business and government, it embodies the collaborative ethos essential for building effective hazard-impact pathways.

In all the studies discussed above, care needs to be taken to allow for confounding variables affecting the impact. In some cases, these may have larger magnitude than the hazard. For instance, the day of the week and public holidays are dominant influences on most health statistics. These can be incorporated in the analysis using auxiliary variables. It is also important to include all potentially relevant environmental variables, not just those hypothesised to be dominant. Thus, temperature, humidity, wind and radiation may act together in cases of heat or cold health impacts. Where causal connections can be established, it may be possible to identify specific groups of people who are vulnerable, and to target warnings at these groups, enabling specific protective responses (e.g. thunderstorm asthma; Dabrera et al. 2013).

Examples of Methods: Impacts on Services Services such as power, water, transport and telecommunications rely on extensive networks of infrastructure that are vulnerable to damage by hazards, either at the nodes of the network (e.g. water treatment plants, power switching stations) or by cutting the connectors (e.g. roads, pipes, cables). Some networks have built in connection redundancy, enabling rerouting to take place if one connector is broken. However, if a key node is taken out of service, it may affect tens of thousands of people, as seen in England in 2007 when a water treatment plant was flooded resulting in the loss of piped water to 420,000 people, several orders of magnitude more than those whose houses were flooded (Pitt 2008).

Reliable data on service impacts are difficult to access, often being commercially sensitive. This results in most work to determine relationships between hazard and service impacts being led by service providers, within their own organisations. Examples of publicly published documents that explore the root causes of such failures are rare, and this has resulted in two very different approaches being used. For public use, the aggregate impact can be analysed statistically, relating media and emergency services reports of impact to the hazard and some aggregate measures of exposure and vulnerability, such as population. The ability to conduct this type of analysis is dependent on the type and consistency of monitoring undertaken by the service provider. This, in turn, can be related to whether regulatory bodies enforce standards of delivery that require monitoring and reporting of service provision (e.g. the number of customer minutes lost, in the case of the energy sector, or fines related to excess sewage leaks, in the case of water companies). For the infrastructure owner, a much more detailed forensic analysis can be undertaken that relates the loss of service to system characteristics such as redundancy and design, operational characteristics (e.g. maintenance schedules) and management characteristics (e.g. availability of technicians on call). This analysis is likely to be mainly process-based, incorporating engineering models of structural failure and computer models of network failure. Ideally, there should be a connection between the statistical aggregated approach and the process-based forensic approach, and this may be possible where infrastructure is in full public ownership.

Impacts to services frequently encompass two impact components within close temporal proximity: the first being immediate (e.g. a broken node, vehicle accident, loss of track due to earthworks failure) and the second being elongated (e.g. loss of service for a period of time; congestion and extended travel times; closures and diversion increasing pressure on the broader network). This means that impact assessments for services need to consider both the drivers that lead to the initial impact and the controls that exacerbate or reduce the secondary/tertiary impacts, to enable a complete 'event' impact assessment. Complex system modelling can provide insights into this impact cascade and support identification of vulnerable or highly exposed hot spots across the network. This can be particularly relevant when trying to understand damage to ecosystem services such as destruction of mangroves, leading to loss of storm protection, or destruction of habitat leading to a reduction in pollinators for crops. However, it is important to consider how this information might be incorporated in forward modelling, for example, should the model focus on forecasting the likelihood of the initial impact (short-chain impact) or look to forecast the total impact of an event (long-chain impact). Similarly, when looking at total impact, it is important to be able to ascertain the range of different impact severities that can be observed so that forward modelling can effectively delineate different scales of impact. We will come back to this in Sect. 5.2.3.

As highlighted previously, service providers may only monitor one of these two impact components or use proxies that partially capture these components, and therefore understanding the full scope of impacts associated with a hazardous event can be very challenging. In addition, as with impacts on people, confounding variables must also be considered. For example, immediate and elongated impacts on road networks can have a number of causes beyond adverse weather, including drivers' abilities, responsiveness, health and behavioural traits (e.g. an audacious individual versus a careful or timid individual); the network's resilience and capacity to absorb shocks; and proximity (in time and space) to available adjacent services that can support recovery. Careful statistical analysis using the same approaches as for epidemiology is needed to identify the part played by the weather. Call et al. (2018) used a cross-sectional approach to identify the contribution of hazardous weather to multi-vehicle traffic accidents on US highways and identified visibility obstruction (due to snow, intense rain or fog) as the primary cause on high-speed roads.

Methods for Financial Impacts Financial impacts of weather-related hazards can be both direct and indirect. It is also often convenient to incorporate financial valuations of intangible impacts representing the human and service losses. Thus, a full analysis of the costs arising from a severe weather event can be very complex even where impact chains are short. Figure 5.2 summarises the main headings under which impacts should be identified. In the top half of the diagram, we have the direct impacts for which financial data should exist, albeit they may be difficult to access. The lower part of the diagram deals with the intangible impacts, whose magnitude first needs to be obtained from data sources such as those discussed under human and service impacts, which then need to be monetised. The tangible and intangible often overlap, and care needs to be taken to avoid double counting. For instance, the direct cost of a traffic accident will include recovery and repair of the vehicle, attendance of the emergency services, recovery of the occupants, any required treatment for injuries and any loss of earnings to the occupants. Indirect impacts will include the effects of shock, discomfort from injuries and the opportunity costs of time lost due to road disruption. The cost of having medical and emergency services available to attend the accident may be part of the direct costs but may also be part of what society pays to reduce the impact of traffic accidents. It must also be noted that 'value' is perception orientated, varying depending on reporting level (e.g. individual, organisation, community or nation). The value reported is therefore closely related to the reporter type and purpose.

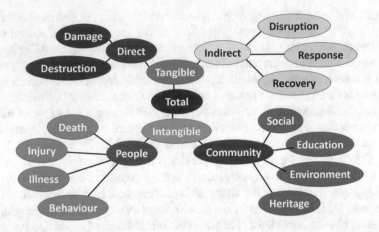

Fig. 5.2 Sources of economic costs of natural hazards. (Adapted from Deloitte, 2016)

The study of financial impacts is part of the science of economics. Financial costs to the individual or business are studied in Microeconomics (see, e.g. Kolmar 2017), while aggregate impacts on national finances are studied in Macroeconomics. Here we are mostly concerned with microeconomic methods, but it should be borne in mind that a disaster produces economic winners as well as losers and that the total cost of a disaster to the nation will be less than the sum of the losses borne by individuals and businesses.

Studies of the cost of weather impacts usually focus on the benefit of an intervention, such as a warning, rather than on the impact itself (e.g. Perrels et al. 2013). However, the methods used are often the same. These methods are summarised in WMO (2015). They may be divided into methods based on historical costs and methods based on people's perception of value. Like epidemiology, economic analysis relies on the application of standard statistical methods (see, e.g. Grant 2018, Cleff 2019).

Data on the financial impacts of severe weather are not systematically collected. Reporting for the Sendai framework (UNDRR 2015) includes regional and sectorial breakdowns of costs, but these are not yet widely available. The main disaster databases, such as EMDAT (EMDAT 2021) and DesInventar (UNISDR 2015), start from insured costs, adding estimates of uninsured and indirect costs when reported, e.g. in the media. Unfortunately, as shown by Panwar and Sen (2019), there is considerable uncertainty in these figures. Apart from insurance payouts, sources of economic data include reported production figures, tax returns, company reports and stock market valuations. Lazo et al. (2011) used state- and sector-level gross domestic product (GDP) data to relate changes in macroeconomic activity to weather anomalies. In doing so, they accounted for external changes in technology and for changes in the level of economic inputs (i.e. capital, labour and energy). An indirect method of assessing loss of business was investigated by Eyre et al. (2020) using the reduced level of social media posts on Facebook to indicate the period of business closure. Direct costs of responding to an emergency can also be obtained from government spending on health and emergency services, on emergency grants to local administrations and on increased social support funding. Panwar and Sen (2020) found a clear signal of increased Indian government spending, increased debt and decreased tax revenue in the 2 years following major flooding events.

Having obtained economic data, analysis of the impact due to the hazard is usually achieved using time-series analysis, relating a change in the impact data to the time of a hazardous event, usually with an allowance for the economic impact to be delayed and to occur over a period. However, cross-sectional analysis may also be used, comparing the changes over the period of the hazard between areas affected and areas not affected. As with the application of these techniques in epidemiology, care must be taken to make the data consistent and to exclude confounding factors. For instance, if monthly earnings are used, it is necessary to adjust for the number of working days. Where impacts in different locations are being compared or aggregated, the composition of the affected populations in terms of demography, gender and economic status needs to be allowed for.

Intangible losses such as deaths and injuries can be given a value, based on loss of potential earnings. Legal liability is usually based on the 'pecuniary' costs associated with the loss – loss of potential income, in the case of death, or costs of treatment and loss of income associated with injury. However, intangible losses are more usually estimated using contingent valuation methods, such as willingness to pay, as described below. Similarly, service losses can be given a value based either on the cost of recovery or on the price that would have been paid for the missed services. However, for critical services such as water and power, where the price is often highly regulated, it is again more normal to value them using contingent valuation methods. This approach is also normally used for valuing ecosystem services such as clean air and water.

Contingent valuations may be estimated using a variety of survey techniques. The most commonly used is willingness to pay (WTP). For instance, it is widely used to place a notional average value on a life (VSL, the Value of a Standard Life) or injury, for economic applications. Its use requires expert input, or the results may not be credible. The US Department of Transportation (DOT; Moran and Monje 2016) established a VSL of $9.6 million in 2015 based on WTP. They also monetised the value of injuries as fractions of VSL. Cho and Kurdzo (2019) used their data to estimate the economic value of the US radar network in reduced injuries and deaths from tornadoes. This approach requires that the estimate given when people are surveyed is consistent with their actual behaviour, at least in an average sense across the surveyed sample. An alternative approach seeks to estimate more directly how people value intangible losses by their behaviour and the costs they voluntarily bear. For instance, people will pay a premium for a more expensive car with extra safety features; or a worker may look for a premium for working in a job that is vulnerable to the weather. Again, expert design is needed to disentangle different influences as, for instance in the case of waterfront properties that are both more vulnerable and more desirable.

The value of weather-related losses may also be estimated from the costs that society bears to reduce or prevent losses, for example, the cost of aviation safety features such as the airport wind-shear radar systems (Hallowell & Cho 2010) or the cost of winter road maintenance (Venäläinen & Kangas 2003). However, there is a danger of creating a circular argument if WTP was used to justify the installation in the first place. The cost of ambulance and emergency medical services may similarly be associated with a perceived valuation of the deaths that they prevent and the injuries that they treat.

It is sometimes possible to transfer valuations from one context to another, for instance between regions of a country. There are many studies investigating the value placed on particular medical treatments. Where those treatments are associated with recovering from a weather-related injury or disease, the valuation may be transferred. More generally, where similar impacts have different causes, the loss estimates should be similar. However, caution should be taken when transferring studies between countries, as valuations may be strongly influenced by country-specific economic and cultural factors.

All these approaches have the weakness that they depend strongly on the wealth of society. This may not matter too much for studies internal to one country, but when making international comparisons, it is not satisfactory. One way of circumventing this is to relate all financial costs to the household income of the individual, the turnover of the business or the GDP of the country.

5.2.3 Forward Modelling of Impacts

Having identified a cause and effect association that is relevant to the end user's decisions, it must be turned into a predictive tool to be of use. Approaches to impact modelling range from simple overlaying of hydrometeorological information with vulnerability and exposure datasets to produce qualitative statements about potential impacts (e.g. Robbins & Titley 2018), via statistically linking hazard magnitude (e.g. weather parameters or flood depths) to observed impacts, to formally quantifying the risk and impacts of events as a function of hazard likelihood, vulnerability and exposure. The decision on which approach is appropriate to implement depends on the strength and completeness of the hazard-impact relationship, the needs of the user, the data available for forward modelling and the required resolution and timeliness of model output. Often, statistical analysis can determine at what magnitude of hazard impacts may start to occur. This is particularly the case where the collected impact data used in the analysis were binary (impact or no impact). Where impact data are continuous in nature, it may be possible for the statistical analysis to identify break points or step changes where a change in hazard magnitude results in a different severity of observed impact (e.g. health impacts associating with increasing or decreasing temperature). This can allow thresholds to be established which can be used by hazard modellers to produce more informative impact-orientated forecasts. For some users, it may be enough that they know where and when to expect impacts (of any type). Others may need to understand the spatial and severity variability of potential impacts over time and have these described in terms of the different types of impacts that may occur (e.g. Aldridge et al. 2020). For yet others, a specific quantification of impact (e.g. the number of homes that could be flooded) may be required for them to prioritise and make appropriate decisions. To obtain this detail, impact modellers can employ a range of techniques, but the main underpinning requirements include (1) a reliable description of the hazard and (2) a way to describe the 'consequences' of the interacting environmental hazard. The former will be discussed in Sect. 5.3, while the latter will be outlined below.

Where thresholds cannot be determined to address future potential risk, or where thresholds only address part of the risk assessment, additional information in the form of vulnerability of individuals, properties or infrastructure can be combined with their exposure to determine likely consequences (impacts) of the hazardous event. Vulnerability and exposure are often discussed as though they are well-defined characteristics. Exposure as defined by the UNDRR (2017) is arguably the easier to describe as it represents a measure (number) of people or tangible assets that are in a hazard-prone area. For physical exposure, this is often considered a fixed problem that can be solved by obtaining appropriate spatial data (e.g. satellite, Lidar, mapping surveys, traffic count point data), either by physically surveying an area or by purchasing commercial datasets. While such an approach is an important building block, it is by no means sufficient for its accurate representation in impact modelling. Firstly, it must be kept up to date. The optimum update frequency of exposure data is challenging to determine and varies depending on the type of data being used and the decisions that need to be taken. For building stock, road networks and agricultural elements (e.g. crop types), the datasets may need to be updated at least annually. For livelihood data, updates may only be needed on five-yearly timescales; however, large-scale shocks to the area where the data were obtained (e.g. conflict or mass migration) might radically change the data and mean that immediate update would be needed for the exposure to still be representative. Secondly, it is necessary to know how the population exposure varies with time. This is most easily illustrated by considering the exposure of children to a tornado. At night they are at home, so their exposure is the same as the other members of their household. During school hours they will be at school perhaps many miles from home in a building of different construction with different shelter possibilities. Before and after school, they may travel by car or bus on a public highway to reach a third location, perhaps playing sport. At weekends or during the summer break, they may visit relatives or undertake other recreational activities, possibly leaving the area altogether to be replaced by visitors who are unfamiliar with local hazards. Each different location has different exposures and vulnerabilities.

In the absence of real-time exposure data, impact modellers can use existing trends, if identifiable from historical data, to model the dynamic behaviour of people and assets so that this temporal variability can be captured. By way of example, the vehicle over-turning (VOT) model (Hemingway and Robbins 2020) forecasts the risk of disruption due to vehicles overturning in strong winds. Exposed elements are the vehicles on the road, counted through manual and automated count points, across the transport network. These data are used to map the average temporal variability of traffic flows by vehicle and road type on an hourly basis. Using this information, a time-varying, average definition of exposure on the road network is used within a risk calculation. However, this is based on historical average usage of the network and does not account for changes in exposure due to road closures associated with road maintenance activities, or short-term shocks to the system that may dramatically reduce traffic flow. It is therefore important that the assumptions and caveats used in the model are clearly documented and transparent for downstream users, so that risk forecasts can be interpreted effectively. Another way to model drivers' dynamic exposure, especially when also interested in the socio-demographic characteristics of those exposed, is to use recent advances in mobility models following an activity-based approach. This framework, used to micro-simulate individual travel-activity patterns, considers travel behaviour as derived from the demand of activity participation and aims to predict the sequence of activities undertaken by individuals (McNally 1995). Activity-based models are of increasing interest for dynamic exposure assessment, as seen in air pollution exposure studies (Beckx et al. 2008, Beckx et al. 2009). Flood exposure studies can also benefit from the rich information provided by this kind of mobility modelling. Indeed, combining individual travel-activity simulations with road flood forecasts enables a thorough assessment of motorists' exposure and its evolution in time and space, relative to the flood hazard (Shabou et al. 2017, 2020).

Vulnerability is harder to define. Ways of defining, measuring and assessing vulnerability vary considerably across research disciplines (Wisner 2016). One reason for this is that vulnerability is often the result of numerous interrelated factors. Several studies use composite metrics or indices which pull together proxy indicators to provide an operational representation of a characteristic or quality of a system (Birkmann 2006, Fuchs et al. 2018), and describe the individual aspects of an asset that increase or reduce vulnerability to a particular hazard. The choice of number, type, weighting and integration method of the indicators is dependent on available data and also the complexity of the risk being modelled. These decisions have large implications on the resulting risk and impact assessment and ultimately on the downstream utility of the information for decision-making. As with exposure, vulnerability is temporally and spatially varying, and identifying ways to express this for forward modelling is important. Terti et al. (2019) used a supervised machine learning technique to link historic impact observations of flash flood human losses with social exposure and vulnerability proxy data in order to predict the outbreak of impact (e.g. fatalities, injuries) within a flash flood or fast-evolving weather event. This type of approach relies on a large set of reliable and precise impact data which, when available, allows the critical interplay of flood water and human mobility to be accounted for at hourly time steps. Alternatively, rapid vulnerability assessment in the wake of humanitarian crises (WFP 2018) can support a better understanding of changing vulnerabilities, as can the use of earth observation data and, potentially, social media. Updates to the vulnerability and exposure then need to be pulled into the impact model so that it has the best representation of current conditions and enables a more accurate impact assessment. Approaches to do this effectively are still being developed for short-term, routinely run impact models, but such updates are critical in instances where modellers are interested in capturing multi-hazards and long-chain impacts, and in fragile and conflict-affected regions.

Most physical science impact models focus on describing physical vulnerability (i.e. the potential for physical impact). This is because most of these models look to identify direct and tangible impacts. However, in cases where the cascade of impacts is important (e.g. social protection), other forms of vulnerability (economic, social, cultural/environmental, psychological) become important (see, e.g. Cutter et al. 2003, Babcicky & Seebauer 2021). For instance, in large cities, especially those growing rapidly in developing countries, migrants and poor people often create informal settlements in open areas that have not been developed because of high exposure to hazards such as flooding, landslides or land contamination. Without money they are likely to be poorly nourished, increasing their vulnerability to disease from flood water. Without power and communication, they will likely not receive any warning of an imminent flood, and in any case, they may not understand the language of the warning. Without transport, they may not be able to respond to the warning even if they receive it. These issues are not restricted to developing countries as highlighted by Wolshon (2006) who identifies the lack of transport of over 100,000 poorer people in New Orleans as the major failing of the evacuation when Hurricane Katrina hit in 2005.

As illustrated in the context of exposure data, where the hazard-impact relationship is based purely on historical evidence, the predictive model needs to be constrained to behave sensibly when outside the range of historical data, and to incorporate any anticipated extension of historical trends (both in terms of possible hazard magnitudes and impacts). The probability distribution from the historical analysis needs to be combined with uncertainty information from the hazard prediction and the results scrutinised in terms of the ability of the model to distinguish between occasions when action should and should not be taken, to see if they provide useful information for decision-making. As with any model, developing an impact model involves making choices and assumptions about the relevant pathways to impacts, the relevant aspects of vulnerability, the appropriate scale or level of detail or the value of some threshold or parameter. The impact model itself therefore adds to the overall uncertainty in the impact predictions, over and above the uncertainty in the hazard. Generally speaking, modelling uncertainty can be attributed to two main causes: (1) parameter uncertainty, arising from the impossibility to find exact parameter values due to lack of data, imperfect process understanding and the need to use approximations, and (2) structural uncertainty, related to how processes are being represented, aspects that were omitted or computational limitations on, for example, resolution. Note there is also a residual element of unpredictability arising from inadequacies of the models, limitations to the validity and completeness of our knowledge or simply inherent unpredictability in the process being modelled. Techniques exist to explore, understand and quantify model uncertainty, for example, through global or one-at-a-time sensitivity analysis of the key parameter values, or by exploring alternative model structures. Impact forecasts should always be accompanied by estimates of their uncertainty so that the warning can be based on a realistic risk assessment.

5.3   Capabilities of the Hazard Forecast

Chapter 6 will describe hazard prediction more broadly, but here we are concerned with understanding the ability of forecasts to capture those aspects of the hazard that are most relevant to its impact. The level of detail available in an operational warning system will depend on whether the hazard forecast is produced as part of a general portfolio of information or if it is a bespoke service. Accuracy depends on the quality of the underlying meteorological and environmental models and their ability to assimilate observations. Reliability requires a probabilistic approach, but also depends on effective feedback of verification to forecast improvement.

Prediction of impact requires relevant information about the hazard. In general, the restrictions on hazard data access are less than those on socio-economic data. However, while the analysis may show a strong relationship between hazard and impact, the specific predictors required may be much more difficult to access than standard weather variables. A simple example is lightning, which kills many people worldwide, but which is not generally predicted by weather models and for which forecasts tend to be very general. This is not helped by the fact that a lightning bolt can travel 10 km or more between cloud and ground.

Hazard forecast capabilities vary significantly among different hazards. The most damaging impacts are flood- and wind-related. For major river floods, the meteorological input may be predictable for a week or more ahead, and the travel time for the flood can add to this. However, details of the flood depth may be critical for impact and are dependent on highly uncertain knowledge of the river, including vegetation and sediment, and the state of repair of levées. While storm surges are often predictable days in advance, their inshore growth is extremely sensitive to the shape of the bathymetry, and thus to the track of storm winds. Flash floods are sensitive to errors in both location and intensity of the causal rainstorms, while urban surface water flooding occurs on space scales too small for proper resolution in current models. Wind hazard predictions have corresponding limitations due to the influence of topography and the built environment. Damage is often caused by gusts of a few seconds duration that are not directly predicted by models. The limitations of winter hazard forecasts are particularly associated with their sensitivity to the proximity of the freezing point, both at the surface, for frost, ice and the accumulation of snow, and above the surface, for freezing rain and ice storms. Hazards associated with severe convection, such as tornadoes and large hail, are inherently unpredictable given their small spatial and temporal scales, and the rapid development of the parent storms. Wildfire growth and movement is sensitively dependent on the interaction of the local wind and topography. While temperature is generally a well-predicted variable, its detailed distribution around and within buildings is not currently predictable. The same is true for pollution, exacerbated by a lack of real-time knowledge of emissions.

This brief summary of the limitations of hazard forecasts emphasises the dangers of a mechanistic approach to taking hazard data and turning it into a deterministic impact prediction tool for use in a warning. However, for each of these hazards, there is a degree of predictability present that enables a probabilistic approach to provide usable impact information. Since the hazard is only predictable in a probabilistic sense and the impact is related to the hazard statistically, proper assessment of risk requires the appropriate combining of these sources of uncertainty. Whereas impact probabilities are based on statistics of historical association, time-dependent hazard probabilities should be obtained from ensemble prediction systems, when these are available. Careful analysis is required to ensure that the resulting risk assessment includes all relevant sources of uncertainty while avoiding double counting.

Lead times for accurate hazard predictions are important for warnings and vary widely according to the hazard. Prediction of the location of a tornado is only possible a few minutes ahead, whereas a major river flood may be predictable a week or more in advance. For very fast response, provision of impact forecasts fully automatically from the hazard inputs can be very attractive. However, care must be taken with quality control of the hazard inputs. This should start with ensuring that spurious hazard values are not used – for instance due to spurious echoes from radar data in a precipitation nowcast. Empirically based impact models have a limited range of application, due to constraints with the training data, so outputs should not be used automatically for hazard values beyond or even near those limits. As discussed above, outputs should be probabilistic. Where a fully automated system is in use, outputs with large uncertainty or that exceed historical norms should automatically be flagged for inspection before issue.

For impact prediction, hazard forecasts need to be evaluated in user-relevant terms. This places demands on the availability of hazard observations, which will be addressed in more detail in Chap. 6. The examples above hint at some of the challenges in selecting an appropriate variable and range of values to include in any evaluation. Where the hazard is very local, the model may not be representative of the same area as the observation, requiring downscaling of the forecast or upscaling of the observation before a meaningful comparison can be made.

5.4   Bridging the Gap Between Impact Forecaster and Hazard Forecaster

A traditional epidemiologic, economic or engineering study is often undertaken as a one-off project by an academic or consultant using a conveniently accessible impact dataset. They obtain hazard and exposure data from the easiest (or cheapest) available source, then perform the detailed statistical analysis, draw conclusions, publish the results and move on to the next, potentially unrelated, study. Subsequent application to warnings may be undertaken, independently, by a public or private hazard forecasting organisation, which selects the most accessible published impact relationship for translation into a predictive tool. In this process, the initial epidemiology suffers from a lack of understanding of the possibilities and limitations of hazard science, while the warning application misrepresents the impact data and the limitations of the statistical analysis.

Development of effective operational impact warning services requires sustainable multidisciplinary partnerships to overcome these challenges. Each collaborator in the partnership builds awareness of data issues in partner disciplines leading to new best practices within and between collaborating partners.

Experience has shown that co-design of hazard/impact studies benefits from governance structures, such as:

- Steering groups – responsible for ethics, sanctioned analysis techniques that will deliver statistical confidence and outputs suitable for operational use.
- Working groups – responsible for negotiating adjustment and supply of data and execution of analysis.
- Stakeholder reference groups – responsible for feedback on how analysis results can be deployed in operational impact warning environments.

Co-design between the groups ensures meteorological hazard data are structured to match exposure and vulnerability data structures, enabling statistical analysis to be executed within the ethical and procedural constraints of the impact sector (e.g. health or engineering sector) to produce outputs at the highest statistically significant resolution possible to address user requirements.

5.4.1 Matching Data Needs

As we have seen, impact data are generally not openly available except in highly aggregated forms. This is true for engineering impacts, health impacts, infrastructure service impacts and business impacts. While data confidentiality is the primary barrier, there are also technical barriers arising from standards and formats, especially when the measured or observed quantity is highly specialist. Health practitioners use standardised disease, illness and cause of death (impact) codes to categorise illness and disease, while codes of practice support quality control and enshrine ethical practices for sustainment of life and privacy. These approaches mean that the release of health data to external researchers can be slow, degrading the value of subsequent impact studies (as described in Sect. 5.2.2).

Hazard data are available from a bewildering variety of sources with different characteristics. High-quality in situ observations are sparse in space, while remotely sensed observations only indirectly capture the variable of interest. Models provide ideal datasets, but even reanalyses have inaccuracies and biases that may distort the analysis. Hazard models often generate gridded data which are more easily adapted to match less flexible health and social data constraints. Hazard modelling centres often only archive a small subset of output data. Since impact models require to be fitted to long time series, that can be a key determinant in what data are best to use – with corresponding constraints on the resulting prediction system. Early identification of archive issues should be on the agenda for new partnerships, so that required data can be retained. Care must be taken when impacts are dominated by small scales that corrections are made for altitude, shelter, proximity to water, etc. where necessary. The differences between these various data sources can significantly affect subsequent statistical analysis and must be consistent with the anticipated operational usage.

While national-scale impact modelling may be carried out by generalist statistical modellers and used for advice to governments or international organisations, detailed modelling for warnings at individual or community level requires the involvement of the data owner. Except where the impact is very direct, they may be sceptical of any link with the hazard, or of any value in identifying such a link. In such cases, establishing a mutually beneficial relationship is necessary before attempting any modelling.

Building successful impact models requires matching of hazard data and impact data: their temporal and spatial specificity and the variables and regimes of relevance. Models are often built using thresholds, usually in the hazard, but often driven by the significant thresholds in the impacts. It is essential to determine the abilities of models to predict these thresholds before making use of the results. Threshold exceedance should always be predicted probabilistically, both for hazards and for impacts. At the same time, some hazard prediction errors may be unimportant for the impact, and it may be possible to use a simpler, faster forecasting tool to produce the required information.

Impact modelling partnerships can produce results more suited to implementation of operational hazard-impact warning systems when partners co-design the hazard exposure/impact analysis research. Human health impact data custodians retain ethical management of experiment design and how the results are released. Social data custodians can equally address their data management requirements during the co-design phase. The ability to include social support, income, housing and census among similar data types is a very powerful determinant for health and economic outcomes. Similarly, physical spatial data allow the use of building stock age, quality, density and percentage of green areas for natural cooling as exposure controls for impact studies.

5.4.2   Evaluation

Model evaluation should be carried out in terms relevant to the end user and on independent data from those used in the historical analysis. For rare hazards, this may pose a challenge of achieving significance, especially if there are confidentiality constraints on use of the data, as in the health sphere. The development of effective partnerships can enable and enhance the robustness of evaluation techniques, while the sharing of such information can also enhance the partnership by providing a harmonisation of scientific understanding and focus direction. Impact model evaluation requires the verification of multiple components (e.g. hazard forecasts, exposure, vulnerability, thresholds, impact and risk forecasts, warnings), and the decision on appropriate metrics is therefore critical to obtain a complete picture of model performance. This is particularly important when statistical models are being used, as changes in verification metrics may indicate the need for recalibration, either of the hazard inputs or of the impact model itself. Both input and output verifications are important, and both make heavy demands on data. Input verification needs to be relevant to the thresholds of significance for the impact model. So, for instance, where a hazard severity index is the input, it may be that only performance in the upper 5% is relevant. Co-designing the appropriate ranges, metrics and methods for verification is an important part of the partnership. Verification of the impact outputs involves access to routine impact data, possibly requiring regular post-event surveys (e.g. as used by Taylor et al. 2019) or routine extraction and processing of social media (Spruce et al. 2021) and other data sources (e.g. as used by Robbins & Titley 2018). The data acquisition methods should be standardised so that successive verification gives consistent results. This is by no means a straightforward process for impact data, particularly where data sources may be available intermittently, in non-direct formats and where classification of impacts (based on severity or type) depends on the aggregation of multiple sources. Careful consideration is needed in how data are classified and how this classification is processed for the purposes of evaluation. The nature of impact databases (often skewed towards larger-magnitude events), the style of classification or standardisation of the impact data and the time it takes to produce consistent and reproducible data may influence when evaluation and verification can take place, and partnerships need to be aware of this as it has implications for project longevity. It is important therefore that verification approaches are co-designed with end users, so that outputs are relevant to their needs and inform best use of the model, and with hazard modellers, so that sensitivities of the model can be assessed and guide improvements to the hazard prediction.

5.5 Examples of Partnership

Box 5.1 Joint Australian Tsunami Warning Centre (JATWC)
John Nairn

The Australian Tsunami Warning Service (http://www.bom.gov.au/tsunami/about/jatwc.shtml) has been sustainably developed and delivered through the Joint Australian Tsunami Warning Centre (JATWC). Plate tectonic and seismic monitoring skills within Geoscience Australia have been virtually coupled with the Bureau of Meteorology's ocean monitoring, modelling and continuous, non-stop message creation and dissemination systems. Curiously, JATWC's greatest success has been the overwhelming number of 'no threat' warning messages issued to the Australian community, countering the

(continued)

proliferation of warnings issued by numerous international warning centres. This service has very successfully curtailed inappropriate community reactions to no-threat seismic events and built community confidence for very rare tsunami warnings.

The bridge between hazard and impact is built upon the foundations of the data paradigms that underpin each discipline. The hazard gap is not very large between two geophysical disciplines in the example of the JATWC, although it is very clear that the best way to build the service was deemed to be through a joint facility where trust is codified, operationally tested and refined.

Coastal inundation from a tsunami is a public health threat. Flooding can result in drownings, and risks from exposed electricity hazards and infection sources including raw sewage and contaminated food and water supplies. Infrequent and widely distributed events generate a wide range of data sources which usually require social science practitioners to locate and document the nature and scale of inundation impacts. The JATWC hazard partners require these data to verify warnings issued. The opportunity to evaluate the service is naturally limited by the infrequency of coastal inundation attributable to tsunamis. It is extremely hard to build trust where health communities are so infrequently exposed to warnings, although it might be argued that confidence may be growing through the issuance of 'no-threat warnings'.

Box 5.2 JAM Partnership
Sarah Millington

JAM (Joint Agency Modelling) is a tool for UK national emergency response to an atmospheric radiological release anywhere in the world (Millington et al. 2019), initially developed in response to the radiological atmospheric release from the Fukushima Daiichi Nuclear Power Plant in 2011. A partnership among the Office for Nuclear Regulation (ONR), Met Office, Public Health England, Environment Agency, Food Standards Agency, Scottish Environment Protection Agency, Northern Ireland Environment Agency, Food Standards Scotland and Natural Resources Wales provides an operational modelling system and delivery of agreed guidance using expertise and scientific software supported by training and routine tests. It is funded by UK government to provide input to the Scientific Advisory Group for Emergencies (SAGE) on the impacts of a radiological release from a nuclear facility. SAGE is a group of experts chaired by the Government Chief Scientific Advisor to deliver coordinated scientific advice to aid central government on forming the strategic emergency response.

(continued)

The process begins with notification of an incident at a nuclear facility to the JAM partners. The next step requires a source term – information about what has and what might be released, how much, when and where. This is provided by the nuclear facility operator or, if unavailable, by ONR using an agreed fixed format pro forma. The source information is fed into the modelling system, with NWP data and receptor parameters (e.g. assumed age for the dose calculations: infant, 10-year-old or adult). Several JAM partners have contributed to the modelling, but, for efficiency, all modelling is carried out at the Met Office.

At the heart is the Numerical Atmospheric-dispersion Modelling Environment (NAME), a Lagrangian particle-trajectory model used to model the atmospheric transport and dispersion of gases and particulates (Jones et al. 2007). Emissions are simulated as computational particles, advected by the NWP three-dimensional wind field with turbulent dispersion simulated using random-walk methods. The particles evolve with time, e.g. to simulate radioactive decay, dry and wet deposition. Time-integrated activity concentrations in air and in material deposited on the ground are estimated to provide health impact assessments.

A dose model provided by Public Health England calculates effective doses that are used to show potential areas where authorities should take action, such as sheltering indoors. The Food Standards Agency provides deposited values of radionuclides that would result in exceeding European standards for cow's milk and leafy green vegetables; these are compared with the estimated deposited values to indicate areas where restrictions should be considered. The Environment Agency applies a similar approach to indicate the geographical area of potential impact on surface water abstraction for drinking water.

Partners scrutinise model outputs and form a consensus brief for SAGE by teleconference, incorporating health impacts, food impacts, surface water impacts and uncertainties in the modelling. SAGE and government agencies use the brief with other scientific information to form guidance for those leading the national emergency response, typically led by a senior government minister. Updates to source terms and NWP data or questions from SAGE would initiate further cycles of the JAM process for as long as required (Fig. 5.3).

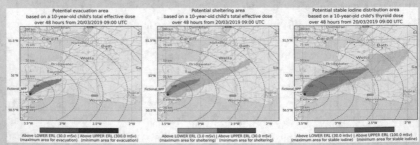

Fig. 5.3 Display of potential areas for protective health actions: (**a**) evacuation area, (**b**) sheltering area (**c**) distribution area for stable iodine in a fictional case based on the Chernobyl release using km-scale regional NWP data. (© Crown Copyright 2021, Met Office)

Box 5.3 Australian Heat-Health Partnership
John Nairn

Extreme heat is best understood as the accumulation of heat from persistent high temperatures. Unfortunately, media demands for simple headlines have focused on maximum temperature as the single determinant of heatwave severity. The minimum temperature is more important. As a result, heat-impacted sectors of the community have little credible experience of the impact of extreme heat. This is further confounded by the inconsistent inclusion of humidity, which is applied inconsistently and has been unnecessary for the development of response interventions as has been shown in Australia's heatwave service.

The Bureau of Meteorology's national heatwave service is based upon a location-based percentile treatment of 3-day average maximum and minimum temperatures (excess heat factor, EHF; Nairn and Fawcett 2014). This heatwave severity classification index determines the presence of low-intensity, severe or extreme heatwaves for each 3-day period. Several epidemiological studies have determined that rising EHF is a generally good determinant of escalating heatwave impacts across Australia. It would be highly misleading to suggest this arose from execution of a well-planned health and meteorological collaboration. Development of the EHF provided new insight into the physical nature of heatwaves and how they have changed within the climate record, past and projected (Nairn & Fawcett 2013). Public Health researchers found EHF attractive and invited collaboration with the author, leading to publication of a sequence of Public Health research articles demonstrating the efficacy of EHF. International support for an Australian heatwave service model was investigated with the support of a Churchill Trust Fellowship, which then led to the launch of the Bureau's pilot heatwave service in January 2014. This service was launched with the assistance of a heatwave stakeholder reference group, recruiting health and emergency service agency participation from across the nation. Pre- and post-season meetings shared performance data and gathered suggestions for how the service could improve.

Epidemiological studies have shown that EHF is an effective predictor of heatwave impact (Scalley et al. 2015, Jegosarthy et al. 2017, Williams et al. 2018). It has also been shown to be effective across climate (Perkins et al. 2012, Nairn & Fawcett 2017), multi-day (Nairn et al. 2018), multi-week forecasts (Hudson & Marshall 2016) and CMIP projection timescales (Wang et al. 2018). Deeper collaboration is required with health institutions to develop rapid verification of heatwave impact if the community is to build confidence in extreme heat warnings. Australia is still growing effective multi-agency partnerships, with significant progress undertaken through the recent (2020) completion of the Reducing Illness and Lives Lost from Heatwaves (RILLH)

(continued)

project. The Bureau of Meteorology led this collaboration of the Australian Bureau of Statistics (ABS), Department of Health (DOH) and Geoscience Australia (GA) to develop individual, suburban, city and regional heatwave vulnerabilities across Australia at regional, city and neighbourhood scales. Further work is required to understand how the results can inform behavioural recommendations for impact-based warnings that are sensitive to location.

5.6 Summary

- Data are fundamental to all impact forecasting. High-quality data are typically only available through partnership with data owners. Mapping data are required for exposure, socio-economic data for vulnerability and health, economic and service data for the impacts themselves. Data need to be accurate and consistent, requiring that they are collected, processed and updated to defined standards.
- Relevant impacts differ between end users. Whereas the public are interested in impacts on individuals, government, businesses and humanitarian organisations may be more interested in aggregate impacts. The methods and the data need to be selected according to the application.
- Hazards produce both tangible and intangible impacts, cascading from direct impacts to several levels of indirect impacts. They may be categorised into human, financial and service impacts, each of which can be translated to a financial value. When comparing financial impacts, it can be helpful to normalise according to household income, business turnover or country GDP.
- Impacts can be forecast by modelling failure processes or by fitting statistical models to historical data. Selecting the appropriate model requires an understanding of the information required by the end user as well as consideration of the available data.
- Hazard forecasts need to be carefully processed for use in impact models with any biases removed. They should also be probabilistic – with unbiased probabilities. For many applications they may need to be site-specific.
- To avoid the pitfalls in impact forecasting, it is essential that the provider of the hazard information and the impact model developer work closely together. Their understanding of the end user's problem will be very different, so prior to model development, it is essential that these differences are shared openly and the approach to be adopted is mutually agreed.
- Once a model is in use, it is important that the information user (e.g. the warner or emergency manager) has access to the expertise behind the model in order to query performance in critical situations. Ideally the relationship should be long term so that updates can be incorporated to use recent data or improved modelling.

References

Aldridge, T., O. Gunawan, R. J. Moore, S. J. Cole, G. Boyce and R. Cowling, 2020. Developing an impact library for forecasting surface water flood risk. J. Flood Risk Man, **13,** e12641. https://doi.org/10.1111/jfr3.12641

Armitage, P., G. Berry and J. N. S. Matthews, 2002. *Statistical Methods in Medical Research.* Blackwell Science Ltd. DOI:https://doi.org/10.1002/9780470773666

Armstrong, B., F. Sera, A. M. Vicedo-Cabrera, R. Abrutzky, D. OudinÅström, M. L. Bell, B-Y. Chen, M. S. Z. S. Coelho, P. M. Correa, T. N. Dang, M. H. Diaz, D. V. Dung, B. Forsberg, P. Goodman, Y-L. L. Guo, Y. Guo, M. Hashizume, Y. Honda, E. Indermitte, C. Íñiguez, H. Kan, H. Kim, J. Kyselý, E. Lavigne, P. Michelozzi, H. Orru, N. V. Ortega, M. Pascal, M. S. Ragettli, P. H. N. Saldiva, J. Schwartz, M. Scortichini, X. Seposo, A. Tobias, S. Tong, A. Urban, C. D. C. Valencia, A. Zanobetti, A. Zeka and A. Gasparrini, 2019. The Role of Humidity in Associations of High Temperature with Mortality: A Multicountry, Multicity Study. *Environ. Health Perspectives,* **127(9).** https://doi.org/10.1289/EHP5430

Babcicky, P. and S. Seebauer, 2021. People, not just places: Expanding physical and social vulnerability indices by psychological indicators. J. Flood Risk Man., 2021;e12752. https://doi.org/10.1111/jfr3.12752

Beckx, C., R. Torfs, T. Arentze, L. Panis, D. Janssens and G. Wets, 2008. Establishing a dynamic exposure assessment with an activity-based modeling approach: methodology and results for the Dutch case study. *Epidemiology,* **19,** S378–9.

Beckx. C., L. Panis, T. Arentze, D. Janssens, R. Torfs and S. Broekx, 2009. A dynamic activity-based population modelling approach to evaluate exposure to air pollution: methods and application to a Dutch urban area. *Environmental Impact Assessment,* **29,** 179–185. DOI:https://doi.org/10.1016/J.EIAR.2008.10.001

Birkmann J., 2006. Measuring vulnerability to natural hazards. UNU Press 55–77

Brown S. K., S. F. Jenkins, R. S. J. Sparks, H. Odbert and M. R. Auker, 2017. Volcanic fatalities database: analysis of volcanic threat with distance and victim classification. *J. Applied Volcanology,* **6,** 15 https://doi.org/10.1186/s13617-017-0067-4

Call, D. A., C. S. Wilson and K. N. Shourd, 2018. Hazardous weather conditions and multiple-vehicle chain-reaction crashes in the United States. *Meteorol. Appl.,* **25,** 466–471. DOI: https://doi.org/10.1002/met.1714

Campbell, S. L., P. J. Jones, G. J. Williamson, A. J. Wheeler, C. Lucani, D. M. J. S. Bowman, F. H. Johnston, 2020. Using Digital Technology to Protect Health in Prolonged Poor Air Quality Episodes: A Case Study of the AirRater App during the Australian 2019–20 Fires. *Fire,* **3,** 40. https://doi.org/10.3390/fire3030040

Cavalière, C., P. A. Davoine, C. Lutoff and I. Ruin, 2020. Geolocated tweets as a means of observing extreme natural events. first specifications. In: *Mobility in the Face of Extreme Hydrometeorological Events* 2. ISTE, 149–175.

Cho, J. Y. N. and J. M. Kurdzo, 2019. Weather Radar Network Benefit Model for Tornadoes. *J. Appl. Meteorol. Clim.,* **58,** 971–987, DOI: https://doi.org/10.1175/JAMC-D-18-0205.1

Cleff, T., 2019. *Applied Statistics and Multivariate Data Analysis for Business and Economics: A Modern Approach Using SPSS, Stata, and Excel.* Springer.

Cutter, S. L., Boruff, B. J. and Shirley, W. L. (2003). Social vulnerability to environmental hazards. Social Science Quarterly, 84(2), 242–261. https://doi.org/10.1111/1540-6237.8402002

Dabrera, G., V. Murray, J. Emberlin, J. G. Ayres, C. Collier, Y. Clewlow and P. Sachon, 2013. Thunderstorm asthma: An overview of the evidence base and implications for public health advice. *Int. J. Med.,* **106,** 207–217. DOI: https://doi.org/10.1093/qjmed/hcs234

Davison C. R. and T. A. Rutke, 2014. Assessment and characterization of volcanic ash threat to gas turbine engine performance. *J. Engineering Gas Turbines Power,* **136.** https://doi.org/10.1115/GT2013-94079

Deloitte Access Economics, 2016. *The economic cost of the social impact of natural disasters.* Australian Business Roundtable for Disaster Resilience & Safer Communities,

Diakakis, M., G. Deligiannakis, E. Andreadakis, K. N. Katsetsiadou, N. I. Spyrou and M. E. Gogou, 2020. How different surrounding environments influence the characteristics of flash flood-mortality: The case of the 2017 extreme flood in Mandra, Greece. *J. Flood Risk Man.*, **13**, e12613. DOI: https://doi.org/10.1111/jfr3.12613

EM-DAT, 2021. CRED / UCLouvain, Brussels, Belgium – www.emdat.be (D. Guha-Sapir)

Eyre, R., F. De Luca and F. Simini, 2020. Social media usage reveals recovery of small businesses after natural hazard events. *Nature Communications*, **11**, 1629. https://doi.org/10.1038/s41467-020-15405-7

Fiorino L., V. Macillo and R. Landolfo, 2019. Shake table tests of a full-scale two-story sheathing-braced cold-formed steel building. *Engineering Structures*, **151**, 633–647. DOI:https://doi.org/10.1016/J.ENGSTRUCT.2017.08.056

Fleming, L. E., A. Haines, B. Golding, A. Kessel, A. Cichowska, C. E. Sabel, M. H. Depledge, C. Sarran, N. J. Osborne, C. Whitmore, N. Cocksedge and D. Bloomfield, 2014. Data Mashups: Potential Contribution to Decision Support on Climate Change and Health. *Int. J. Environ. Res. Public Health*, *11*, 1725-1746. https://doi.org/10.3390/ijerph110201725

Fuchs S., T. Frazier and L. Siebeneck, 2018. *Vulnerability and resilience to natural hazards*. Cambridge University Press 32–52. ISBN: 9781107154896

Garske, S. I., S. Elayan, M. Sykora, T. Edry, L. B. Grabenhenrich, S. Galea, S. R. Lowe and O. Gruebner, 2021. Space-Time Dependence of Emotions on Twitter after a Natural Disaster. *Int. J. Environ. Res. Public Health*, **18**, 5292. https://doi.org/10.3390/ijerph18105292

Grant, D., 2018. *Methods of Economic Research: Craftsmanship and Credibility in Applied Microeconomics*. Springer. ISBN: 978-3-030-01734-7

Grossi, P. and H. Kunreuther, (Eds.) 2005. *Catastrophe Modelling: A new approach to managing risk*, Springer. ISBN 978-0-387-23129-7

Hallowell, R. G. and J. Y. N. Cho, 2010. Wind-shear system cost-benefit analysis. *Lincoln Lab. J.*, **18**, 47–68.

Hemingway, R. and J. Robbins, 2020. Developing a hazard impact model to support impact-based forecasts and warnings: The Vehicle OverTurning (VOT) Model. *Meteorol. Appl.*, **27**, e1819. https://doi.org/10.1002/met.1819

Hudson D. and A. G. Marshall, 2016. Extending the Bureau of Meteorology's heatwave forecast to multi-week timescales. *Bureau Research Report* - BRR016 ISBN: 978-0-642-70681-2

Jegasothy, E., R. McGuire, J. Nairn, R. Fawcett and B. Scalley., 2017. Extreme climatic conditions and health service utilisation across rural and metropolitan New South Wales. *Int. J. Biometeorol.*, **61**, 1359–1370. doi: https://doi.org/10.1007/s00484-017-1313-5.

Jones, A. R., D. J. Thomson, M. Hort and B. Devenish, 2007. The U.K. Met Office's next-generation atmospheric dispersion model, NAME III. In: Borrego, C., Norman, A.L.(Eds.), *Air Pollution Modeling and its Application* XVII (Proceedings of the 27th NATO/CCMS International Technical Meeting on Air Pollution Modelling and its Application. Springer, 580–589.

Jonkman, S. N., 2005. Global perspectives on loss of human life caused by floods. *Nat. Hazards*, **34**, 151–175.

Kolmar, M. 2017. *Principles of Microeconomics: An Integrative Approach*. Springer

Lazo, J. K., M. Lawson, P. H. Larsen and D. M. Waldman, 2011. U.S. Economic sensitivity to weather variability. *Bull. Amer. Meteor. S.,* **92**, 709-720. DOI:https://doi.org/10.1175/2011BAMS2928.1

Lombardi, D. A., 2010, The case-crossover study: a novel design in evaluating transient fatigue as a risk factor in road traffic accidents. *Sleep,* **33**, 283–284. doi: https://doi.org/10.1093/sleep/33.3.283

McNally, M. G. 1995, An Activity-based Microsimulation Model for Travel Demand Forecasting", in D. Ettema and H. Timmermans, eds. *Activity-based Approaches to Transportation Modeling*, Elsevier

Millington, S., M. Richardson, L. Huggett, L. Milazzo, K. Mortimer, C. Attwood, C. Thomas, D. Edwards and D. Cummings, 2019. Joint Agency Modelling – A process to deliver emergency response national guidance for a radiological atmospheric release. In *Extended Abstracts, 19th*

International Conference on Harmonisation within Atmospheric Dispersion Modelling for Regulatory Purposes, 3-6 June 2019, Bruges, Belgium.

Moran, M. J. and C. Monje, 2016. Guidance on treatment of the economic value of a statistical life (VSL) in U.S. Department of Transportation Analyses-2016 adjustment. Department of Transportation Memo., 13 pp., https://cms.dot.gov/sites/dot.gov/files/docs/2016%20Revised%20Value%20of%20a%20Statistical%20Life%20Guidance.pdf

Mulchandani, R., M. Smith and B. Armstrong, English National Study of Flooding and Health Study Group, C. R. Beck and I. Oliver, 2019. Effect of Insurance-Related Factors on the Association between Flooding and Mental Health Outcomes. *Int. J. Environ. Res. Public Health*, **16**, 1174. doi:https://doi.org/10.3390/ijerph16071174

Munro, A., R. S. Kovats, G. J. Rubin, T. D. Waite, A. Bone, B. Armstrong and the English National Study of Flooding and Health Study Group, 2017. Effect of evacuation and displacement on the association between flooding and mental health outcomes: a cross-sectional analysis of UK survey data. *Lancet Planet Health*, **1**, e134–41. https://doi.org/10.1016/S2542-5196(17)30047-5

Nairn, J. and R. Fawcett, 2013. Defining heatwaves: heatwave defined as a heat-impact event servicing all community and business sectors in Australia. *CAWCR technical report* 60. ISBN: 9781922173126. http://www.bom.gov.au/research/cawcr-reports.shtml

Nairn, J. R. and R. J. B. Fawcett, 2014. The excess heat factor: A metric for heatwave intensity and its use in classifying heatwave severity. *Int. J. Environ. Res. Public Health*, **12(1)**. doi: https://doi.org/10.3390/ijerph120100227.

Nairn J. and R. J. B. Fawcett, 2017. Heatwaves in Queensland, *Aust. J. Emerg. Manag.*, 32(1)

Nairn J, B. Ostendorf and P. Bi, 2018. Performance of Excess Heat Factor Severity as a Global Heatwave Health Impact Index, *Int. J. Environ. Res. Public Health*, **15(11)**.

Panwar, V. and S. Sen, 2019. Disaster Damage Records of EM-DAT and DesInventar: A Systematic Comparison. In *Economics of disasters and climate change*. Springer. DOI https://doi.org/10.1007/s41885-019-00052-0

Panwar, V. and S. Sen, 2020. Fiscal Repercussions of Natural Disasters: Stylized Facts and Panel Data Evidences from India. *Nat. Hazards Rev.*, **21(2)**. DOI: https://doi.org/10.1061/(ASCE)NH.1527-6996.0000369

Perrels, A., Th. Frei, F. Espejo, L. Jamin and A. Thomalla, 2013. Socio-economic benefits of weather and climate services in Europe. *Adv. Sci. Res.*, **10**, 65–70. doi:https://doi.org/10.5194/asr-10-65-2013

Perkins, S. E., L. V. Alexander and J. R. Nairn, 2012. Increasing frequency, intensity and duration of observed global heatwaves and warm spells. *Geophys. Res. Letters*, **39(20)**. doi: 10.1029/2012GL053361.

Pitt, M., 2008. *Learning Lessons from the 2007 Floods*. Cabinet Office, London. http://webarchive.nationalarchives.gov.uk/20100807034701/http:/archive.cabinetoffice.gov.uk/pittreview/thepittreview/final_report.html.

Raaschou-Nielsen, O., Z. J. Andersen, R. Beelen, E. Samoli, M. Stafoggia, G. Weinmayr, B. Homann, P. Fischer, M. J. Nieuwenhuijsen, B. Brunekreef, W. Xun, K. Katsouyanni, K. Dimakopoulou, J. Sommar, B. Forsberg, L. Modig, A. Oudin, B. Oftedal, P. E. Schwarze, P. Nafstad, U. De Faire, N. L. Pedersen, C.-G. Östenson, L. Fratiglioni, J. Penell, M. Korek, G. Pershagen, K. T. Eriksen, M. Sørensen, A. Tjønneland, T. Ellermann, M. Eeftens, P. H. Peeters, K. Meliefste, M. Wang, B. Bueno-de-Mesquita, T. J. Key, K. de Hoogh, H. Concin, G. Nagel, A. Vilier, S. Grioni, V. Krogh, M.-Y. Tsai, F. Ricceri, C. Sacerdote, C. Galassi, E. Migliore, A. Ranzi, G. Cesaroni, C. Badaloni, F. Forastiere, I. Tamayo, P. Amiano, M. Dorronsoro, A. Trichopoulou, C. Bamia, P. Vineis and G. Hoek, 2013. Air pollution and lung cancer incidence in 17 European cohorts: Prospective analyses from the European Study of Cohorts for Air Pollution Effects (ESCAPE). *Lancet Oncol.*, **14**, 813–822. DOI:https://doi.org/10.1016/S1470-2045(13)70279-1

Robbins J.C., J. Nairn, G. Williamson, A. Wheeler, S. Campbell, D. Bowman and F. Johnston, 2017. Challenges for verifying global heatwave and coldwave forecasts: Can emerging technology help? In: *21st International Congress of Biometeorology* 3-7 Sept, 2017, Durham, UK

Robbins J.C. and H. A. Titley, 2018. Evaluating high-impact precipitation forecasts from the Met Office Global Hazard Map using a global impact database. *Meteorol. Appl.*, **25**(4), 548–560.

Ruin, I., J.-D. Creutin, S. Anquetin and C. Lutoff, 2008. Human exposure to flash- floods - Relation between flood parameters and human vulnerability during a storm of September 2002 in southern France. *J. Hydrol.*, **361**, 199–213.

Ruin, I., J.-D. Creutin, S. Anquetin, E. Gruntfest and C. Lutoff, 2009. Human vulnerability to flash floods: Addressing physical exposure and behavioral questions. In Samuels, P. et al. (Eds) *Flood Risk Management: Research and Practice*. Taylor & Francis Group, London, 1005–1012.

Ruin, I., C. Lutoff, B. Boudevillain, J. Creutin, S. Anquetin, M. Rojo, L. Boissier, L. Bonnifait, M. Borga, L. Colbeau-Justin, L. Créton-Cazanave, G. Delrieu, J. Douvinet, E. Gaume, E. Gruntfest, J-P. Naulin, O. Payrastre and O. Vannier, 2014. Social and Hydrological Responses to Extreme Precipitations: An Interdisciplinary Strategy for Postflood Investigation. *Wea. Clim. Soc.*, **6**, 135–153. DOI: https://doi.org/10.1175/WCAS-D-13-00009.1

Scalley, B. D., T. Spicer, L. Jian, J. Xiao, J. Nairn, A. Robertson and T. Weeramanthri, 2015. Responding to heatwave intensity: Excess Heat Factor is a superior predictor of health service utilisation and a trigger for heatwave plans. *Australian and New Zealand J. Public Health*, **39**, 582–587. doi: https://doi.org/10.1111/1753-6405.12421.

Schwartz, R. M., C. N. Gillezeau, B. Liu, W. Lieberman-Cribbin and E. Taioli, 2017. Longitudinal Impact of Hurricane Sandy Exposure on Mental Health Symptoms. *Int. J. Environ. Res. Public Health*, **14**, 957. doi:https://doi.org/10.3390/ijerph14090957

Shabou, S., I. Ruin, C. Lutoff, S. Debionne, S. Anquetin, J-D. Creutin and X. Beaufils, 2017. MobRISK, a model for assessing the exposure of road users to flash flood events, *Nat. Hazards Earth Syst. Sci.*, **17**, 1631–1651. https://doi.org/10.5194/nhess-17-1631-2017

Shabou, S., I. Ruin, C. Lutoff, S. Chardonnel and S. Debionne, 2020. Assigning travel-activity patterns based on socio-demographics for flood risk assessment. In Lutoff et Durand. (Eds*), Mobilité face aux événements hydrométéorologiques extrêmes 2. Rythmes d'adaptation*. ISTE Editions Ltd, London, 135–164.

Smith, K. and D. Petley, 2009. *Environmental Hazards: Assessing risk and reducing disaster*, Routledge

Spruce, M. D., R. Arthur, J. Robbins and H. T. P. Williams, 2021. Social sensing of high-impact rainfall events worldwide: A benchmark comparison against manually curated impact observations, Nat. Hazards Earth Syst. Sci., 21(8), 2407–2425. DOI: https://doi.org/10.5194/nhess-21-2407-2021

Taylor, A. L., A. Kause, B. Summers and M. Harrowsmith, 2019. Preparing for Doris: Exploring Public Responses to Impact-Based Weather Warnings in the United Kingdom. *Wea. Clim. Soc.*, **11**, 713–729. DOI: https://doi.org/10.1175/WCAS-D-18-0132.1

Terti, G., I. Ruin, S. Anquetin and J. J. Gourley, 2015. Dynamic vulnerability factors for impact-based flash flood prediction. *Nat. Hazards*, **79**(3), 1481–1497.

Terti, G., I. Ruin, S. Anquetin and J. J. Gourley, 2017. A Situation-based Analysis of Flash Flood Fatalities in the United States. *Bull. Amer. Meteorol. S.*, **98**(2), 333–345.

Terti, G., I. Ruin, J. J. Gourley, P. Kirstetter, Z. Flaming, J. Blanchet, Z. Arthur and S. Anquetin, 2019. Towards Probabilistic Prediction of Flash Flood Human Impacts. *Risk analysis*, **39**(1), 140–161. https://doi.org/10.1111/risa.12921.

UNDRR, 2017. Terminology: Online Glossary. https://www.undrr.org/terminology

UNDRR, 2015. Sendai Framework for Disaster Risk Reduction 2015–2030. Geneva, Switzerland: United Nations Office for Disaster Risk Reduction, 37. https://www.preventionweb.net/files/43291_sendaiframeworkfordrren.pdf. (Accessed 2/9/2021)

UNISDR, 2021. *DesInventar*. https://www.desinventar.net

U.S. Department of health and human services, 2012. *Principles of Epidemiology in Public Health Practice: An Introduction to Applied Epidemiology and Biostatistics*. https://www.cdc.gov/csels/dsepd/ss1978/SS1978.pdf

Venäläinen, A. and M. Kangas, 2003. Estimation of winter road maintenance costs using climate data. *Meteorol. Appl.*, **10**, 69–73. DOI:https://doi.org/10.1017/S1350482703005073

Wang, Y., F. Nordio, J. Nairn, A. Zanobetti and J. D. Schwartz, 2018. Accounting for adaptation and intensity in projecting heat wave-related mortality. *Environ. Res.*, **161**. doi: https://doi.org/10.1016/j.envres.2017.11.049

WFP, 2018. Vulnerability analysis and mapping: Food security analysis at the World Food Programme. https://docs.wfp.org/api/documents/WFP-0000040024/download/

Williams, S., K. Venugopal, M. Nitschke, J. Nairn, R. Fawcett, C. Beattie, G. Wynwood and P. Bi, 2018. Regional morbidity and mortality during heatwaves in South Australia. *Int. J. Biometeorol.*, **62**, 1911–1926. doi: https://doi.org/10.1007/s00484-018-1593-4

Wisner B., 2016. *Vulnerability as Concept, Model, Metric and Tool.* https://pdfs.semanticscholar.org/f324/e211e4b03b1bee082c6f1288ed0121e6c804.pdf

WMO, 2015. Valuing weather and climate: economic assessment of meteorological and hydrological services.

Wolshon, B., 2006, Evacuation planning and engineering for Hurricane Katrina. *The Bridge*, **36**, 27–34

Xie Q., L. Wang, L. Zhang, W. Hu and T. Zhou, 2019. Seismic behaviour of a traditional timber structure: shaking table tests, energy dissipation mechanism and damage assessment model. *Bull. Earthquake Engineering*, **17**, 1689–1714

Chapter 6
Connecting Weather and Hazard: A Partnership of Physical Scientists in Connected Disciplines

Brian Golding, Jenny Sun, Michael Riemer, Nusrat Yussouf, Helen Titley, Joanne Robbins, Beth Ebert, Tom Pagano, Huw Lewis, Claire Dashwood, Graeme Boyce, and Mika Peace

B. Golding (✉) · H. Titley · J. Robbins
Met Office, Exeter, UK

WMO/WWRP HIWeather project, Geneva, Switzerland
e-mail: brian.golding@metoffice.gov.uk

J. Sun
National Center for Atmospheric Research, Boulder, CO, USA

WMO/WWRP HIWeather project, Geneva, Switzerland

M. Riemer
University of Mainz, Mainz, Germany

WMO/WWRP HIWeather project, Geneva, Switzerland

N. Yussouf
Cooperative Institute for Mesoscale Meteorological Studies, University of Oklahoma,
Norman, OK, USA

NOAA/OAR/National Severe Storms Laboratory, Norman, OK, USA

School of Meteorology, University of Oklahoma, Norman, OK, USA

WMO/WWRP HIWeather project, Geneva, Switzerland

B. Ebert
Bureau of Meteorology, Bushfire and Natural Hazards Cooperative Research Centre,
Melbourne, VIC, Australia

WMO/WWRP HIWeather project, Geneva, Switzerland

T. Pagano
Bureau of Meteorology, Melbourne, VIC, Australia

H. Lewis
Met Office, Exeter, UK

C. Dashwood
British Geological Survey, Nottingham, UK

© The Author(s) 2022
B. Golding (ed.), *Towards the "Perfect" Weather Warning*,
https://doi.org/10.1007/978-3-030-98989-7_6

Abstract Achieving consistency in the prediction of the atmosphere and related environmental hazards requires careful design of forecasting systems. In this chapter, we identify the benefits of seamless approaches to hazard prediction and the challenges of achieving them in a multi-institution situation. We see that different modelling structures are adopted in different disciplines and that these often relate to the user requirements for those hazards. We then explore the abilities of weather prediction to meet the requirements of these different disciplines. We find that differences in requirement and language can be major challenges to seamless data processing and look at some ways in which these can be resolved. We conclude with examples of partnerships in flood forecasting in the UK and wildfire forecasting in Australia.

Keywords Ocean · Flood · Hydrology · Fire · Heat · Cold · Winter · Seamless · Coupled model · Forecaster · Evaluation

6.1 Introduction

There are many weather-related hazards, of which only a few occur in the atmosphere itself. Others arise from the interaction of the atmosphere with other components of the natural environment, including oceans, rivers and the land surface. In this chapter, we show that:

- A wide variety of environmental hazards result from the weather
- Hazards are predicted using process models or statistical models, which may be coupled with or integrated into NWP models
- Some weather systems give rise to multiple hazards, which must be predicted consistently
- Hazard forecasting methods and terminology have evolved separately to meet the needs of science and users, sometimes in very different ways from those used in weather forecasting
- When linking hazard models to NWP models, care must be taken that variables represent the same things in each model, that space and timescales match and that biases have been removed
- Observations of hazards are fundamental to understanding and for verification but are not widely available or easily accessible
- Successful partnerships come from a shared understanding of the different methods and viewpoints of the different hazard sciences and the agreement of shared objectives

G. Boyce
Flood Forecasting Centre, Exeter, UK

M. Peace
Bureau of Meteorology and Bushfire and Natural Hazards Cooperative Research Centre, Melbourne, Australia

6.2 Hazard Forecasting

In this section, we look at a variety of weather-related hazards for which forecasts might be needed. We start with a brief section covering some general issues in hazard forecasting: model building, availability of observations, consistency, timeliness and interfaces between models. Then, we look in turn at river flood, coastal marine hazards, surface water flood, wet landslide, winds in cyclones, orographic windstorm, severe convective storm, wildfire, extreme heat, air pollution, fog and winter weather. Table 6.1 provides an overview of aspects of forecasting in each area. We conclude with a consideration of multi-hazard forecasting and evaluation of hazard forecasts.

6.2.1 General Aspects of Hazard Prediction

Prediction of hydrometeorological hazards starts by identifying states of the environment that cause significant socio-economic impact. Observation and prediction of these states are the basis of monitoring and warning. While observations of the atmosphere are widely available, observations of land surface conditions, the oceans and atmospheric pollution are much sparser and more difficult to access. Observations of extreme conditions are especially rare. Process models are often developed and calibrated using data from field experiments, but these rarely include extremes. Remote sensing offers high spatial and temporal resolution and can capture these rare events, but with less detail and lower accuracy. Future improvements in hazard prediction, particularly using data-driven techniques, require clearer definition of the required hazard variables, development of sensing methods, reporting to agreed standards, open exchange and accessible archives (Fig. 6.1).

From observations, a description of the physical processes involved in creating the hazardous states is developed, aimed at generating a mathematical model. Process models embody our understanding of how the hazard develops and can allow for complex interactions of multiple forcings. They may be used directly for prediction or may be used as simulators to generate data for use in training statistical models. When used for prediction, outputs often require further post-processing and transformation for them to be useful in warnings, including calibration to remove biases, downscaling to specific locations and translation into variables that relate to socio-economic impact.

Where the processes are not fully known or are too complex or uncertain to model, the process-based description is complemented or replaced by a statistical or empirical model derived from past or simulated data. Ranging from simple linear regression to convolutional neural networks operating in a deep learning framework, these techniques bypass the need to model the processes. They are unbiased, by design, and may be tuned for individual locations or areas. Statistical models require a large training dataset spanning the observed space of the predictand and an

Table 6.1 Contributing weather variables, methods of prediction and ancillary data inputs for a selection of hazards

Hazard	Primary weather inputs	Forecasting approach	Other inputs
Riverine/fluvial flood	Rainfall time/space distribution in river basin Soil moisture distribution Snowmelt distribution	Rainfall-runoff model Hydrodynamic model	River topography Soil hydraulic properties Dams, weirs, etc. Height and condition of riverbanks/leveés Status of vegetation and sediment in river
Surface water/ pluvial flood	Rainfall rate	Inundation model Drainage model	Surface topography Drain topography Drain condition
Coastal flood	Pressure and wind space/ time distribution in ocean basin	2-D or 3-D ocean model Surface wave model	Ocean bathymetry Height and condition of coastal protection Tides and currents Beach sediment status
Extreme wind (gust)	Wind speed and direction Boundary layer stability	Empirical	Local shelter, funnelling, etc.
Fire weather	Rainfall accumulation Wind speed and direction Temperature Soil moisture	Empirical index	Vegetation type and moisture
Fire spread	Wind speed and direction Humidity	Coupled fire model	Vegetation type and moisture
Extreme heat	Temperature Radiation Wind Humidity	Indices based on percentiles or physiological models	Building topography Artificial heat Air conditioning usage Thermal properties of building materials
Surface ice and frost	Surface temperature Humidity	Road surface model	Road structure Sky view Road treatment
Ice accumulation	Temperature Humidity Precipitation rate Wind speed	Ice accumulation model	Thermal properties of base object
Snow	Snow accumulation Snow rate Temperature Wind speed and direction Ground temperature		Shelter obstacles Surface characteristics Road treatment
Poor visibility	Stability Humidity Precipitation rate and type	Empirical	Water sources Aerosol sources

(continued)

Table 6.1   (continued)

Hazard	Primary weather inputs	Forecasting approach	Other inputs
Air pollution	Stability Wind speed and direction	Dispersion model Atmospheric chemistry model	Pollution sources
Landslide	Rainfall accumulation Rainfall rate	Empirical model	Rock/soil structure Recent soil movement Soil moisture

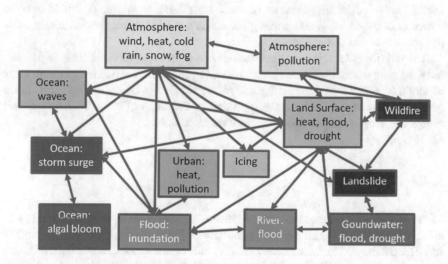

Fig. 6.1 Selection of commonly applied process-based hazard prediction models and some of the linkages between the processes involved. (© Crown Copyright 2021, Met Office)

independent testing dataset for evaluation. Some methods, such as decision trees, Bayesian networks and fuzzy logic, include human reasoning in the design of the model, while data-driven approaches such as machine learning (ML) allow the model to be determined by the relationship between input and output data with only general guidance on model structure. ML is rapidly gaining use in hazard prediction, facilitated by open-source code libraries and fast access to data from multiple sources (e.g. Lagerquist et al. 2020).

The credibility and usefulness of hazard forecasts are dependent on consistency – in time and space – between variables and between products. A flood forecast for a different catchment from the rain forecast and an ice forecast for earlier than the corresponding fall in temperature are inconsistencies that could undermine credibility. While a single forecaster is unlikely to make mistakes of this sort, products issued from different locations on different schedules and with different spatial/temporal resolutions can easily fall into these traps. On the other hand, some apparent inconsistencies are real. In this case, they need to be justified, e.g. a river flood occurring days after the rain. Inconsistency between experts is inevitable but must be resolved before a warning is issued, so as to avoid confusion, e.g. if the meteorological forecaster says there will be flooding but the hydrological forecaster says there will not.

Time is critical for warning and response. Most hazard forecasting methods must wait for weather forecast inputs. Minimising this delay is important – but poor-quality inputs will produce misleading warnings however quickly they are available. Frequent rapid updates can be helpful if skill improves with each update, and fast post-processing for multiple output locations and requirements is an advantage. For some hazards, such as tornadoes or flash floods, minutes of lead time can be critical, but a misleading forecast is worse than none, so warnings should only be issued or changed when consistent guidance is available – from observations, multiple forecasts and ensembles.

The interfaces between process models can be sources of error. This is particularly the case where biases can accumulate. For instance, a small temperature error of say 0.5 °C may be ignored by the meteorologist as being unimportant, but a persistent bias of 0.5 °C in the evaporation calculations of a soil moisture deficit model can grossly distort flood and drought calculations.

6.2.2   River Flood

Floods are the most frequent type of natural hazard associated with disasters and affect more people than all other types combined (55% of the global total from 1994 to 2013, CRED 2015). When heavy rainfall falls on the ground, some will infiltrate. The excess may directly inundate the land surface or drain into streams and rivers that then swell beyond their banks. When heavy rainfall infiltrates pervious rock strata, it may also create excess groundwater, resurfacing in distant locations to cause flooding. Flooding may also result from melting of snow: seasonally in snow-dominated landscapes or episodically, such as in rain-on-snow events (Fig. 6.2).

The speed of the flood wave down a river depends on the steepness of the catchment and on the magnitude of the flood. Large floods can be faster or slower, depending on the cross-section profile of the channel and the friction of the water passing over different materials in the riverbed. When the river overflows its banks, flooding spreads across adjacent low-lying areas. In the flattest regions of some continents, floods may spread thousands of square kilometres and persist for weeks (Sajjad et al. 2019).

Natural systems interact with human systems throughout the landscape (Sene 2008). Built environments are often made of impervious materials that prevent infiltration, resulting in increased runoff which is then channelled through canals and pipes. Gates and leveés are used to control the path of the flood, and reservoirs may allocate space for flood control, while in exceptional circumstances, dam failures can exacerbate floods. Deforestation and wildfire also affect the balance between runoff, evapotranspiration and infiltration.

Five primary modelling challenges for flood forecasters (Pagano et al. 2014; Adams and Pagano 2016) are:

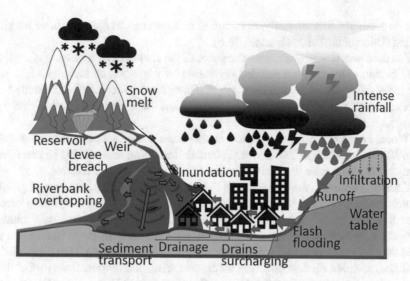

Fig. 6.2 Processes involved in modelling terrestrial flooding. (© Crown Copyright 2021, Met Office)

(a) Estimating antecedent conditions of the catchment/watershed, often through accounting of historical precipitation
(b) Predicting future precipitation, often using Numerical Weather Prediction models
(c) Partitioning precipitation into runoff and infiltration
(d) Tracking the flood wave as it travels downstream across the landscape
(e) Relating the river flow to river depth and/or extent at key sites of interest

The simplest approach for large rivers has been to observe the river flow (or level) at upstream locations and then to relate these, statistically, to the later observed flow or level at the location of concern. Given adequate observations, this approach can provide accurate flood forecasts. However, it can only be applied to measured locations and requires that the upstream-downstream relationship is recalibrated frequently.

For the upper reaches and rapidly responding rivers, rainfall-runoff models relate the river flow to the rainfall in the catchment or watershed with simple representations of rainfall infiltration and evapotranspiration of soil moisture. Traditionally, such models have been basic, representing the system as a collection of "leaky buckets" (e.g. Perrin et al. 2003), with parameters tuned to get a good fit between historical simulations and observations of river flow (Duan et al. 1993). The spatial dimension can be represented in one of three ways (Khakbaz et al. 2012): the catchment may be lumped, where a single time series of catchment average rainfall forces the model; it may be semi-distributed, with the landscape represented by irregularly shaped but hydrologically homogeneous areas; or it may be distributed, with the landscape represented by a regular grid (like Numerical Weather Prediction models). In the latter two cases, runoff is aggregated to the catchment outlet using routing methods. Traditionally, both runoff and routing components have required tuning (Duan et al. 1993; Overton 1966).

For the lower reaches of rivers, the shape of the channel and its surroundings, the composition of the channel bed and vegetation are important characteristics, requiring the application of hydraulic models. These have historically represented the river as a series of segments, with a cross-section shape and bed friction that vary along its length.

Where appropriate, flood forecasters also make use of water operations models (e.g. Klipsch and Hurst 2007), which simulate the filling and spilling of reservoirs, many of which operate to fixed rules, sometimes set by laws or treaties.

Current research is moving towards the application of three-dimensional gridded models that use process-based descriptions of the flow of water, integrating runoff, river flow and inundation (e.g. Yamazaki et al. 2011), with some also including subsurface hydrogeological flows. Satellite measurements of flood inundation are used to calibrate, update and verify these models (Wu et al. 2014). These approaches facilitate extended models covering whole countries, and outputs from global flood models (e.g. the ECMWF GLOFAS model, Alfieri et al. 2013) are now freely available to flood warning authorities.

6.2.3   Coastal Marine Hazards

Coastal erosion and flooding are hazards that affect many major cities. They may result from local wind-driven ocean waves, remotely generated swell waves or storm surges (storm tides) (Fig. 6.3). Here, we do not consider tsunamis, which are geologically initiated. The energy in storm waves can destroy coastal flood defences, both natural and man-made, while the combined elevation of storm tide and storm waves can project large amounts of water over remaining defences, inundating the land behind. The destruction of coastal defences may allow further flooding from subsequent regular high tides if no action is taken to repair the damage.

In situ observations of ocean waves are made by ships and buoys – the latter moored to fixed locations. Basic measurements are the average wave height (usually of the one third highest waves) and the average time between waves. Modern buoys can analyse the wave structure into a frequency and direction spectrum. Storm surges are measured by tide gauges after removal of the astronomical tide. In both cases, the measurements are very sparse compared to the variability of the phenomena being measured. Satellites are also able to provide information on waves and tides, but there are considerable challenges in using it.

The main meteorological inputs are the distribution of wind and pressure across upstream ocean areas, but accurate flood prediction also depends on knowledge of detailed bathymetry, especially near the shore, and of the coastal defences, whether natural or engineered.

Simple deep-water ocean wave predictions can be obtained using statistical relationships of wave height and wavelength/wave period to local wind speed and the duration and/or fetch of ocean that it is blowing over. Outside the influence of a

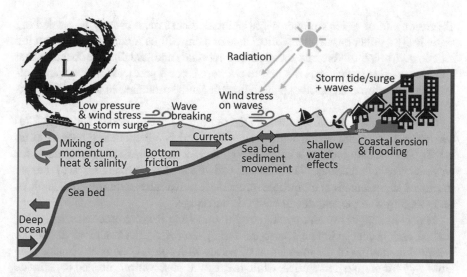

Fig. 6.3 Processes involved in modelling coastal flood hazards. (© Crown Copyright 2021, Met Office)

storm, swell waves propagate along great circles with little loss of energy until they reach shallow water.

Current best practice forecasts use third-generation ocean wave models (Komen et al. 1994), which incorporate energy transfers between different wavelengths. While the processes that transfer energy from wind to waves are broadly understood, the details are too complex for wave prediction models which, instead, rely on simplified relationships between mean wind at a defined height and the spectrum of wave energy. Wave breaking is also represented in a simplified way. The most advanced of these models represent the influences of ocean currents and of refraction and diffraction in shallow water (Booij et al. 1999).

Storm surges (storm tides) are generated by low pressure and wind acting on the ocean, producing a large-scale wave. When this propagates into shallow water, its height is magnified, producing an enhanced high tide which can overtop coastal defences producing serious flooding (Pugh and Woodworth 2014). In some locations, storm surges propagate along the coast, and simple predictions can then be obtained by upstream observation, using statistical relationships between surge heights at different locations, derived either from observations or from modelling. Simple predictions may also be obtained from statistical relationships between prior pressure and wind speed/direction patterns in specific areas of the ocean and observed surges. However, such relationships have limited application, and current best practice is to use a vertically integrated 2-D ocean model (Flather 2000). Since the surge and the astronomical tide interact, it is important that the ocean model reproduces the astronomical tides. Inshore amplification of the surge is sensitive to small-scale bathymetry, so requires a high-resolution local model.

Having computed mean water depth using the surge model and wave height using the wave model, whether offshore or using an inshore model, the estimation of erosion and of water volumes overtopping defences must usually be carried out statistically – either based on historical data or using offline models tuned to specific locations. Choice of forecast location is important, and usually, the location that local knowledge indicates is the first to overtop. Once a predicted overtopping volume has been forecast, flood extent and depth can be modelled in the same way as for a river breach.

Current research is leading towards the use of 3-D ocean models for surge forecasting and their integration with ocean wave and NWP models (Cavaleri et al. 2018). The benefits include a better energy budget and consistency, especially for wave-current interactions. The use of variable inshore resolution will enable improved representation of inshore processes, potentially requiring inclusion of time-varying water extent due to both tide and surge.

Major challenges remain, particularly in modelling inshore processes, including seabed and beach sediment transport during storms (e.g. Carniel et al. 2011). Detailed modelling of the interaction of waves with coastal defences remains possible only for simple geometries, while the regular observations needed to estimate failure likelihood are simply not available, particularly for natural defences. It therefore seems likely that the final step of computing defence failure and overtopping volume will continue to be based on statistical relationships for the foreseeable future.

Other coastal hazards for which warnings may be required include dangers to bathers from rip tides/currents and the growth of potentially toxic algal blooms (red tides). Prediction of algal blooms requires a much wider interaction of ocean, land and atmospheric processes, with temperature, nutrient runoff and biological processes as key components (Pettersson and Pozdnyakov 2012; Zohdi and Abbaspour 2019).

6.2.4   Surface Water Flood

Whereas a flood wave in a river can be tracked downstream, the prediction of flooding from intense rain that has yet to enter a watercourse or that overflows from drains, ditches and other minor watercourses is much more challenging, requiring detailed knowledge of the rainfall intensity distribution. Once the water is on the ground, predicting its movement requires an extremely high-resolution representation of the surface and how it might change as the flowing water picks up, transports and deposits material or erodes new channels. Predictions must also account for absorption of water into the ground, requiring knowledge of soil moisture for natural surfaces and drainage capacity in urban areas (Bach et al. 2014).

Simple approaches to surface water flood prediction rely on statistical analysis of thresholds in rainfall depth and duration beyond which flooding is observed to occur in particular locations. When using these, the rainfall amount should be adjusted for absorption into the ground, to give an "effective rainfall" threshold for flood occurrence. More sophisticated approaches have been developed and may be suitable for specific applications (Shaw et al. 2011). For small rural and upland catchments, rainfall-runoff models can be used to route water into minor watercourses for flash flood prediction. For urban areas, hydrodynamic models of various complexities can be used to model surface and/or subsurface drainage networks (Bach et al. 2014). More generally, distributed inundation models are now available with the capability to use gridded rainfall time series as input and to model the flow of water across the land surface and in water courses (Bates et al. 2004). These approaches are all highly sensitive to the land surface specification, with metre-scale horizontal resolution and centimetre-scale vertical accuracy necessary in urban areas. Given the limited predictability of intense local rainfall, and the consequent need for speed, approaches using pre-computed flood scenarios are being adopted as an alternative to real-time computation (Aldridge et al. 2020; Birch et al. 2021).

In the future, real-time computation will be coupled with real-time updating of flow paths in critical urban areas.

6.2.5 Wet Landslide

The spatial distribution of landslides in an area reflects variations in the underlying geological, geomorphological and hydrological conditions. Landslides can be triggered by intense rainfall, snowmelt and ground vibration such as earthquakes or human-induced changes in stress conditions, e.g. road construction and quarrying. Landslide inventories provide an overview of the extent of landsliding in an area, spatially and temporally, and can include valuable information on the types of processes occurring. Inventories can be populated using direct mapping, either on the ground or using remote sensing, or by accessing archive material and harvesting social media data and news reports.

Rainfall is one of the most common triggers of landsliding with global fatalities focused in south, southeast and eastern Asia linked to the summer monsoon, typhoons and La Niña events (Froude and Petley 2018; Petley 2009). The impact of these landslides, not restricted to loss of life, is wide-ranging, and many communities are affected by loss of livelihoods and damage to transport links and infrastructure.

Landslide type is strongly influenced by the intensity, frequency and duration of the rainfall as well as by antecedent conditions. Large, deeper-seated failures respond more slowly to hydrological changes, while shallow slides and debris flows are most commonly triggered by short-duration, high-intensity rainfall events (Martelloni et al. 2012). Shallow landslides, the subject of most early warning systems, can be extremely rapid and destructive. They commonly occur due to the rapid infiltration of rainfall leading to a rise in pore water pressures and shear failure or due to sediment entrainment in surface water runoff (Baum et al. 2010; Godt and Coe 2007; Wieczorek 1996) (Fig. 6.4).

Landslides are forecast at scales ranging from national to slope scale. At slope scale, monitoring of deformation alongside hydrogeological and meteorological

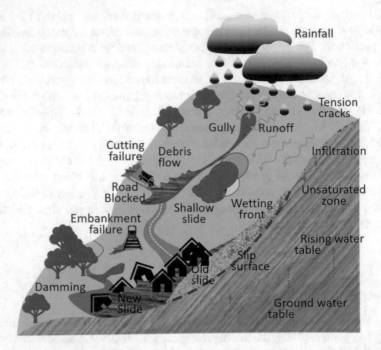

Fig. 6.4 Processes involved in modelling landslide hazards. (© Crown Copyright 2021, Met Office with input from BGS © UKRI 2021)

parameters can be used to produce slope-scale thresholds related to rates of deformation and changes in groundwater level, as well as highlight precursors to failure including the build-up of destabilising groundwater pressures.

At the regional to local scale, landslide forecasting is commonly carried out through the estimation of rainfall thresholds that lead to failure. Empirically derived thresholds are widely used, based on statistical analysis of the local historical rainfall record alongside a detailed, dated inventory of landslides (Caine 1980; Guzzetti et al. 2008; Brunetti et al. 2010). The most widely used rainfall variables are cumulated event rainfall-duration, intensity-duration or antecedent rainfall. Rainfall data are mostly obtained from rain gauges but also from radar and satellite. The quality and reliability of the threshold depend on the rainfall data quality, network density, temporal resolution (hourly, daily or coarser) and accuracy as well as on the landslide record (Gariano et al. 2020; Nikolopoulos et al. 2015).

Antecedent conditions may be incorporated by setting hydrological rather than rainfall thresholds (Reichenbach et al. 1998). The Norwegian Water Resources and Energy Directorate (NVE) has developed a forecasting system for rainfall- and snowmelt-induced landslides which combines real-time data (discharge, groundwater levels, soil water content) with modelled hydrometeorological conditions (Krøgli et al. 2018).

Physically based models which couple slope stability and hydrological models, to produce a spatial and temporal forecast of landsliding, have also been developed, most commonly at a slope or catchment scale (Montgomery and Dietrich 1994; Baum et al. 2002; Salvatici et al. 2018; Guzzetti et al. 2020). However, this approach requires significant amounts of geotechnical, mechanical and hydrological data.

6.2.6 Extreme Winds in Cyclones

Damaging winds are associated with both tropical and extra-tropical cyclones. These travelling storms typically form or intensify over oceans, which provide their main energy source, but may then move over land.

Tropical cyclones gain their energy primarily from the condensation of water vapour, evaporated from tropical seas, as it is lifted in the updraughts of storm clouds. They are associated with the most destructive winds which may reach speeds up to 300 km/h in the eyewall, a ring of intense precipitation and high winds wrapped around the centre of circulation. Strong winds and tornadoes can also occur far from the centre of the storm, associated with spiral bands of heavy rain and thunderstorms. Forecasts and warnings of winds are generally based on surface observations (including ships and buoys), Doppler radar (when in range), satellite data (both imagery and scatterometer winds) and aircraft reconnaissance (where available) and using model outputs from statistical models, statistical-dynamic models and NWP models (Cangialosi et al. 2020). For the nowcasting of tropical cyclone impacts, observations of tropical cyclone structure, track and intensity are crucial (Leroux et al. 2018), with satellite and aircraft reconnaissance flights in particular providing timely information on changes to the locations and intensity of the strongest winds, e.g. changes in the eyewall structure.

There has been a large improvement in tropical cyclone track prediction by NWP models, but it has proved more challenging until recently to improve intensity forecasts (Yamaguchi et al. 2017). However, both regional higher-resolution models and global NWP models are now providing useful guidance, although the forecasting of rapid intensification remains challenging (Short and Petch 2018; Magnusson et al. 2019; Knaff et al. 2020). Satellites provide critical observations of the storm's initial conditions, together with information on the wider environment affecting its evolution. Predictions from a variety of NWP models are widely shared and synthesised into the official advisories issued for each ocean basin.

Improvements to model physics and dynamics are required, along with representation of model uncertainties, to reliably simulate the possible developments. There is also great potential to improve the use of ensemble-based uncertainty information in tropical cyclone forecasts and warnings (Titley et al. 2019). Improvements in model resolution, in atmosphere-ocean coupling and in the representation of momentum exchange at the sea surface should lead to further improvements in intensity prediction.

Extra-tropical cyclones acquire their energy from the temperature gradient between tropics and poles, supplemented by the release of latent heat of condensation of water vapour as it cools. These storms cover a much larger area than tropical cyclones, so it is particularly important to identify which areas will be subject to damaging winds. NWP models are generally very accurate at predicting the intensity and track of such cyclones, but it is only in recent years that details of the wind structure, such as the sting jet (Clark and Gray 2018), have been adequately captured. Damaging winds associated with embedded convection, in urban areas, and in areas of complex topography, are generally predicted using empirical rules and forecaster experience, though very-high-resolution models are showing some promise at direct prediction.

6.2.7 Orographic Windstorms

Damaging winds can occur when the wind blows across a mountain range and the upper air temperature structure creates a barrier to upward motion, resulting in either large amplitude atmospheric waves or a hydraulic jump (Whiteman 2000). Such windstorms have a variety of names around the world (Fig. 6.5).

Traditional forecasting approaches use analysis of observed wind direction and vertical temperature soundings to identify conditions favourable for storm development. NWP models provide good guidance in many cases, but the requirements for vertical resolution of the temperature structure and horizontal resolution of the mountains make accurate predictions of the marginal conditions for onset and cessation difficult, so the forecaster often uses a combination of methods to refine the timing and intensity of the storm. Improvements to NWP resolution will continue to improve the skill of predictions, many of which are freely available from global and regional NWP centres.

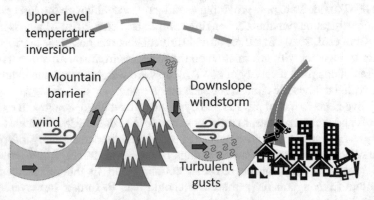

Fig. 6.5 Processes involved in modelling orographic windstorms. (© Crown Copyright 2021, Met Office)

6.2.8   Extreme Winds, Lightning and Hail in Severe Convective Storms

Hazards associated with severe convective storms include intense rainfall, damaging straight-line winds and tornadoes, large hail and lightning. Vertical motion in convective clouds is fuelled by the release of latent heat of condensation, producing a deep cloud composed of water at lower levels and ice higher up (Yau and Rogers 1996), characterised by severe turbulence and icing. Most of the damaging winds are a result of outflow generated by thunderstorm downdrafts or tornadoes (Church et al. 1993). Hail and lightning result from the processes of freezing and melting of raindrops (Mason and Mason 2003) (Fig. 6.6).

Forecasters monitor Doppler radar scans for indicators of severe wind or hail, other remote sensing tools such as satellite and lightning networks and/or spotter input. They also look for indices in the storm environment such as convective available potential energy, convective inhibition or helicity that are statistically related to the occurrence of severe thunderstorm phenomena.

Convective-scale NWP models provide useful guidance in issuing warnings of extreme wind, lightning and hail threat. Part of the challenge is the accurate analysis of storms and their surrounding environment in the initial conditions, for which the assimilation of radar and satellite data is essential. The rapid growth of uncertainty in these forecasts requires the use of ensemble systems to generate probabilistic forecasts (Wheatley et al. 2015). As resolution, data assimilation and ensemble prediction improve at these convection-permitting scales, the contribution of NWP to forecasting these hazards will become dominant.

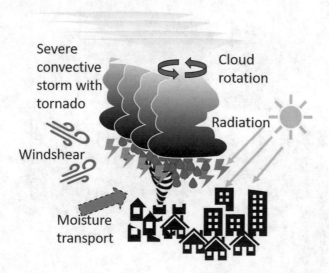

Fig. 6.6 Processes involved in modelling convectively generated severe weather. (© Crown Copyright 2021, Met Office)

6.2.9   Wildfires

Predicting wildfire is becoming increasingly important with climate change and the move of people into peri-urban spaces (NAS 2020; Ubeda and Sarricolea 2016) (Fig. 6.7). The primary meteorological inputs for wildfire prediction are temperature, wind speed and direction, relative humidity, lightning and precipitation, including antecedent precipitation which affects fuel moisture. Environmental inputs include vegetation type, fuel load and moisture content. The recent burn history of an area has a strong influence on the vegetation available to be burned. Topography influences fire spread through meteorological effects such as wind channelling, boundary layer structure and rain enhancement/shadow. Large, intense fires can modify the surrounding meteorological environment, which may lead to unpredictable and dangerous fire behaviour. The passage of a cold front, accompanied by a substantial shift in wind direction, can turn a head fire into a much broader flank fire with a much longer fire front. A significant uncertainty in wildfire prediction is ignition, which is often caused naturally by lighting but can also be caused by inadvertent or deliberate human activity.

Fire weather prediction occurs on multiple timescales. Coupled modelling at seasonal and multi-week timescales gives outlooks for anomalous fire weather conditions based on indices that combine multiple weather and environmental variables (e.g. Bedia et al. 2018; Dowdy 2020). In the medium and short range, ensemble and deterministic NWP is routinely used to predict fire weather conditions and is particularly important for forecasting significant wind changes. When fires are occurring or expected, forecasters need to make forecasts of detailed weather conditions at locations specified by the agencies responsible for firefighting.

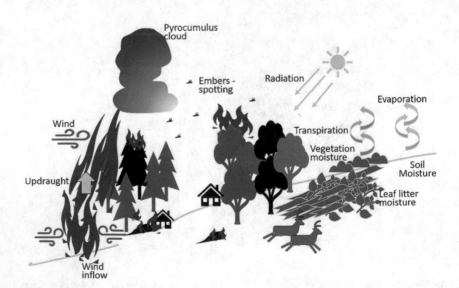

Fig. 6.7 Processes involved in modelling wildfire hazards. (© Crown Copyright 2021, Met Office)

Simple statistical models may be used to represent linear (downwind) fire spread in simple conditions, for example, grass fires in flat terrain. Fire spread models take meteorological and environmental inputs and compute the speed and direction of the fire spread and may provide other parameters, such as flame height or fire radiative power. These models can represent changes associated with up-slope and down-slope direction and may include spotting. The outputs are very sensitive to wind direction. When run in ensemble mode, the probabilistic outputs can represent uncertainty in weather and fuel inputs and fire behaviour processes.

Very-high-resolution coupled fire-atmosphere models are now becoming available that simulate the feedback processes causing fires to grow explosively into extremely dangerous and unpredictable fires (e.g. Filippi et al. 2018; Jiménez et al. 2018). These models can simulate the transport of embers to start new fires and the acceleration of winds in hydraulic jumps in down-slope flow. Some models can also represent fire-generated thunderstorms (pyrocumulonimbus), generated by convergence of air towards the fire, that can generate lightning and downbursts with gusty winds. In general, these models are still too costly to run operationally (Peace et al. 2020).

Successful application of the best fire risk and fire behaviour models has the potential to enable more effective fire prevention and firefighting, with reduced loss of life and damage to property. However, remaining challenges include reliable observation of both fuel state and wind at sub-kilometre scales; accurate forecasting of weather and environmental inputs, including fuel state, on these scales; effective modelling of complex fire behaviour; and more complete observations of fire behaviour to validate and improve the models. In the light of recent evidence of increased wildfire activity due to climate change, more emphasis on addressing these challenges can be expected in the next few years (Dowdy 2020).

6.2.10   Extreme Heat

Extreme heat is generally associated with meteorological conditions that vary slowly in space and time. Temperature is the main variable of interest, but humidity, wind speed and radiation are also relevant for predicting the impacts of heat stress on health. Cities are particularly vulnerable to extreme heat because of the urban heat island (UHI) effect, where the absorption and re-emission of radiation by asphalt and concrete surfaces and the concentration of industry and other heat sources cause heat to be retained (Fig. 6.8). The relationship between outside and inside temperature in buildings is also important for some health impacts. Thresholds for defining heat waves are often defined relative to the local climate (e.g. Nairn and Fawcett 2014), requiring both observed and model climatologies.

Coupled numerical modelling systems are now capable of predicting heat wave conditions more than a week in advance and the likelihood of weather regimes associated with heat waves several weeks in advance (e.g. Marshall et al. 2014). UHI effects can be estimated from satellite remote sensing and urban sensor networks

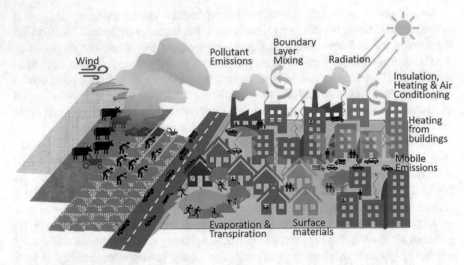

Fig. 6.8 Processes involved in modelling heat and pollution hazards. (© Crown Copyright 2021, Met Office)

and used in post-processing to predict heat stress conditions on sub-kilometre grids. Pinpointing local heat stress requires detailed understanding of neighbourhood and street-scale conditions, including sun/shadows, green canopy, wind flow around buildings and ventilation within buildings. This level of detailed modelling is currently beyond the routine prediction capability of operational weather services. Statistical models can be used to relate local effects to the larger-scale meteorological and environmental conditions, perhaps guided by urban canopy models.

The future "smart city" will exploit new sensors, communication technology and the internet of things to predict heat stress and its impact using data-driven forecasting approaches.

6.2.11 Extreme Pollution

Hazardous pollution results when industrial and transport emissions are trapped in a stable boundary layer and when wildfire smoke and dust are transported into populated areas. Poor air quality from primary pollutants may be exacerbated by secondary pollution from photochemical reactions.

Quantitative measurement and forecasting of air pollution is a relatively young science, borrowing much from weather prediction. Variations of exposure at ground level within the urban fabric are important. Air quality monitoring relies on sparse surface networks of (primarily urban) air monitoring sites supplemented by limited satellite remote sensing capabilities (EEA 2020). The recent deployment of inexpensive fine particle ($PM_{2.5}$) sensors offers opportunities to improve monitoring, forecast initialisation and post-processing (Lewis et al. 2018b).

For directly emitted chemicals such as the oxides of sulphur and nitrogen, the principal uncertainty is in specifying the emissions and how they diffuse close to source. Static emissions inventories from industry and other anthropogenic sources require a large effort to update, so often become outdated as the industrial landscape and regulatory practices evolve. Simple passive or chemical transport and diffusion modelling approaches can provide useful predictions of high concentration levels for use in warnings (WMO 2020).

State-of-the-art predictions of hazardous near-surface concentrations of $PM_{2.5}$, ozone and other pollutants use high-resolution limited area numerical air quality models. These can be characterised as "offline", meaning gas and aerosol chemistry is computed in a chemical transport model using meteorological conditions as input, or "inline/online", meaning the gases and aerosols can influence the radiation, temperature and cloud microphysical properties of the weather model (e.g. Wang et al. 2020). Inline models, while more accurate, are more complex and expensive.

All approaches rely on the accuracy of the emissions and the wind, and for limited area models, the inward transport of pollutants at the boundaries must be accurately specified. Observations of aerosol optical depth from satellites and ground-level data can enhance forecast accuracy.

6.2.12 Fog

Fog is primarily a hazard to people using transport networks. The critical visibilities for which warnings are required vary significantly between users, depending on their speed of motion and the complexity of the landscape being navigated. For a pilot of a fast military jet in mountainous terrain, safe flight may require several kilometres of clear visibility, whereas for a car driver on a straight road, 100 m is normally sufficient (Call et al. 2018). Distinguishing such visibility thresholds is challenging both to observe and to forecast. Variations occur at very small scales and may be related to local water and/or pollution sources. Over the ocean, minor changes in sea surface temperature can produce the same effect (Isaac et al. 2020; Fallmann et al. 2019). Once formed, a thick fog bank may persist until there is an air mass change. However, it is the time of clearance that the user requires, and this can be as hard to predict as formation.

Current forecasting methods mostly use representative vertical temperature soundings, from observations or model predictions, and apply detailed site-specific modelling of local heat and humidity budgets in the vertical column, allowing for any change in mixing due to the wind (see, e.g. Gultepe et al. 2007). However, very-high-resolution NWP models (of 100s of metres grid length) are showing useful accuracy and will likely become the preferred approach in the next decade (Price et al. 2018).

6.2.13   Frost, Ice, Snow and Freezing Rain

Prediction of ice-covered surfaces is required for warnings of hazardous road and footway conditions, for anti-icing treatment of roads and aircraft and for warnings of ice accumulation on structures and cables. Ice can form when pre-existing water freezes, by frost deposition, from freezing rain or from compaction of snow or other frozen hydrometeors. Freezing rain (Changnon and Creech 2003) is a major hazard for road users and for trees and towers (Fig. 6.9).

While regular products from NWP models can provide useful guidance, warnings need to be based on careful analysis of the thermal conditions of the lower atmosphere and the exposure of the road or other surface (Karsisto et al. 2017). Road icing is generally forecast by 1-D heat balance models that incorporate influences on the local short- and long-wave radiation, including local shading from daytime sun and inhibition of nocturnal cooling by local heat sources such as walls and trees, together with a detailed representation of the thermal properties of the road. Large-scale conditions are taken from a NWP forecast. Accumulation of ice on structures and cables requires models of water availability and the heat budget (Fikke et al. 2006).

NWP can also provide guidance on frozen precipitation, but accuracy remains low, due to limitations in cloud physics and inadequate resolution, so forecasters commonly rely on interpretation of the low-level vertical temperature structure, both observed and predicted. Even small amounts of snow or ice on untreated roads can be hazardous, so a judgement is needed as to the amount of snow that will settle on the road, bearing in mind the effects of wind, shelter by trees, buildings and hedges, local convective enhancements and variability due to orography.

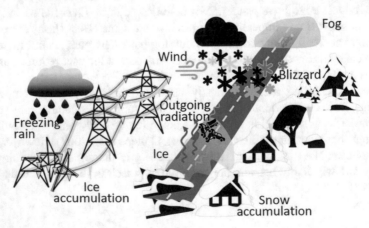

Fig. 6.9 Processes involved in modelling winter hazards. (© Crown Copyright 2021, Met Office)

6.2.14 Multi-hazards

A weather system may give rise to several hazards, each with distinct warning requirements, potentially leading to very complex warnings. To enable the most useful advice to be provided, hazard forecasters need precise information about the different hazards and their interactions. Table 6.2 gives some examples of multi-hazards in weather systems, which we illustrate by looking in more detail at the most of extreme of these: the tropical cyclone.

Tropical cyclones are one of the most destructive meteorological phenomena and are associated with several different hazards that can cause significant impacts on life and property (WMO 2017). In addition to destructive winds, the combination of wind-driven waves and low pressure can produce a coastal storm surge, destroying coastal defences and causing coastal flooding. Extremely heavy rainfall associated with tropical cyclones can lead to landslides and serious pluvial and fluvial flooding. Tornadoes and lightning are also commonly associated with tropical cyclones. Different hazards may occur together in the same or neighbouring regions, leading to difficulties for emergency preparedness, response and recovery programmes. Impacts from the strongest winds are greatest close to landfall location in the eye-wall of the storm, but the highest storm surge is displaced to the side experiencing onshore winds, while precipitation and fluvial flooding can extend far inland from the landfall location. Warnings based only on track and peak intensity do not provide a complete picture of multiple and cascading hazards. For example, when planning evacuations from areas of extreme winds, it is important to ensure that evacuation centres are not in a flood area. The impact of a tropical cyclone also depends on its translation speed, which affects the duration of strong winds and the accumulated rainfall, and the land characteristics, vulnerability and exposure of the area being impacted. Accurate predictions of all hazards, and their associated uncertainty, across the entire multi-hazard event can provide the basis for more effective communication of the multiple risks to life and property.

Table 6.2 Multiple hazards associated with some weather systems

Weather system	Principal hazards
Tropical cyclone	Extreme wind; coastal, river and surface water flood; landslide; tornado; lightning
Mid-latitude cyclone	Extreme wind; coastal, river and surface water flood; landslide; snow and blizzard; ice storm
Anticyclone	Heat wave/cold wave; air pollution; fog
Severe convective storm	Extreme wind; tornado; lightning; hail; intense rain; river and surface water flood; landslide

6.2.15   Evaluation

We started this section with the challenge of obtaining credible hazard observations. When evaluating the performance of hazard forecasting systems, two different approaches are used to overcome the lack of observations. For a process model, it may be assumed that accurate prediction of normal conditions is a necessary, if not sufficient, condition for accurate prediction of extremes. For some hazards, such as ocean waves and storm surges, river levels or visibility, sufficient observations are available for statistical verification in non-hazardous conditions. Where the extreme distribution is sufficiently well defined by the observations, it may also be possible to establish whether the asymptotic behaviour of the predictions is consistent with that observed. Where a hazardous threshold is exceeded sufficiently often for statistical significance, the ability to forecast exceedance or non-exceedance of the threshold can be tested using a variety of binary scores (Jolliffe and Stephenson 2012). The second approach uses case studies to evaluate performance. Given several of these, it may be possible to infer conditions under which predictions are more, or less, realistic.

6.3   Capabilities of the Weather Forecast

The weather forecasting process is described in detail in Chap. 7. Here, we briefly introduce the three main prediction approaches of Numerical Weather Prediction (NWP), rapid update nowcasting and statistical forecasting, together with the role of the forecaster in relating these sources of guidance to the hazard of concern. We then summarise the capabilities of weather forecasts in general and for each of the main hazard-related weather variables. We conclude with a brief section on evaluation.

6.3.1   NWP Models/Ensembles

The primary source of quantitative meteorological input for hazard prediction comes from Numerical Weather Prediction (NWP) models, run routinely by weather services, which form the basis for public weather forecasts. The World Meteorological Organization's Global Data-processing and Forecasting System (WMO 2019) supports an information cascade in which a small number of countries share information from global ensemble NWP systems and several countries in each region share information from regional NWP systems, providing every weather service with access to forecast guidance.

As will be described further in Chap. 7, NWP models encapsulate a complex set of processes in the atmosphere (Coiffier 2011) and in the interaction of the atmosphere with the land and ocean surfaces. The mathematical description of each process is an approximation to reality containing uncertain parameters. When integrated in a forecast model, these parameters must be mutually adjusted over extensive trial periods to reproduce the observed weather accurately.

Weather prediction is an initial value problem, which means that the resulting forecast depends on the initial state. Since that initial state is uncertain, so is the forecast, and as the forecast evolves, uncertainty eventually swamps the result as the limit of predictability is reached. Before that state is reached, the quality of the forecast depends critically on the initial state. Thus, the availability of observations and their incorporation into the model through data assimilation (see Chap. 7) are of critical importance. Many hazards are associated with parts of the atmosphere where energy is being released rapidly, and it is in these areas that the growth of uncertainty is also greatest. An important part of the NWP forecast is an assessment of this uncertainty, obtained using an ensemble of predictions from slightly different initial conditions.

Physical constraints also limit the accuracy of high-impact weather forecasts. Computer power has increased dramatically, but the accuracy of forecasts, especially of hazardous weather, remains limited by resolution, and every doubling of spatial resolution requires more than a tenfold increase in computer power and a fourfold increase in communication bandwidth. The best current NWP models have global grid spacings of ~10 km and regional grid spacings of ~1 km. Ensemble prediction systems tend to have slightly coarser resolution. Since models with a coarser grid spacing than about 4 km are unable to represent deep convection, except in a statistical sense, they are generally less good at predicting weather systems in which convection is a major energy source – including those in the tropics and summer mid-latitudes. Even models with a 1 km grid spacing are unable to represent the detailed near-surface processes that occur in fog and wildfire evolution and that determine the variation in heat and pollution between urban buildings. Speed of forecast production also impacts on the time between observation of the initial state and availability of the forecast. For global models, gathering observations from around the world takes time, while longer forecasts increase the computer time required, so it can be as much as 6 hours before a forecast is disseminated.

Communication restrictions mean that only a limited range of variables from an ensemble NWP forecast can be shared, often at much reduced time and space resolution and with limited probabilistic information. It is therefore important that the information required to meet users' needs is selected carefully. In many cases, the variables of interest are not those that govern the atmospheric evolution, so they must be derived from the model output in a post-processing step (Vannitsem et al. 2021). State-of-the-art systems enable users to define the processing required remotely and to receive just the results required. This is increasingly being realised through hosting of the full database in "the cloud" (i.e. in shared storage facilities connected through the internet).

A state-of-the-art warning chain will carefully consider the trade-offs between timeliness and accuracy. Resolution must be good enough to represent the weather feature that is the source of the hazard, but given a significant degree of uncertainty, probabilistic information is crucial.

Challenges for the future are to increase the availability of information to hazard forecasters, including more relevant variables, at higher resolution, with a better description of uncertainty, but targeted to the areas and thresholds of concern.

6.3.2 Nowcasting Tools

We have seen that using ensemble NWP as the basis for hazard prediction takes time. For very rapidly developing hazards, a faster response to new information from local observations is required. The collection of resulting techniques is referred to as nowcasting (Browning 1982) and may be used for lead times of a few minutes up to a few hours, depending on the hazard.

To achieve maximum speed of response, nowcasting systems typically use a limited source of observations and a simplified predictive model, focused on a single variable, such as rainfall, strong wind or large hail. The first systems were based on radar reflectivity observations and used linear extrapolation of the position of areas exceeding a critical threshold (Wilson et al. 1998). Most current systems continue to be radar and/or satellite based, since these observing systems give detailed spatial coverage in a single data stream. Prediction also continues to be based on extrapolation, but with an increasingly sophisticated choice of variables and methods.

Nowcasts depend critically on the initial state. Processing of the observational input to remove errors is particularly important. Ensembles of nowcasts are also used, particularly where it is useful to identify the sensitivity of the output to different estimates of trend, whether in position or intensity. Since nowcasting tools generally produce a forecast of a single variable, it is important to avoid inconsistencies, either between different nowcasting tools or between the nowcast and NWP guidance. For example, a cloud nowcast based on satellite imagery, and a rain nowcast based on radar, may easily produce an intense rain forecast in the same location as clear skies. Such differences must be avoided if trust is to be built and maintained; effective methodologies to blend nowcasts and NWP are currently the subject of much research (e.g. Atencia et al. 2020).

6.3.3 Statistical Models and Machine Learning Algorithms

Statistical models contribute to weather forecasting, both to correct biases in NWP outputs (Vannitsem et al. 2021) and for processes that are too complex or time-consuming to incorporate in the NWP model, including very-short-range forecasting of the boundary layer and severe convective storms. Statistical methods are designed to be unbiased and may be tuned for individual locations, so are ideal for translating NWP outputs to site-specific forecasts. However, if spatio-temporal correlations are important, relationship-preserving methods such as analogue ensembles (using observations from historic cases similar to the current situation, e.g. Clark et al. 2004) may be needed.

As noted earlier, statistical models are limited by the availability of training and testing data that span the full range of required outputs. Since hazards are often associated with extremes, particular care is needed to ensure that the model gives sensible results in these conditions. Non-stationarity in the data (e.g. due to climate change) is also a challenge, requiring either frequent recalibration or the provision of an auxiliary model. These difficulties may be overcome if data can be generated with a sufficiently realistic simulation model.

Many different statistical approaches are available, ranging from decision trees involving multiple human inputs to purely data-driven approaches, such as multi-layer neural networks. However, machine learning techniques are rapidly gaining use, facilitated by open-source code libraries (e.g. Lagerquist et al. 2020). Performance assessments show these approaches can be competitive with human judgement and physical modelling.

6.3.4 The Professional Weather Forecaster

A forecaster combines outputs from a range of tools with experience and professional knowledge to reach a judgement on the future occurrence of weather, especially hazardous weather (Pagano et al. 2016). Forecast outputs are limited by the processing capability of the forecaster, which may limit them to focusing on areas of prior expected hazard or of a particular vulnerability. Where the response depends on fine judgements of cost and benefit, the ability to estimate the distribution of hazard probability reliably is key and is only achieved by the very best forecasters.

Private sector weather forecasters provide a paid-for service, which may be part of a general media information service, funded by advertising, or a consultancy service to a specific industry that has a weather-related vulnerability. In the media, the forecaster is primarily a communicator, using their expert knowledge and judgement to interpret the general forecast from the model and/or weather service professional, to create actionable messages for their audience. The challenges for this sector are in understanding the strengths and weaknesses of their inputs and in relating the information received to the concerns of their audience. This role was considered in more detail in Chap. 4.

The consultant forecaster is generally focused on specific clients at specific locations, with particular needs and vulnerabilities. Translating the general information coming from models and forecasters into the advice needed by their clients requires them to select relevant data and to apply hazard-specific prediction techniques. They may themselves predict the hazard and its impact, or they may produce bespoke weather information for others to use. In achieving this, they will use now-cast and machine learning tools, tuned specifically to the needs of their customers. Key challenges for this sector are the trade-off between the accuracy and cost of the

NWP data they source and maintaining the reliability and accuracy of their tools. Tools often originate from academia, but their maintenance requires either the regular purchase of upgrades or significant effort from the consultant. Users of multiple consultants may see consistency problems due to use of different tools, different base data sources or different judgements.

6.3.5 Evaluating Weather Forecasts

Using meteorological forecasts effectively for hazard forecasting requires thorough understanding of their performance. For medium-range forecasts, standard scores for probabilistic and deterministic forecasts provide a useful assessment of the prediction of large weather systems. Since the variables concerned change smoothly on these scales, the statistics are well behaved. As the event gets closer and the details become better resolved and predicted, the forecast timing and location of synoptic-scale storms, cold fronts, tropical cyclones and other features can be evaluated using object-based verification approaches. Standard observations of surface and upper air variables can also be used to verify forecasts of environments conducive to hazardous weather. In addition to surface-level variables such as rainfall and temperature, vertical elements such as stability, wind shear and boundary layer depth should be evaluated when they are direct inputs to hazard predictions (e.g. air pollution, fog, freezing rain).

Direct evaluation of weather forecasts at hazard scale is more difficult as the standard observation network is rarely dense enough to capture the important details and there may be few observations of extremes. Remotely sensed observations from radar and satellite are spatially complete but only approximate the variables of interest such as rainfall and wind. At high resolution, traditional verification scores are hampered by the "double penalty" where small timing or location errors in a forecast feature cause the event to be predicted where it didn't occur and missed where it did occur. Spatial verification approaches accommodate this situation (Mittermaier and Roberts 2010; Raynaud et al. 2019), but for some hazards, the spatial context is important (e.g. a river basin, a coastal city), so correctly predicting the location is crucial.

6.3.5.1 Resolution

Many hazards occur on small spatial and short temporal scales, e.g. a flash flood or a wind squall. It is a formidable challenge for weather forecast models to resolve these. The latest local-area NWP models use grid spacings of around 1 km, while research models use grid spacings down to 100 m (e.g. Lean et al. 2019). Such high-resolution models greatly improve forecast precision, but with considerable uncertainty in the small-scale detail.

Evaluation of operational forecasts indicates that reducing the grid length improves model performance at large scales as well as making it possible to resolve small scales. Improved accuracy in the 50–200 km scale range enables forecasters to better interpret observations and finer-scale predictions.

High-resolution models improve forecast accuracy most prominently for weather phenomena that are influenced by the improved representation of the atmosphere's lower boundary – orography, the urban fabric, variability in land use, etc. For small-scale weather phenomena that are sensitive to the larger-scale flow, the benefits of high resolution may be masked by uncertainties at larger scales.

6.3.5.2 Precision and Accuracy

For use in hazard prediction, weather forecasts may need to be both precise and accurate. Precision in forecasting refers to its "fineness" in space, time and other attributes dictated by the hazard. For instance, if some threshold in heat stress was reached at a forecast of 38.4 °C, the hazard forecaster would want to know where, when and whether this would happen, not just if it would be "extremely hot". Similarly, distinguishing between intense rain during and after an outdoor festival could be very important.

Accuracy refers to how well a forecast matches the observation. When measuring accuracy, the spatial and temporal scales of forecast and observations must be matched by upscaling or downscaling. The choice of whether to verify at the finer or coarser scale depends on the precision required by the downstream hazard model. As well as verifying at specific locations, some verification methods can measure errors in the location and timing of meteorological features such as storm systems and fronts (e.g. Dorninger et al. 2018).

Forecasts have systematic error (bias) and random error components. Biased forecasts are particularly damaging when input to hazard models that were developed using observations, so it is advisable to remove biases if possible. Random errors can be reduced through aggregation or averaging, e.g. spatially by catchment or fetch averaging or temporally by accumulation or dose averaging, at the loss of some forecast precision. When observations have significant uncertainty associated with them due to instrument error or representativeness (e.g. rain gauge measurements of convective precipitation), aggregation and averaging of observations may also be needed.

6.3.5.3 Reliability

When a risk assessment is being made, the likelihood is as important as the intensity. Ensemble forecasting systems are used to provide probability forecasts, e.g. of rainfall accumulation exceeding a threshold in a particular location and over a certain time period. The reliability of probability forecasts from ensemble prediction systems has improved enormously over the last 25 years (Bauer et al. 2015), although post-processing and calibration of probabilities are still needed (Williams et al. 2014). A probability forecast must be verified as part of a collection of forecasts, not alone. Probability verification measures, such as the Brier score (Jolliffe and Stephenson 2012), assess the following qualities: (i) reliability, agreement between forecast probability and the observed frequency; (ii) sharpness, tendency to forecast probabilities near 0% or 100%, as opposed to values clustered around the mean; and (iii) resolution, ability of the forecast to resolve events into subsets with characteristically different outcomes.

6.3.6 Predictability of Hazard-Relevant Variables

Due to scale interactions and the chaotic nature of the atmosphere, there are *intrinsic* limits to predictability that even an optimal (yet physically reasonable) forecast system could not overcome. These limits vary substantially between hazard-relevant variables and are a function of the weather system that is associated with the hazard. The intrinsic limit of predictability is a hypothetical concept because an "optimal" forecast system does not exist. Yet the concept is important because it underpins the use of probabilistic forecast frameworks while also guiding improvements of state-of-the-art forecast systems. Conceptually, predictability is most severely limited in the presence of potential "bifurcations" (e.g. Keller et al. 2019) such as are seen in the tracks of tropical cyclones. Bifurcations may occur in a more general sense when atmospheric conditions are close to specific thresholds, e.g. for freezing rain a temperature near 0 °C at the ground, for convective initiation a forcing that is close to the convective inhibition. Below, we provide an overview of *practical* limits of predictability, i.e. predictability limits as observed in current state-of-the-art systems, for several hazard-relevant variables and for typical weather situations.

Rain Extreme rain is a function of duration and the area of interest. For large areas, extremes over long periods may dominate. Global ensemble NWP has considerable capability in predicting the persistent regimes that produce such long period extremes, but usage is hampered by biases in the modelled rainfall and lack of adequate datasets to recalibrate with. At shorter durations, we may identify three main types of rainfall extremes: interaction of atmospheric rivers or conveyor belts with orography, typically over multi-day periods (Shearer et al. 2020); organised rainbands, often with embedded convection, typically over periods up to a day; and intense convective rain over periods of an hour or so. Atmospheric rivers are predictable for few days ahead, but with limited spatial accuracy until lead times of a few hours. Organised rainbands are predictable for a day or so ahead, but details of intensity and duration are uncertain until shorter lead times. Intense convection is typically only predictable in a regime sense for a few hours, and individual storms are currently unpredictable except by nowcasting methods at less than an hour's lead time (Wang et al. 2019). Future developments in the assimilation of storm-related data in kilometre-scale models should lead to improvements.

Wind While the mean wind is well predicted by current forecasting methods up to several days ahead, extreme local winds are an unresolved challenge. Tropical cyclone winds are beginning to be captured skilfully by the latest generation of kilometre-scale models, at least for forecasts up to a day ahead. Prediction of tornadoes and other wind extremes related to severe storms is largely possible only through statistical inference using predicted indices of atmospheric structure, though diagnosis of predicted cloud structure and rotation in kilometre-scale forecasts are taking us closer to direct prediction (Wang and Wang 2020) up to timescales of a few hours. Orographic wind prediction has some skill up to a day ahead in models that adequately resolve the orography, provided the vertical resolution is able to capture the vertical structure of the atmosphere. However, extreme winds in the vicinity of steep gradients and buildings are not currently predictable, except in a statistical sense, because of their scale. Improvements in resolution can be expected to provide significant progress in short-range prediction.

Winter precipitation The snow/rain boundary is diffuse and difficult to define, yet the impact of crossing it at the surface is profound. The same is true of other varieties of freezing and frozen precipitation. The extents of liquid and frozen precipitation can often be predicted a day ahead, but accurate positioning and timing of the boundary, especially in slow-moving weather systems, may not be achieved until a few hours ahead. Prediction needs to be site-specific, because of the sensitive dependence on height, so requires post-processing of model gridded outputs.

Temperature The general structure of temperature in the atmosphere is highly predictable up to several days ahead. However, models struggle with the detail of boundary layer structure, especially during the transitions from day to night and vice versa (Lapworth 2006, Papadopoulos and Helmis 1999). In low turbulence situations, such as under nocturnal inversions, details of the land surface may be significant. The unpredictability of these flows may be such that deterministic prediction is not possible, even at very short lead times, with implications for fog and frost warnings. In urban areas, the crude representation of urban structures limits predictability.

Atmospheric boundary layer The principal meteorological variables relevant for boundary layer hazards such as air pollution and fog are mean wind (for transport) and wind variance or turbulence (for diffusion). Since turbulence is not a primary variable, pollution models often infer it indirectly from gross boundary layer characteristics, such as boundary layer depth and mean temperature gradient. These are problematic for stable boundary layers when elevated pollution levels are a particular problem. NWP models have some capability for prediction of widespread, persistent fog, though with significant uncertainty in density and timing, even at lead times of only a few hours. Patchy diurnal fog is currently unpredictable except in a very general sense, due to limitations in humidity, in the turbulent structure of stable boundary layers and in the resolution of significant features of the land surface (McCabe et al. 2016; Fallmann et al. 2019; Ducongé et al. 2020).

6.4   The Bridge Between Weather and Hazard

In this section, we explore the challenges of connecting the disciplinary languages, processes, timescales and cultures, the organisational hierarchies, the different mindsets and the technical capabilities that impede communication between hazard forecasters and weather forecasters.

6.4.1   Institutional Barriers

It is increasingly recognised that partnerships between expert bodies, for example, national meteorological services and flood or other hazard agencies, are necessary for effective hazard prediction (e.g. Demeritt et al. 2013). For such partnerships to grow and flourish, the barriers that separate institutions must be overcome. Some of these barriers arise from political and economic decisions of government that, for instance, promote competition amongst public bodies for funding or power. Barriers may also arise from entrenched institutional procedures, which may be embedded in legislation, especially in institutions having a long history (Pagano et al. 2001). Such procedures may be tuned to the needs of particular customers, with their own history and governance structure, especially when these customers are the dominant funding source (e.g. civil aviation or the military). These barriers need to be recognised and strategies developed for overcoming them before proceeding with partnership building.

In building an institutional partnership, each party brings their scientific and technical expertise which, when integrated, can be enormously powerful. Successful communication between partners needs to start by translating the goal of the partnership into each partner's language and then identifying a mutually beneficial objective. Although partners share the common goal of enhancing community safety, their differing mandates and areas of responsibility can lead to different priorities. National meteorological services typically operate at national scale, while hazard authorities often operate at state, region or city scale. If operational practices and requirements for meteorological information differ amongst hazard authorities in different regions, then the complexities of serving multiple users with similar but not identical data can slow down effective integration of weather and hazard prediction. Standardisation of service levels and practices can lead to improved consistency and facilitate broader and stronger partnerships.

When developing forecasting systems, meteorological agencies can choose to develop their own or to import systems developed elsewhere. While NWP models had crude representations of the land and ocean surfaces, it was normal for meteorological scientists to incorporate the available knowledge. With increasing complexity, the choice now is to import expertise (e.g. a team of hydrologists), to import a model (e.g. an ocean wave model) or to develop a partnership with a centre of expertise. For example, ECMWF developed its global flood forecasting system, GLOFAS (Alfieri et al. 2013), using an imported model, which it has further developed by employing hydrologists in-house.

Embedding meteorologists within operational agencies that are responsible for hazard prediction or, conversely, embedding hazard forecasters within meteorological agencies is becoming more common practice (Uccellini and Ten Hoeve 2019; see also Wildfire Case Study, below). This complements the integration of hazard models. Experts working in partnership can interpret and integrate important details of a hazardous situation and form a consensus view on the evolution (or possible trajectories) of the hazard. This more united view usually leads to better decisions as discussed in Chap. 4 and builds valuable trust between the partners.

6.4.2   Shared Situational Awareness

In real-time operational response, one person, however skilled, cannot provide expert interpretation for every hazard. However, as soon as responsibilities are split up, there is the possibility of inconsistency, even when the same model guidance is shared. Mechanisms for ensuring consistency of message need to be built into the operational structure of the partnership. This can involve sharing of observational data and guidance statements or of a multi-hazard dashboard (e.g. NOAA 2021a). It should also include frequent conferences between forecasters to enable different interpretations of the forecast guidance to be discussed and a common version agreed. With modern technology, the time between such conferences can be bridged using informal messaging tools, such as "chat rooms" (e.g. NOAA 2021b). All of this is greatly facilitated when all partners use a common set of tools and view the same data. This will increasingly be achieved by placing data in the "cloud".

6.4.3   Connecting Disciplinary Cultures

6.4.3.1   Hydrology and Meteorology

Accurate weather forecasts are essential to most flood forecasting systems. While air temperature forecasts are needed for some applications (such as determining evaporation or snowmelt), the primary variable of interest is precipitation.

Precipitation is notoriously difficult to forecast, with NWP having relatively coarse resolution, substantial biases and limited skill (Cuo et al. 2011). However, recent advances in model resolution and the use of ensembles have made the outputs more relevant to flood forecasting applications. Historically, most rainfall-runoff models have been based on parametrised conceptualisations developed in the 1970s (Pagano et al. 2014). Much effort continues to be directed to improving the calibration of such models. Furthermore, much of the NWP improvement stems from better assimilation of remotely sensed observations, whereas research in hydrologic data assimilation (e.g. Chen et al. 2013) is little used.

These points reflect a cultural difference in the use of models by meteorologists and hydrologists (Pagano et al. 2016). Generally, NWP systems are run on super-computers. After automated post-processing, the results are reviewed by the meteorologist to either accept, adjust or replace. Depending on the context, river forecasting may be more iterative, with lightweight models being run iteratively until the hydrologist is satisfied. Although this builds confidence, it prevents the use of more objective techniques, such as data assimilation and statistical post-processing. In response, there is an increasing operational trend towards side-by-side river forecasting systems, one complex and objective and another simple and adjustable.

Meteorologists are increasingly aiming to generate precipitation forecast products in probabilistic form (e.g. 25% chance of exceeding 15 mm). Although probabilities better represent the uncertainty in the forecast, they lack the spatial covariances and correlations of observations. Given that the relationship between rainfall and runoff is highly non-linear, such spatial information is essential to accurate runoff forecasting. In response to this, hydrologists have developed methods to convert probabilistic rainfall forecasts (including forecasts from different models at different lead times) into seamless, physically realistic ensembles, primarily through sampling patterns in historical observations (Clark et al. 2004; Bennett et al. 2017). Some of these approaches require objective hindcasts of NWP models, consistent with the current operational versions, which are expensive to generate.

6.4.3.2 Oceanography and Meteorology

On the face of it, oceanographers and meteorologists should communicate easily, since both are physics-based sciences of geophysical fluids on the earth's surface. The history of the two sciences has, however, resulted in quite different approaches to some aspects of their science. Oceanography is predominantly focused on research rather than operations and on ship-based experimental research rather than modelling, whereas meteorology has been focused on operational prediction since the nineteenth century. Indeed, the history of meteorology is dominated by the synoptic map – an analysis of conditions simultaneously sampled at multiple locations over a large area. Oceanographers rarely study the global ocean as a single entity, focusing on individual ocean basins, whereas meteorologists naturally take a global view. On the other hand, the object of a weather forecast is often a single point, whereas points in the ocean are rarely of interest, except on continental shelves where offshore production facilities have been constructed. The coastline is of tremendous importance to an oceanographer, as processes are very compressed in the inshore zone, and the water edge is a model boundary. While coasts are also important to meteorologists, they tend not to be considered in any greater detail than elsewhere over land. Until recently, the oceanographer has always had to work with minimal observational data, whereas the meteorologist is much better supplied – even over the oceans. Mathematically, the large-scale behaviour of the ocean is strongly constrained by boundaries – laterally at the coasts and vertically at the seabed – while its motion is driven by momentum transfer from the wind. On the other hand, small-scale motions are internally driven and much smaller than typical weather disturbances. In the atmosphere, internal dynamics drive much of the large-scale motion, while local forcing may be more important at small scales. Buoyancy is important in both but is driven by temperature and humidity in the atmosphere, as opposed to temperature and salinity in the ocean.

There is a long history of interaction in marine weather forecasts. However, the resulting interdisciplinary science has tended to be isolated from core ocean science. Genuine partnership has grown more recently with the development of coupled global models for climate studies. Such partnerships tended not to focus on coastal hazards. However, the coupled modelling approach is gaining increased use in weather forecasting (Pullen et al. 2017), and so the number of meteorologists and oceanographers working across this interface has grown.

6.4.3.3 Meteorology and Other Disciplines

Meteorological models run on a global grid in the medium and long range, often moving to limited area at shorter ranges to allow high-resolution grids. They are updated at regular intervals. Hazard models tend to be run on demand by the user, and the meteorological input may not be as fresh as desired. Hazard modellers frequently operate at jurisdictional (state, county or local government) level, so it is necessary to interpolate or "cookie cut" weather model output to obtain appropriate input for their hazard models. Post-processing of ensemble forecasts can provide the "worst-case", "best-case" and "expected" weather inputs for downstream models. However, sophisticated users increasingly ingest individual ensemble members to generate ensemble hazard forecasts of flood, fire, air quality, etc.

Different hazard models use weather information quite differently so weather modellers and forecasters must be flexible in their capability to serve hazard models with the necessary weather information. Weather models offer much richer information than many hazard models have been designed to ingest. As weather and hazard models become more tightly coupled, some physical attributes and processes that are critical to accurate hazard forecasting, such as vegetation, soil moisture and aerosol, will need to be treated more carefully by meteorologists.

6.4.4 Technical Constraints

6.4.4.1 Data and Standards

The data communication bandwidth within institutions is often orders of magnitude greater than that between institutions. As a result, the downstream partner may receive highly degraded input data, e.g. in spatial or temporal resolution, in domain or in the resolution of the probability distribution. Increasing bandwidth not only requires faster datalinks, but bigger databases, more expensive processing and more sophisticated interfaces, so choosing the optimum is important.

Weather model outputs conform to standards set by the World Meteorological Organization (WMO 2019). Hazard modelling communities operate to different, often local, standards, so effort is required to make weather and hazard models "talk to each other". State-of-the-art NWP models are computationally intensive and require high-performance computers (WMO 2013). Hazard models generally have less intensive compute requirements. When coupling hazard models more closely to weather models, hazard models could be transferred to meteorological centres where the NWP outputs can feed directly into the hazard model as in the ECMWF GLOFAS system (Alfieri et al. 2013) or NWP outputs can be transferred to the hazard modellers' environments for local use or one or both models can operate in a shared cloud environment. Each has challenges in terms of computational efficiency, control of model upgrades and speed of operation.

6.4.4.2 Spatial and Temporal Scales

The scales required to assess the impact of the hazard must be the driver for all parties. The resolution used by the hazard agency to meet these demands may require unachievably fine resolutions for the input weather data. In this case, downscaling of the weather forecasts may be needed. The benefits of doing this need to be clearly evaluated.

Hazard impacts often occur on small scales, and as a result, hazard prediction frequently places high demands on the meteorological information supplied as input. In a 1-day forecast, very-high-resolution limited area NWP may be able to meet this demand. For hazard prediction on timescales of days to a week or more, required for some mitigating actions, the influence of the large-scale meteorological flow and uncertainty on the local detail dictates the use of global models, with consequent coarser spatial resolution. The predicted timing of events similarly loses precision for longer-range forecasts. This may not meet the needs of the hazard agency for highly precise information. However, it follows from the inherent unpredictability of the atmosphere and has to be accommodated by the coupled prediction system.

6.4.4.3 Uncertainty and Bias

Weather forecasters and hazard modellers use numerical forecasts in different ways. Meteorologists accommodate errors in the predicted location and timing of high-impact weather from NWP paying particular attention to the large-scale patterns (at medium range) or the mode of convection (at short range) and applying their experience to interpret the forecast. Hazard models are often less able to accommodate errors such as rainfall falling in a different catchment because of a "minor" positional error, a surge occurring at a different state of the tide because of a "minor" timing error or snow failing to reach the ground because of a "minor" temperature error. Ensemble prediction offers a means to transfer uncertainty in weather forecasts into uncertainties in hazard forecasts. However, while ensembles are good at capturing uncertainty, they may still be biased. It is therefore important to correct systematic errors in model outputs by statistical post-processing prior to their ingestion into hazard models (Gascon et al. 2019).

6.4.4.4 Uncertainty

Atmospheric forecasts are essentially uncertain due to the chaotic behaviour of the atmosphere. An effective partnership will recognise that this is not a shortcoming in the data input to the hazard forecast but, rather, a fundamental limitation that must be reflected in the hazard forecasting system. Some hazards, such as those associated with severe thunderstorms, reinforce the uncertainty from the basic meteorology and must be predicted in probabilistic terms, while others, such as flood predictions for a large river system, may reduce it. An essential part of developing a hazard forecasting system is to identify how uncertainty will be incorporated, both in the meteorological inputs and in the hazard outputs.

6.4.4.5 Consistency of Heat, Water, Gas and Momentum Fluxes

Many hazards occur at the interface between atmosphere and land or ocean. The ability to model this interface is currently crude, with fluxes leaving one model often inconsistent with the requirements of the receiving model. Over land, the descriptions both of the land surface, including buildings, trees, rocks, etc., and of the turbulent processes through which interaction occurs are extremely simplified. Over the sea, the surface is generally considered to be horizontal, ignoring the turbulent effects of waves and moisture exchanges due to spray. Advances in understanding these processes require detailed and painstaking research supported by expensive field measurements.

6.4.5 Model Integration

From the first coupled climate models of the 1970s, representations of the long-timescale interactions between the physical and chemical state of the atmosphere along with feedbacks between atmosphere, land surface, ocean and cryosphere were necessary for decadal to centennial simulations. Since then, the scope of earth system coupling has been extended to introduce greater complexity and fidelity and to include processes such as aerosol chemistry, dynamic vegetation and ice sheet dynamics (see, e.g. Jones et al. 2011; Cornell et al. 2012).

The translation of this "whole system" thinking to weather forecast timescales is less mature but is a growing area of research and application (Rabier et al. 2015; Belair 2015). Typically, NWP has made simplifying assumptions that omit or parameterise earth system interactions so as to minimise the computational cost and complexity of forecasting systems, recognising that these processes usually make a second-order contribution to predictive skill. For example, assuming that the analysed sea surface temperature valid at the start of a simulation cycle will persist for the duration of the weather forecast has been considered sufficient at most operational NWP centres. Today, this situation is changing with the application of atmosphere-land-ocean coupled ensemble NWP systems increasingly common for global-scale systems running forecasts on timescales of days to weeks (Harrigan et al. 2020). This has been shown to improve predictive skill in tropical regions including a better representation of tropical cyclone evolution and intensity. Remaining challenges include extending this "whole system" approach to data assimilation and ensemble prediction. While assimilation methods and capability are well advanced for atmosphere and ocean components, they are less developed for other components such as atmospheric chemistry or hydrological state. Challenges also remain in coupled atmosphere-ocean data assimilation, arising from the different timescales required for initialisation of ocean and atmosphere components (Frolov et al. 2016). Further work is also required on the design of representative initial condition and model uncertainties in coupled ensemble systems, so as to capture the impact of interactions on uncertainty.

For shorter-range regional prediction systems, there is similarly a growing recognition of the potential value of increasing model complexity, including regional coupled environmental prediction systems (Lewis et al. 2018a; Fallmann et al. 2019). At kilometre scale, the relevance of earth system processes becomes important for better representing the heterogeneity of the landscape to improve model skill, notably at coastlines and around urban environments. The prospect of a more integrated catchment-resolving approach to hydrometeorological prediction also becomes possible. Critically, the advance towards ensemble numerical environmental prediction provides a framework from which to develop consistent outputs for simulation of multiple hazards. In general, environmental hazards have a strong meteorological driver, e.g. the multiple impacts of a storm in a coastal environment through strong winds producing inland inundation from sea surge and wave overtopping and through heavy rainfall and saturation of the land surface leading to high river flows, overbank inundation and potential for landslide and other linked hazards. The goal is to represent multi-hazard probabilities, accounting for uncertainty propagation through a connected system. At these scales of interest, interactions of physical and biogeochemical systems with the built environment and human systems also become increasingly relevant and offer a further frontier for bridging across communities, science disciplines and modelling capability.

Integrated modelling requires attention to be paid to the scales of interest in each domain. Integration of NWP with land surface hydrology requires recognition of the much finer horizontal resolution required for accurate hydrological prediction, especially in urban areas (Cuo et al. 2011), as well as the sensitivity to heat fluxes and evapotranspiration. Ocean models need to run with fine horizontal resolution to represent the nearshore ocean and especially the inter-tidal zone, and vertical fluxes of heat and momentum depend on modelling of waves and currents. Estuaries are particularly complex, requiring interactions amongst inshore ocean, river and flood inundation, often in an area of complex meteorology, bearing in mind that the temperature and composition of river water may influence the temperature and biology of the coastal ocean and hence any coastal atmospheric circulations. Coupling of air composition into weather models requires both the radiative and cloud microphysical impacts of aerosols to be considered simultaneously with getting the ground-level pollutant concentrations right. Composition models typically contain large numbers of species and the chemical reactions between them, resulting in much expanded prediction codes with many parameters (Freitas et al. 2011). They also require specification of pollutant sources – which may change for a variety of reasons, some of which may have regular patterns in time and space, some may be weather dependent and others may be associated with specific events such as festivals.

6.4.6 User-Oriented Verification

To generate confidence across the partnership, the quality of the inputs delivered by each partner should be measured in terms that reflect their use by the other partner, as well as in terms that support their internal development (Ebert et al. 2018). For hazard forecasting, aspects of weather forecast quality that are important may include location and timing of features such as storms and fronts, structure and variability, and magnitude and extremity. For example, verifying rainfall for flood prediction requires assessing whether it was located over the catchment of interest and whether it had the right intensity distribution to produce the observed runoff and flood height. Temperature verification for heat wave forecasting assesses whether the predicted temperatures were sufficiently extreme and of sufficient duration to lead to health impacts. This sort of diagnostic evaluation complements more traditional metrics of forecast accuracy. The discipline of routine objective forecast verification practised in operational meteorology can be extended to hazard prediction, providing suitable hazard observations are available.

Identifying the root cause of a deficiency in hazard forecasts is important and requires collaboration. Where errors can be related to a bias in the meteorological input, this may be straightforward. However, where processes are complex, it may be necessary to explore them in detail to establish whether the cause is in the process representation, in the input meteorology or elsewhere. It may, indeed, be a mixture, and model tuning often leads to error cancellation that only becomes apparent in extreme conditions. Joint field programmes can be a valuable opportunity for exploring such issues.

Verification should be oriented to the aspects of the prediction system that are most relevant to the decision-maker at the end of the warning chain. Knowing how forecasts become less accurate at longer lead times helps the decision-maker understand the risks of acting (or not) on a forecast that may turn out to be a false alarm or a missed event. Verifying forecasts and warnings of socio-economic impact (if they have been made) is extremely difficult, as discussed in Chap. 4. Visual comparison of forecasts overlaid with evidence of hazard impact can be informative and helps tell the story. Measuring the performance of the forecast elements that can be objectively verified is also important. When decision-makers have thresholds for taking action based on the forecast, then verifying forecasts of threshold exceedance at the location of interest gives the user the quality information required to develop an appropriate level of confidence.

6.5 Examples of Partnerships

Box 6.1 Flood Forecasting Centre Case Study
Graeme Boyce, Flood Forecasting Centre, UK

Following devastating floods during the summer of 2007, the UK government was determined to develop a more "joined-up" approach to both preparing for and responding to flooding. The Pitt Review (Pitt 2008) identified both the need for all organisations involved to be willing to work together and share information and the importance of forecasting and prediction in enabling emergency planners and responders to reduce the risk and impact of flooding. Its recommendation was clear – the Environment Agency (EA), as the lead flood risk management authority for England and Wales (subsequently responsibility for Wales was devolved to Natural Resources Wales (NRW)), and the Met Office, the UK national meteorological service, should work together, through a joint centre, to improve the technical capability to forecast, model and warn against all sources of flooding. In April 2009, the Flood Forecasting Centre (FFC) became operational, creating a national capability, for England and Wales, to provide advanced notice of potential flood risk from all natural sources of flooding (river, coastal, surface water and groundwater) through a daily flood guidance service delivered to all organisations with a statutory responsibility to respond to flooding. The most important role for the FFC was, and still is, to provide flood guidance to the response community; however, with commendable foresight, the scope of the centre also included the remit to engage directly with customers/users and to deliver ongoing service improvements based on feedback from those using the service.

From its inception, the FFC has placed the science of hydrometeorology at its core. A small team of meteorologists from the Met Office, and hydrologists from the EA, was recruited and cross-trained to gain a deeper understanding

(continued)

of each other's disciplines and customer needs – creating a cadre of professionally accredited operational hydrometeorologists. With different training, institutional backgrounds and employment terms, there was an initial challenge of creating trust, which was overcome by a combination of openness and establishing a common purpose. However, the complications of dual IT systems linked to the parent institutions remain and will not be easily solved.

The centre was set up with a goal to forecast the impact of floods from natural sources, with as long a lead time as possible. To do this, it was recognised that concepts of likelihood and uncertainty would need to be incorporated into guidance information and this resulted in a risk matrix taking a central role in presenting the likelihood of flooding, over the next 5-day period, within the FFC's primary product – the Flood Guidance Statement (Fig. 6.10).

The risk matrix was co-designed with the Met Office National Severe Weather Warning Service (NSWWS) ensuring both used consistent concepts and terminology, helping to promote a joined-up service with our response community. Partnership and collaboration were key to its initial ability to become embedded within the flood risk incident management structure and remain vital to its success. The Flood Guidance Statement is co-produced with local forecast teams from the EA/NRW and operational meteorologists from the Met Office. At times of heightened flood risk, FFC duty managers routinely brief senior officials within central government and the EA on the flood risk at a national scale to support strategic decisions. Considerable effort is made to maintain an authoritative and consistent flood risk message during periods of heightened flood risk across the FFC partnership to support flood incident management decision-making. This level of collaboration is maintained when planning improvements to forecasting capabilities of customer facing products. The default position is to maintain common forecasting and visualisation systems where possible and work in partnership with Met Office and EA/NRW colleagues to improve these for mutual benefit. The "bridge" that the FFC provides from the Met Office to the EA/NRW flood management authorities has improved the pull-through of science into operational use and has reduced the time taken for this to happen. Prior to 2009, the lead time generally provided by flood forecasts across England and Wales was measured in hours. Over the past 10 years, flood risk guidance has routinely been provided for the next 5 days, and now the Centre is expanding the user base for its 30-day Flood Outlook service (Fig. 6.11).

Engagement with the flood responder community is also coordinated across the partnership, and this allows the forecasting authorities to present a more coordinated approach and increase the benefit gained from forecasting and warning information. All these improvements have been overseen by a Joint Steering Group, with representation from the Met Office, EA and NRW, and guided by a User Group, with a wide membership from the flood response

(continued)

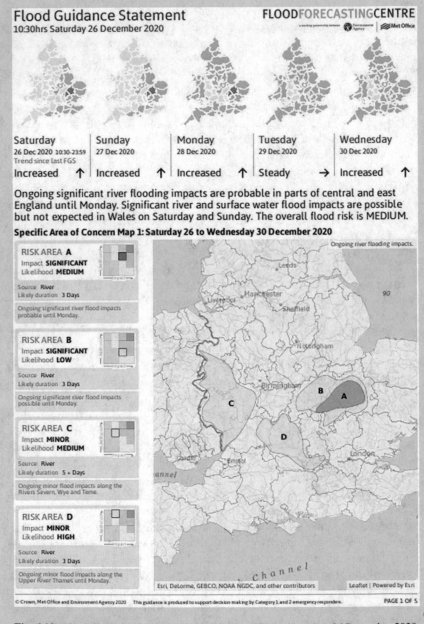

Fig. 6.10 Flood Guidance Statement for England and Wales issued on 26 December 2020. (© Crown Copyright 2020, Flood Forecasting Centre)

(continued)

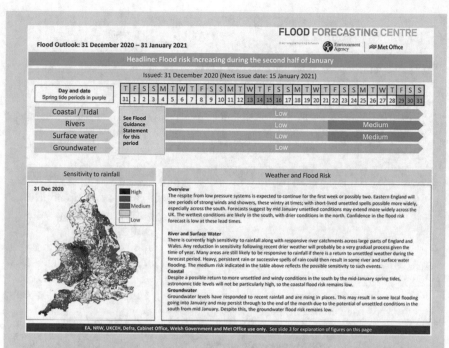

Fig. 6.11 Flood Outlook for England and Wales issued on 30 December 2020. (© Crown Copyright 2020, Flood Forecasting Centre)

community, that has enabled this unprecedented partnership and collaborative approach to continue.

With over 10 years of operational experience, providing a flood guidance service, including periods of significant flooding (e.g. winter 2013/2014, winter 2015/2016 and February 2020), the Flood Forecasting Centre can confidently claim that it has become a very successful partnership bringing together world leading meteorological and hydrological science. With customers from the emergency responder community asked to rate the service provided by the FFC every 2 years since its inception, overall satisfaction rose to 91% in 2019, and 92% were satisfied with the daily Flood Guidance Statement. Trust is critical, both in terms of maintaining a successful partnership and in continuing to deliver a forecasting service that is acted upon and provides value to its user base. Perhaps the most visible example of this trust is the investment by the Environment Agency in over 40 km of temporary barriers to help defend communities at risk of flooding where no permanent defences exist. Their successful deployment is dependent on good, advance notice of flooding which is delivered by the FFC in partnership with forecasting colleagues from the Met Office and Environment Agency. This has only been possible through continued collaboration at all levels of governance and leadership, scientific/technological development and operational delivery over the past 12 years.

Box 6.2 Reflections on Working in Partnership with Fire Agencies During Extreme Fires
Mika Peace, Bureau of Meteorology and Bushfire and Natural Hazards Cooperative Research Centre, Australia

In recent fire seasons, Australia has experienced unprecedented fire events. Through many of these, I have worked inside the state operations centres (SOCs) of fire agencies, providing an enhanced briefing and interpretive role.

Unlike other severe weather phenomena, high fire risk doesn't always translate to impacts; it depends on whether ignition occurs. When extreme fires are active, they happen fast; therefore, a deep appreciation of complexity of the situation and rapid response is required. My role involves working closely with the fire behaviour analysts, as fire prediction crosses the disciplines of fire science and meteorology, requiring cohesive teams with multi-disciplinary knowledge and an ongoing exchange of information. I need to have an evolving narrative as the situation unfolds and new information becomes available through the day. Being embedded in another agency also requires a strong connection and established networks with my home agency, so I can reach out for additional information when required.

I see my role as ensuring there are "no surprises", in the SOC or on the fireground. Extreme fire behaviour will happen, but response can be adapted and risk minimised if everyone has clarity on when and where the fire will run and what fire behaviour will occur. When analysing the data, I'm constantly thinking "what could the weather and fire potentially produce as extreme fire behaviour and what is the likelihood". The process is not as simple as looking at the NWP output; a deeper level of interpretation and pattern recognition is required. Sometimes, communicating with confidence what won't happen is extremely valuable because it focuses energy away from unnecessary concerns.

Inside the state operations centres, there is a prodigious demand for meteorological information, but the value is in interpretation of how the weather will impact fire behaviour. Copious amounts of data are readily available; intelligence is much more difficult to develop and deliver. On bad days, numerous briefings to various audiences are requested, frequently with minimal notice and requiring distillation of complex information into an understandable and immediately relevant message. My experience inside partner organisations is that briefings and conversations are more valued than products.

Emergency management involves political leaders in the decision and response process, and they rely on expert advice from trusted scientists who can communicate clearly. I've been surprised and impressed at how quickly politicians can read a room and determine who has deep expertise and can be

(continued)

trusted for advice as well as their perceptive questions that require comprehensive knowledge to answer.

Extreme fire events are stressful, particularly in a room full of people who have responsibility for making decisions with life and death consequences. So far, I've had an ability to maintain a calm demeanour and clarity of thought during briefings. When I am particularly worried about a day, I'm aware that my concern is projected during briefings, and I've seen the emotion in the delivery of my briefing message being received and responded to just as clearly as the science content. On occasions during a disaster, I've switched from providing science briefings to being a listening ear and providing hugs to colleagues under stress.

In post-event debriefs, I have repeatedly heard our partners emphasise the value they place on trusted relationships with individuals. The counterargument I've heard is that our procedures should be sufficiently robust that relationships don't matter. However, human nature is to value relationships, and emergency management tends to attract empathetic people with altruistic intent, so I believe relationships will continue to be important during extreme events.

Having researchers such as myself in operations has bilateral benefits as, ultimately, stronger research utilisation links will be built, enabling accelerated uptake and adoption of research findings. It will also focus research efforts towards high-impact outcomes addressing real-world issues. I am fortunate to have an extremely rewarding role straddling operations and research. However, what I do is not traditionally a defined career pathway in meteorological agencies. The benefits are intrinsic and therefore difficult to measure and are only fully realised during high-impact events. Long-term investment is required to build cross-disciplinary capability and develop partnerships before events happen so we can be ready to "hit the ground running".

It is not possible to anticipate and plan contingencies for all possible future scenarios. It is probable that the worse-case scenario is beyond what we can imagine, and it is inevitable that cascading and overlapping events will present response challenges that stretch resources beyond capacity. A structure that supports organic response and enables well-connected people to call in any available assistance when faced with predicted and escalating situations will support optimal response to emerging disasters (Fig. 6.12).

Dr. Mika Peace is a fire meteorologist at the Australian Bureau of Meteorology. For 10 years, she worked as an operational forecaster in various locations around Australia, and for the past 10 years, she has held a fire meteorology research role. She is recognised as an expert in fire atmosphere interactions through research on case studies and simulations of extreme fire behaviour using coupled fire atmosphere models to understand the interaction processes between the energy release and the surrounding atmosphere.

(continued)

Fig. 6.12 Mika (left) briefing the New South Wales Rural Fire Service Commissioner (centre) and the New South Wales Premier (right) in the Rural Fire Service Operations Centre ("the room") during the 2019–2020 fire season

6.6 Summary

- Successful hazard predictions require effective application of expertise from each discipline.
- Building partnerships amongst hazard prediction institutions requires time and effort to remove institutional barriers and build shared objectives.
- Hazard disciplines have different languages and cultures. Successful hazard prediction requires members of each discipline to learn the language and culture of their partners.
- Observations of hazards are fundamental to understanding their importance and their causes but are not widely available or easily accessible.
- Linking hazard models to weather models requires care, based on an understanding of the different roles of the relevant variables in each model.
- Linking hazard models to weather models requires choices of data standards, time and space resolution, update frequency, forecast length, representation of uncertainty and measures of quality. Compromises should be driven by user requirements wherever possible.
- Integrated models will increasingly be the basis of hazard prediction in the future. Their implementation should be based on clear evidence of benefit to users.
- Hazard forecasts should be verified against observations using methods that reflect the use of the predictions in warnings.

- Hazard forecasts will be used alongside weather forecasts and should be consistent with them. Shared situational awareness tools can facilitate consistency.

References

Adams, T. E. and T. C. Pagano, (eds)., 2016. *Flood Forecasting: A Global Perspective.* Academic Press.

Aldridge, T., O. Gunawan, R. J. Moore, S. J. Cole, G. Boyce and R. Cowling, 2020. Developing an impact library for forecasting surface water flood risk. *J. Flood Risk Man*, **13**, e12641. doi:https://doi.org/10.1111/jfr3.12641

Alfieri, L., P. Burek, E. Dutra, B. Krzeminski, D. Muraro, J. Thielen and F. Pappenberger, 2013. GloFAS - global ensemble streamflow forecasting and flood early warning. *Hydrol. Earth Syst. Sci.,* **17**, 1161–1175, doi:https://doi.org/10.5194/hess-17-1161-2013.

Atencia, A., A. Kann, Y. Wang and F. Meier, 2020. Localized variational blending for nowcasting purposes. *Meteorol. Zeit.*, **29**, 247–261.

Bach, P. M., W. Rauch, P. S. Mikkelsen, D. T. McCarthy and A. Deletic, 2014. A critical review of integrated urban water modelling Urban drainage and beyond. *Env. Modell. & Software*, **54**, 88–107. doi: https://doi.org/10.1016/j.envsoft.2013.12.018

Bates, P. D., D. Mason, S. Neelz, G. Pender, G. Villaneuva and M. D. Wilson, 2004. A framework for flood inundation modelling, in *Flood risk assessment*. Reeve, D. (ed.). Institute of Mathematics and Applications, Southend-on-Sea, 169–178

Bauer, P., A. Thorpe and G. Brunet, 2015. The quiet revolution of numerical weather prediction. *Nature*, **525**, 47–55.

Baum, R. L., W. Z. Savage and J. W. Godt, 2002. TRIGRS—A FORTRAN program for transient rainfall infiltration and grid-based regional slope-stability analysis, U.S. Geol. Surv. Open-File Rcp. 02-0424, 35pp.

Baum, R. L., J. W. Godt and W. Z. Savage, 2010. Estimating the timing and location of shallow rainfall-induced landslides using a model for transient, unsaturated infiltration, *J. Geophys. Res.,* 115, F03013, doi:https://doi.org/10.1029/2009JF001321.

Bedia, J., N. Golding, A. Casanueva, M. Iturbide, C. Buontempo and J. M. Gutiérrez, 2018. Seasonal predictions of Fire Weather Index: Paving the way for their operational applicability in Mcditcrranean Europe. *Clim. Serv.*, **9**, 101–110.

Belair, S., 2015. Regional Environmental Prediction Systems. In Seamless Prediction of the Earth System: from minutes to months, Eds G. Brunet, S. Jones & P. M. Ruti. Geneva: World Meteorological Organization. WMO-No. 1156. ISBN 978-92-63-11156-2.

Bennett, J. C., Q. J. Wang, D. E. Robertson, A. Schepen, M. Li and K. Michael, 2017. Assessment of an ensemble seasonal streamflow forecasting system for Australia. *Hydrol. Earth Sys. Sci.,* **21**, 6007–6030.

Birch, C. E., B. L. Rabb, S. J. Böing, K. L. Shelton, R. Lamb, N. Hunter, M. A. Trigg, A. Hines, A. L. Taylor, C. Pilling and M. Dale, 2021. Enhanced surface water flood forecasts: User-led development and testing. *J. Flood Risk Man.*, **14**, e12691. doi:https://doi.org/10.1111/jfr3.12691

Booij, N., R. C. Ris and L. H. Holthuijsen, 1999. A third-generation wave model for coastal regions – 1. Model description and validation. *J. Geophys. Res.-Oceans*, **104**, 7649–7666

Browning, K. A., 1982. *Nowcasting*. Academic Press Inc (1 Aug. 1982). 256pp, ISBN-10: 0121377601 ISBN-13: 978-0121377601.

Brunetti, M. T., S. Peruccacci, M. Rossi, S. Luciani, D. Valigi and F. Guzzetti, 2010. Rainfall thresholds for the possible occurrence of landslides in Italy. *Nat. Hazards Earth Syst. Sci.,* **10**, 447–458. doi:https://doi.org/10.5194/nhess-10-447-2010.

Caine, N., 1980. The rainfall intensity – duration control of shallow landslides and debris flows. *Geografiska Annaler,* 62, 23–27. doi:https://doi.org/10.1080/04353676.1980.11879996.

Call, D. A., C. S. Wilson and K. N. Shourd, 2018. Hazardous weather conditions and multiple-vehicle chain-reaction crashes in the United States. *Meteorol. Appl.,* 25, 466–471. doi:https://doi.org/10.1002/met.1714

Cangialosi, J. P., E. Blake, M. DeMaria, A. Penny, A. Latto, E. Rappaport and V. Tallapragada, 2020. Recent Progress in Tropical Cyclone Intensity Forecasting at the National Hurricane Center, *Wea. Forecast.*, 35, 1913–1922.

Carniel, S., M. Sclavo and R. Archetti, 2011. Towards validating a last generation, integrated wave-current-sediment numerical model in coastal regions using video measurements. *Ocean. Hydrobiol. Studies,* 40, 11–20.

Cavaleri, L., S. Abdallab, A. Benetazzoa, L. Bertottia, J.-R. Bidlotb, Ø. Breivikc, S. Carniela, R. E. Jensend, J. Portilla-Yandune, W. E. Rogersf, A. Rolandg, A. Sanchez-Arcillah, J. M. Smithd, J. Stanevai, Y. Toledoj, G.Ph. van Vledderk and A. J. van der Westhuysen, 2018. Wave modelling in coastal and inner seas. *Prog. Ocean.*, 167, 164–233. doi:https://doi.org/10.1016/j.pocean.2018.03.010

Changnon, S. A. and T. G. Creech, 2003. Sources of data on freezing rain and resulting damages. *J. Appl. Meteor.*, 42, 1514–1518, doi:https://doi.org/10.1175/1520-0450(2003)042,1514:SODOFR.2.0.CO;2.

Chen, H., D. Yang, Y. Hong, J. J. Gourley and Y. Zhang, 2013. Hydrological data assimilation with the Ensemble Square-Root-Filter: Use of streamflow observations to update model states for real-time flash flood forecasting. *Adv. Water Resources*, 59, 209–220.

Church, C., D. Burgess, C. Doswell and R. Davies-Jone, (Eds), 1993. *The Tornado: Its Structure, Dynamics, Prediction, and Hazards.* AGU Geophysical Monograph Series Volume 79 Print ISBN:9780875900384 Online ISBN:9781118664148. doi:https://doi.org/10.1029/GM079.

Clark, M., S. Gangopadhyay, L. Hay, B. Rajagopalan and R. Wilby, 2004. The Schaake shuffle: A method for reconstructing space–time variability in forecasted precipitation and temperature fields. *J. Hydrometeorol.*, 5, 243–262.

Clark, P. A. and S. L. Gray, 2018. Sting jets in extratropical cyclones: a review. *Quart. J. Roy. Meteorol. S.,* 144, 943–969

Coiffier, J., 2011. *Fundamentals of Numerical Weather Prediction.* Cambridge University Press. 368pp. ISBN-10: 110700103X ISBN-13: 978-1107001039.

Cornell, S. E., I. C. Prentice, J. I. House and C. J. Downy, 2012. *Understanding the Earth System: Global Change Science for Application.* Cambridge University Press. ISBN 9781139560542.

CRED (Centre for Research on the Epidemiology of Disasters), 2015. *The human cost of natural disasters 2015: A global perspective.* [Online] Available at: https://reliefweb.int/sites/reliefweb.int/files/resources/PAND_report.pdf (Accessed 21 Jan 2021).

Cuo, L., T. C. Pagano and Q. J. Wang, 2011. A review of quantitative precipitation forecasts and their use in short-to medium-range streamflow forecasting. *J. Hydrometeorol.*, 12, 713–728.

Demeritt, D., S. Nobert, H. L. Cloke and F. Pappenberger, 2013. The European Flood Alert System and the communication, perception, and use of ensemble predictions for operational flood risk management. *Hydrol. Proc,* 27, 147–157.

Dorninger, M., E. Gilleland, B. Casati, M. P. Mittermaier, E. E. Ebert, B. G. Brown and L. J. Wilson, 2018. The setup of the MesoVICT project, *Bull. Amer. Meteorol. S.,* 99, 1887–1906. doi:https://doi.org/10.1175/BAMS-D-17-0164.1

Dowdy, A. J., 2020. Seamless climate change projections and seasonal predictions for bushfires in Australia. *J. Southern Hemisphere Earth Sys. Sci.*, 70, 120–138.

Duan, Q. Y., V. K. Gupta and S. Sorooshian, 1993. Shuffled complex evolution approach for effective and efficient global minimization. *J. Optimization Theory Appl.*, 76, 501–521.

Ducongé, L., C. Lac, B. Vié, T. Bergot and J. D. Price, 2020. Fog in heterogeneous environments: the relative importance of local and non-local processes on radiative-advective fog formation. *Quart. J. Roy. Meteorol. S.,* 146, 2522–2546. doi:https://doi.org/10.1002/qj.3783

Ebert, E., B. Brown, M. Göber, T. Haiden, M. Mittermaier, P. Nurmi, L. Wilson, S. Jackson, P. Johnston and D. Schuster, 2018. The WMO Challenge to Develop and Demonstrate the Best New User-Oriented Forecast Verification Metric. *Meteorol. Zeit.*, **27**, 435–440. doi:https://doi.org/10.1127/metz/2018/0892

EEA, 2020. Air quality in Europe – 2020 report. doi:https://doi.org/10.2800/786656

Fallmann, J., H. Lewis, J. C. Sanchez and A. Lock, 2019. Impact of high-resolution ocean–atmosphere coupling on fog formation over the North Sea. *Quart. J. Roy. Meteorol. S.*, **145**, 1180–1201. doi:https://doi.org/10.1002/qj.3488

Fikke S., G. Ronsten, A. Heimo, S. Kunz, M. Ostrozlik, P.-E. Persson, J. Sabata, B. Wareing, B. Wichura, J. Chum, T. Laakso, K. Säntti and L. Makkonen, 2006. *COST-727 - Atmospheric Icing on Structures, Measurements and data collection on icing: State of the Art*. MeteoSwiss, 75, 110pp.

Filippi, J. B., Bosseur, F., Mari, C. and Lac, C., 2018. Simulation of a large wildfire in a coupled fire-atmosphere model. *Atmosphere*, **9**, 218.

Flather, R. A., 2000. Existing operational oceanography. *Coastal Engineering*, **41**, 13–40.

Freitas, S. R., K. M. Longo, M. F. Alonso, M. Pirre, V. Marecal, G. Grell, R. Stockler, R. F. Mello and M. Sánchez Gácita, 2011, PREP-CHEM-SRC – 1.0: a preprocessor of trace gas and aerosol emission fields for regional and global atmospheric chemistry models, *Geosci. Model Dev.*, **4**, 419–433. doi:https://doi.org/10.5194/gmd-4-419-2011.

Frolov, S., C. H. Bishop, T. Holt, J. Cummings and D. Kuhl, 2016. Facilitating Strongly Coupled Ocean–Atmosphere Data Assimilation with an Interface Solver. *Mon. Wea. Rev.*, **144**, 3–20. doi:https://doi.org/10.1175/MWR-D-15-0041.1.

Froude, M. J. and D. N. Petley, 2018. Global fatal landslide occurrence from 2004 to 2016. *Nat. Hazards Earth Syst. Sci.*, **18**, 2161–2181. doi:https://doi.org/10.5194/nhess-18-2161-2018.21612181.

Gariano, S. I., M. Melillo, S. Peruccacci and M. T. Brunetti, 2020. How much does the rainfall temporal resolution affect rainfall thresholds for landslide triggering? *Nat. Hazards*, **100**, 655–670. doi:https://doi.org/10.1007/s11069-019-03830-x

Gascon, E., D. Lavers, T. M. Hamill, D. S. Richardson, Z. B. Bouallegue, M. Leutbecher and F. Pappenberger, 2019. Statistical postprocessing of dual-resolution ensemble precipitation forecasts across Europe. *Quart. J. Roy. Meteorol. S.,* **145**, 3218–3235. doi: https://doi.org/10.1002/qj.3615.

Godt, J. W. and J. A. Coe, 2007. Alpine debris flows triggered by a 28 July 1999 thunderstorm in the central Front Range, Colorado. *Geomorphology*, **84**, 80–97.

Gultepe, I., R. Tardif, S. C. Michaelides, J. Cermak, A. Bott, J. Bendix, M. D. Müller, M. Pagowski, B. Hansen, G. Ellrod, W. Jacobs, G. Toth and S. G. Cober, 2007. Fog Research: A Review of Past Achievements and Future Perspectives. *Pure Appl. Geophys.* **164**, 1121–115. doi:https://doi.org/10.1007/s00024-007-0211-x.

Guzzetti F., S. L. Gariano, S. Peruccacci, M. T. Brunetti, I. Marchesini, M. Rossi and M. Melillo, 2020. Geographical landslide early warning systems. *Earth-Sci. Rev.,* **200**, 102973. doi:https://doi.org/10.1016/j.earscirev.2019.102973.

Guzzetti, F., S. Peruccacci, M. Rossi and C. Stark, 2008. The rainfall intensity-duration control of shallow landslides and debris flows: an update. *Landslides*, **5**, 3–17.

Harrigan, S., H. Cloke and F. Pappenberger, 2020. Innovating global hydrological prediction through an Earth system approach. WMO Bulletin, **69(1)**.

Isaac, G. A., T. Bullock, J. Beale and S. Beale, 2020. Characterizing and Predicting Marine Fog Offshore Newfoundland and Labrador. *Wea. Forecast.*, **35**, 347–365. doi:https://doi.org/10.1175/WAF-D-19-0085.1.

Jiménez, P.A., D. Muñoz-Esparza and B. Kosović, 2018. A high resolution coupled fire–atmosphere forecasting system to minimize the impacts of wildland fires: Applications to the Chimney Tops II wildland event. *Atmosphere*, 9, 197.

Jolliffe I. T. and D. B. Stephenson (eds.), 2012. *Forecast Verification: A Practitioner's Guide in Atmospheric Science*. 292 pp. John Wiley & Sons Ltd: Chichester, UK.

Jones, C. D., J. K. Hughes, N. Bellouin, S. C. Hardiman, G. S. Jones, J. Knight, S. Liddicoat, F. M. O'Connor, R. J. Andres, C. Bell, K-O. Boo, A. Bozzo, N. Butchart, P. Cadule, K.D. Corbin, M. Doutriaux-Boucher, P. Friedlingstein, J. Gornall, L. Gray, P. R. Halloran, G. Hurtt, W. J. Ingram, J-F. Lamarque, R. M. Law, M. Meinshausen, S. Osprey, E. J. Palin, L. Parsons Chini, T. Raddatz, M. G. Sanderson, A. A. Sellar, A. Schurer, P. Valdes, N. Wood, S. Woodward, M. Yoshioka and M. Zerroukat, 2011. The HadGEM2-ES implementation of CMIP5 centennial simulations. *Geosci. Model Dev.*, **4**, 543–570. doi:https://doi.org/10.5194/gmd-4-543-2011.

Karsisto, V., S. Tijm and P. Nurmi, 2017. Comparing the Performance of Two Road Weather Models in the Netherlands. *Wea. Forecast.*, **32**, 991–1006. doi:https://doi.org/10.1175/WAF-D-16-0158.1.

Keller, J. H. C. M. Grams, M. Riemer, H. M. Archambault, L. Bosart, J. D. Doyle, J. L. Evans, T. J. Galarneau Jr, K. Griffin, P. A. Harr, N. Kitabatake, R. McTaggart-Cowan, F. Pantillon, J. F. Quinting, C. A. Reynolds, E. A. Ritchie, R. D. Torn and F. Zhang, 2019. The Extratropical Transition of Tropical Cyclones. Part II: Interaction with the Midlatitude Flow, Downstream Impacts, and Implications for Predictability. *Mon. Wea. Rev.*, **147**, 1077–1105. doi:https://doi.org/10.1175/MWR-D-17-0329.1

Khakbaz, B., B. Imam, K. Hsu and S. Sorooshian, 2012. From lumped to distributed via semi-distributed: Calibration strategies for semi-distributed hydrologic models. *J. Hydrol.*, **418**, 61–77.

Klipsch, J. D. and M. B. Hurst, 2007. *HEC-ResSim reservoir system simulation user's manual version 3.0*. US Army Corps of Engineers, Davis, CA, *512*.

Knaff, J. A., C. R. Sampson and B. R. Strahl, 2020. A Tropical Cyclone Rapid Intensification Prediction Aid for the Joint Typhoon Warning Center's Areas of Responsibility. *Wea. Forecast.*, **35**, 1173–1185.

Komen, G., L. Cavaleri, M. Donelan, K. Hasselmann, H. Hasselmann and P. A. E. M. Janssen, 1994. *Dynamics and Modelling of Ocean Waves*, Cambridge Univ. Press, 532pp.

Krøgli, I. K., G. Devoli, H. Colleuille, M. Sund, S. Boje and I. K. Engen, 2018. The Norwegian forecasting and warning service for rainfall- and snowmelt-induced landslides. *Nat. Hazards Earth Syst. Sci.*, **18**, 1427–1450. doi:https://doi.org/10.5194/nhess-2017-426.

Lagerquist, R., A. McGovern, C. R. Homeyer, D. J. Gagne II and T. Smith, 2020. Deep Learning on Three-Dimensional Multiscale Data for Next-Hour Tornado Prediction. *Mon. Wea. Rev.*, **148**, 2837–2861. doi:https://doi.org/10.1175/MWR-D-19-0372.1.

Lapworth, A., 2006. The morning transition of the nocturnal boundary layer. *Boundary Layer Meteorol.*, 119, 501–526.

Lean, H. W., J. F. Barlow and C. H. Halios, 2019. The impact of spin-up and resolution on the representation of a clear convective boundary layer over London in order 100 m grid-length versions of the Met Office Unified Model. *Quart. J. Roy. Meteorol. S.*, **145**, 1674–1689.

Leroux, M., K. Wood, R. L. Elsberry, E. O. Cayanan, E. Hendricks, M. Kucas, P. Otto, R. Rogers, B. Sampson and Z. Yu, 2018. Recent advances in research and forecasting of tropical cyclone track, intensity, and structure at landfall. *Tropical Cyclone Res. Rev.*, **7**, 85–105.

Lewis, H. W., J. M. C. Sanchez, J. Graham, A. Saulter, J. Bornemann, A. Arnold, J. Fallmann, C. Harris, D. Pearson, S. Ramsdale, A. M. de la Torre, L. Bricheno, E. Blyth, V. A. Bell, H. Davies, T. R. Marthews, C. O'Neill, H. Rumbold, E. O'Dea, A. Brereton, K. Guihou, A. Hines, M. Butenschon, S. J. Dadson, T. Palmer, J. Holt, N. Reynard, M. Best, J. Edwards and J. Siddorn, 2018a. The UKC2 regional coupled environmental prediction system. *Geosci. Model Dev.*, **11**, 1–42. doi:https://doi.org/10.5194/gmd-11-1-2018.

Lewis, A., W. R. Peltier and E. von Schneidemesser, 2018b. *Low-cost sensors for the measurement of atmospheric composition: overview of topic and future applications*. Research Report. World Meteorological Organization, Geneva, Switzerland.

Magnusson, L., J.-R. Bidlot, M. Bonavita, A. R. Brown, P. A. Browne, G. De Chiara, M. Dahoui, S. T. K. Lang, T. McNally, K. S. Mogensen, F. Pappenberger, F. Prates, F. Rabier,

D. S. Richardson, F. Vitart and S. Malardel, 2019. ECMWF Activities for Improved Hurricane Forecasts, *Bull. Amer. Meteorol. S.*, **100**, 445–458.

Marshall, A. G., D. Hudson, M. C. Wheeler, O. Alves, H. H. Hendon, M. J. Pook and J. S. Risbey, 2014. Intra-seasonal drivers of extreme heat over Australia in observations and POAMA-2. *Climate Dynamics*, **43**, 1915–1937.

Martelloni, G., S. Segoni, R. Fanti and F. Catani, 2012. Rainfall thresholds for the forecasting of landslide occurrence at regional scale. Landslides, **9**, 485–495.doi:https://doi.org/10.1007/s10346-011-0308-2.

Mason, J. and N. Mason, 2003. The physics of a thunderstorm. *Eur. J. Phys.* **24**, S99.

McCabe, A., R. Swinbank, W. Tennant and A. Lock, 2016. Representing model uncertainty in the Met Office convection-permitting ensemble prediction system and its impact on fog forecasting. *Quart. J. Roy. Meteorol. S.*, **142**, 2897–2910. doi:https://doi.org/10.1002/qj.2876.

Mittermaier, M. and N. Roberts, 2010. Intercomparison of spatial forecast verification methods: Identifying skillful spatial scales using the fractions skill score. *Wea. Forecast.*, **25**, 343–354. doi:https://doi.org/10.1175/2009WAF2222260.1.

Montgomery, D. and W. Dietrich, 1994. A physically based model for the topographic control of shallow landsliding. *Water Resour. Res.,* **30**, 1153–1171.

Nairn, J. and R. Fawcett, 2014. The Excess Heat Factor: A Metric for Heatwave Intensity and Its Use in Classifying Heatwave Severity. *Int. J. Env. Res. Pub. Health*, **12**, 227–53.

NAS, 2020. *Implications of the California Wildfires for Health, Communities, and Preparedness*: Proc. Workshop. Washington, DC: The National Academies Press. https://doi.org/10.17226/25622.

Nikolopoulos, E. I., M. Borga, J. D. Creutin and F. Marra, 2015. Estimation of debris flow triggering rainfall: influence of rain gauge density and interpolation methods. *Geomorphology,* **243**, 40–50. doi:https://doi.org/10.1016/j.geomorph.2015.04.028.

NOAA, 2021a. *NWS SAFER Hazard Dashboard - Situational Awareness for Emergency Response.* https://www.arcgis.com/apps/MapSeries/index.html?appid=ea8b0eeb2e9c45b790329c0ed2fdc225

NOAA, 2021b. *NWSChat.* https://nwschat.weather.gov/.

Overton, D. E., 1966. Muskingum flood routing of upland streamflow. *J. Hydrol.*, **4**, 185–200.

Pagano, T. C., H. C. Hartmann and S. Sorooshian, 2001. Using climate forecasts for water management: Arizona and the case of the 1997-1998 El Niño. *JAWRA J. Amer. Water Resources Assoc.*, **37**, 1139–1153.

Pagano, T. C., F. Pappenberger, A. W. Wood, M. H. Ramos, A. Persson and B. Anderson, 2016. Automation and human expertise in operational river forecasting. *Wiley Interdisciplinary Reviews: Water*, **3**, 692–705.

Pagano, T. C., A. W. Wood, M. H. Ramos, H. L. Cloke, F. Pappenberger, M. P. Clark, M. Cranston, D. Kavetski, T. Mathevet, S. Sorooshian and J. S. Verkade, 2014. Challenges of operational river forecasting. *J. Hydrometeorol.*, **15**, 1692–1707.

Papadopoulos, K. H. and C. G. Helmis, 1999. Evening and morning transition of katabatic flows, *Boundary Layer Meteorol.,* **92**, 195–227.

Peace, M., J. Charney and J. Bally, 2020. Lessons Learned from Coupled Fire-Atmosphere Research and Implications for Operational Fire Prediction and Meteorological Products Provided by the Bureau of Meteorology to Australian Fire Agencies. *Atmosphere*, **11**, 1380.

Perrin, C., C. Michel and V. Andréassian, 2003. Improvement of a parsimonious model for streamflow simulation. *J. Hydrol.*, **279**, 275–289.

Petley D. N., 2009. On the impact of urban landslides. In: Culshaw M G, Reeves H J, Jefferson I, Spink T (eds) *Engineering geology for tomorrow's cities*, engineering geology special publications, vol 22. Geological Society of London, London, 83–99. doi:https://doi.org/10.1144/EGSP22.6.

Pettersson, L. H. and D. Pozdnyakov, 2012. *Monitoring of Harmful Algal Blooms.* Berlin, Springer-Praxis.

Pitt, M., 2008. *Learning Lessons from the 2007 Floods.* Cabinet Office, London. http://webarchive.nationalarchives.gov.uk/20100807034701/http:/archive.cabinetoffice.gov.uk/pittreview/thepittreview/final_report.html.

Price, J. D., S. Lane, I. A. Boutle, D. K. E. Smith, T. Bergot, C. Lac, L. Duconge, J. McGregor, A. Kerr-Munslow, M. Pickering and R. Clark, 2018. LANFEX: A Field and Modeling Study to Improve Our Understanding and Forecasting of Radiation Fog. *Bull. Amer. Meteorol. S.,* **99**, 2061–2077. doi:https://doi.org/10.1175/BAMS-D-16-0299.1.

Pugh, D. and P. Woodworth, 2014. *Sea-Level Science: Understanding Tides, Surges, Tsunamis and mean sea level changes,* Cambridge University Press. 407pp.

Pullen, J., R. Allard, H. Seo, A. J. Miller, S. Y. Chen, L. P. Pezzi, T. Smith, P. Chu, J. Alves and R. Caldeira, 2017. Coupled ocean-atmosphere forecasting at short and medium time scales. *J. Marine Res.,* **75**, 877–921.

Rabier, F., A. J. Thorpe, A. R. Brown, M. Charron, J. D. Doyle, T. M. Hamill, J. Ishida, B. Lapenta, C. A. Reynolds and M. Satoh, 2015. Global Environmental Prediction. *In Seamless Prediction of the Earth System: from minutes to months,* Eds G. Brunet, S. Jones & P. M. Ruti. WMO-No. 1156. ISBN 978-92-63-11156-2.

Raynaud, L., I. Pechin, P. Arbogast, L. Rottner and M. Destouches, 2019. Object-based verification metrics applied to the evaluation and weighting of convective-scale precipitation forecasts. *Quart. J. Roy. Meteorol. S.,* **145**, 1992–2008. doi:https://doi.org/10.1002/qj.3540.

Reichenbach, P., M. Cardinali, P. De Vita and F. Guzzetti, 1998. Regional hydrological thresholds for landslides and floods in the Tiber River Basin (Central Italy). *Environ. Geol.* **35**, 146–159. doi:https://doi.org/10.1007/s002540050301.

Sajjad, A., J. Z. Lu, X. L. Chen, C. Chisenga and S. Mahmood, 2019. The riverine flood catastrophe in August 2010 in South Punjab, Pakistan: potential causes, extent and damage Assessment. *Appl. Ecology Env. Res.,* **17**, 14121–14142. doi:https://doi.org/10.15666/aeer/1706_1412114142.

Salvatici, T., V. Tofani, G. Rossi, M. D'Ambrosio, C. T. Stefanelli, E. B. Masi, A. Rosi, V. Pazzi, P. Vannocci, M. Petrolo, F. Catani, S. Ratto, H. Stevenin and N. Casagli, 2018. Application of a physically based model to forecast shallow landslides at a regional scale. *Nat. Hazards Earth Syst. Sci.,* **18**, 1919–1935. doi:https://doi.org/10.5194/nhess-18-1919-2018.

Sene, K., 2008. *Flood warning, forecasting and emergency response.* Springer Science & Business Media.

Shaw, E. M., K. J. Beven, N. A. Chappell and R. Lamb, 2011. *Hydrology in Practice*, Routledge ISBN 9780415370424, 560pp.

Shearer, E. J., P. Nguyen, S. L. Sellars, B. Analui, B. Kawzenuk, K. L. Hsu and S. Sorooshian, 2020. Examination of Global Midlatitude Atmospheric River Lifecycles Using an Object-Oriented Methodology. *J. Geophys. Res.- Atmos.* 125. doi:https://doi.org/10.1029/2020JD033425.

Short, C. J. and J. Petch, 2018. How Well Can the Met Office Unified Model Forecast Tropical Cyclones in the Western North Pacific? *Wea. Forecast.,* **33**, 185–201. doi:https://doi.org/10.1175/WAF-D-17-0069.1.

Titley, H. A., M. Yamaguchi and L. Magnusson, 2019. Current and potential use of ensemble forecasts in operational TC forecasting: results from a global forecaster survey. *Tropical Cyclone Res. Rev.,* **8(3)** 166–180.

Ubeda, X. and P. Sarricolea, 2016. Wildfires in Chile: A review. *Global Planet. Change,* **146**, 152–161. doi:https://doi.org/10.1016/j.gloplacha.2016.10.004.

Uccellini, L. W. and J. E. Ten Hoeve, 2019. Evolving the National Weather Service to build a weather-ready nation: Connecting observations, forecasts, and warnings to decision-makers through impact-based decision support services. *Bull. Amer. Meteorol. S.,* **100**, 1923–1942.

Vannitsem, S., J. B. Bremnes, J. Demaeyer, G. R. Evans, J. Flowerdew, S. Hemri, S. Lerch, N. Roberts, S. Theis, A. Atencia, Z. B. Bouallègue, J. Bhend, M. Dabernig, L. De Cruz, L. Hieta, O. Mestre, L. Moret, I. O. Plenković, M. Schmeits, M. Taillardat, J. Van den Bergh, B. Van Schaeybroeck, K. Whan and J. Ylhaisi, 2021. Statistical Postprocessing for Weather

Forecasts 1 – Review, Challenges and Avenues in a Big Data World. *Bull. Amer. Meteorol. S.*, **102**, E681–E699. doi:https://doi.org/10.1175/BAMS-D-19-0308.1

Wang, K., Y. Zhang, S. Yu, D. C. Wong, J. Pleim, R. Mathur, J. T. Kelly and M. Bell, 2020. A Comparative Study of Two-way and Offline Coupled WRF v3. 4 and CMAQ v5. 0.2 over the Contiguous US: Performance Evaluation and Impacts of Chemistry-Meteorology Feedbacks on Air Quality. *Geosci. Model Development Discussions*, in review. doi:https://doi.org/10.5194/gmd-2020-218

Wang, Y., J. Gao, P. S. Skinner, K. Knopfmeier, T. Jones. G. Creager, P. L. Heiselman and L. J. Wicker, 2019. Test of a Weather-Adaptive Dual-Resolution Hybrid Warn-on-Forecast Analysis and Forecast System for Several Severe Weather Events. *Wea. Forecast.*, **34**, 1807–1827. doi:https://doi.org/10.1175/WAF-D-19-0071.1.

Wang, Y. and X. Wang, 2020. Prediction of Tornado-Like Vortex (TLV) Embedded in the 8 May 2003 Oklahoma City Tornadic Supercell Initialized from the Subkilometer Grid Spacing Analysis Produced by the Dual-Resolution GSI-Based EnVar Data Assimilation System. *Mon. Wea. Rev.* **148**, 2909–2934. doi:https://doi.org/10.1175/MWR-D-19-0179.1.

Wheatley, D. M., K. H. Knopfmeier, T. A. Jones and G. J. Creager, 2015. Storm-scale data assimilation and ensemble forecasting with the NSSL Experimental Warn-on-Forecast System. Part I: Radar data experiments. *Wea. Forecast.*, **30**, 1795–1817, doi:https://doi.org/10.1175/WAF-D-15-0043.1.

Whiteman, C. D., 2000. *Mountain Meteorology: Fundamentals and Applications*. Oxford University Press. 376pp.

Wieczorek, G. F., 1996. Landslide Triggering Mechanisms. In: Turner, A.K. and Schuster, R.L., Eds., *Landslides: Investigation and Mitigation*, Transportation Research Board, National Research Council, Special Report, Washington DC, 76–90. Oxford University Press.

Williams, R. M., C. A. T. Ferro and F. Kwasniok, 2014. A comparison of ensemble post-processing methods for extreme events. *Quart. J. Roy. Meteorol. S.*, **140**, 1112–1120.

Wilson, J. W., N. A. Crook, C. K. Mueller, J. Sun and M. Dixon, 1998. Nowcasting thunderstorms: A status report. *Bull. Amer. Meteorol. S.*, **79**, 2079–2100.

WMO, 2013. *Cascading Process to Improve Forecasting and Warning Services*. WMO Bulletin **62 (2)**.

WMO, 2017. *Global Guide to Tropical Cyclone Forecasting*. **WMO No 1194**, 397pp.

WMO, 2019. *Manual on the Global Data-processing and Forecasting System*: Annex IV to the WMO Technical Regulations. **WMO-no. 485**.

WMO, 2020. Training Materials and Best Practices for Chemical Weather/Air Quality Forecasting. **ETR-no. 26**.

Wu, H., R. F. Adler, Y. Tian, G. J. Huffman, H. Li and J. Wang, 2014. Real-time global flood estimation using satellite-based precipitation and a coupled land surface and routing model. *Water Resources Res.*, *50*, 2693–2717.

Yamaguchi, M., J. Ishida, H. Sato and M. Nakagawa, 2017. WGNE Intercomparison of Tropical Cyclone Forecasts by Operational NWP Models: A Quarter Century and Beyond. *Bull. Amer. Meteorol. S.*, **98**, 2337–2349.

Yamazaki, D., S. Kanae, H. Kim and T. Oki, 2011. A physically based description of floodplain inundation dynamics in a global river routing model. *Water Resources Res.*, *47(4)*. doi:https://doi.org/10.1029/2010WR009726

Yau, M. K. and R. R. Rogers, 1996. *A Short Course in Cloud Physics* 3rd edition. Butterworth-Heinemann 304pp. ISBN: 9780750632157 ISBN: 9780080570945.

Zohdi, E. and M. Abbaspour, 2019. Harmful algal blooms (red tide): a review of causes, impacts and approaches to monitoring and prediction. *Int. J. Env. Sci. Technol.* **16**, 1789–1806. doi:https://doi.org/10.1007/s13762-018-2108-x.

Chapter 7
Predicting the Weather: A Partnership of Observation Scientists and Forecasters

Paul Joe, Jenny Sun, Nusrat Yussouf, Steve Goodman, Michael Riemer, Krishna Chandra Gouda, Brian Golding, Robert Rogers, George Isaac, Jim Wilson, Ping Wah Peter Li, Volker Wulfmeyer, Kim Elmore, Jeanette Onvlee, Pei Chong, and James Ladue

---

P. Joe (✉)
Environment and Climate Change Canada (Retired), Gatineau, QC, Canada

WMO/WWRP HIWeather project, Geneva, Switzerland

J. Sun
National Center for Atmospheric Research, Boulder, CO, USA

WMO/WWRP HIWeather project, Geneva, Switzerland

N. Yussouf
Cooperative Institute for Mesoscale Meteorological Studies, University of Oklahoma, Norman, OK, USA

NOAA/OAR/National Severe Storms Laboratory, Norman, OK, USA

School of Meteorology, University of Oklahoma, Norman, OK, USA

WMO/WWRP HIWeather project, Geneva, Switzerland

S. Goodman
Thunderbolt Global Analytics, Owens Cross Roads, AL, USA

M. Riemer
Universität Mainz, Mainz, Germany

WMO/WWRP HIWeather project, Geneva, Switzerland

K. C. Gouda
Council of Scientific and Industrial Research, Fourth Paradigm Institute, New Delhi, India

WMO/WWRP HIWeather project, Geneva, Switzerland

B. Golding
Met Office, Exeter, UK

WMO/WWRP HIWeather project, Geneva, Switzerland

R. Rogers
National Oceanographic and Atmospheric Administration, Washington, DC, USA

WMO/WWRP HIWeather project, Geneva, Switzerland

G. Isaac
Weather Impacts Consulting Incorporated, Trail Barrie, ON, Canada

J. Wilson
National Center for Atmospheric Research, Boulder, CO, USA

© The Author(s) 2022
B. Golding (ed.), *Towards the "Perfect" Weather Warning*,
https://doi.org/10.1007/978-3-030-98989-7_7

Abstract Weather forecasts are the foundation of much of the information needed in the warnings we have been considering. To be useful, they require knowledge of the current atmospheric state as a starting point. In this chapter, we first look at the methods used to predict the weather and the resulting demands for observations. Then, we explore the wide variety of sensors and platforms used to obtain this information. There has been a long history of close working between sensor and platform designers and meteorologists that has produced spectacular advances in forecast accuracy. However, the latest high-resolution models require new approaches to obtaining observations that will require different collaborations. Examples are presented of partnerships in space observing and in aviation, a demonstration system from Canada, and the use of testbeds and observatories as environments for progress.

Keywords Numerical Weather Prediction · Nowcasting · Satellite · Radar · Lidar · Sensor · Observing platform · Third-party data · Evaluation

7.1   Introduction

This chapter addresses the challenge of forecasting hazardous weather, focusing on the collaborations needed to meet the requirements of prediction models and forecasters for observations of the current state of the atmosphere. Figure 7.1 provides an overview of the scope of the chapter, in which we shall describe:

- The role of the forecaster in extracting critical user-relevant information from model predictions.
- Data-driven prediction tools, including nowcasting and statistical post-processing.
- The structure and components of a Numerical Weather Prediction (NWP) system with emphasis on assimilation of observations and ensemble prediction in state-of-the-art kilometre-scale NWP systems.
- Current in situ and remote sensing networks and new capabilities that are under development.
- New sources of data that have the potential to enhance observational coverage and density.

P. W. P. Li
Hong Kong Observatory, Hong Kong, China

V. Wulfmeyer
University of Hohenheim, Stuttgart, Germany

K. Elmore · J. Ladue
Cooperative Institute for Mesoscale Meteorological Studies, National Severe Storms Laboratory, Norman, OK, USA

J. Onvlee
Royal Netherlands Meteorological Institute, De Bilt, Netherlands

P. Chong
China Meteorological Administration, Beijing, China

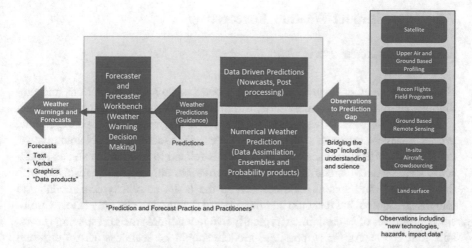

Fig. 7.1 Simplified view of the components of the forecast systems as discussed in this chapter. The orange arrows indicate the interface between weather and hazard warnings/forecast (leftmost orange arrow, discussed in Chap. 6) and the gap between observationalists and the forecaster/prediction system (rightmost orange arrow) that is the subject of this chapter

- Differences of methodology that need to be overcome to build effective partnerships that deliver the observational data needed for prediction.
- Examples of working relationships that have successfully overcome these challenges.

Forecast information is required for a variety of weather variables at different lead times and spatial-temporal resolutions. For warnings, the probability of extreme or unusual conditions relevant to local standards (e.g. infrastructure and construction codes) and expectations are particularly important. Warnings are especially important in densely populated urban environments where hazards can lead to a cascade of impacts (Baklanov et al. 2010, 2018; Grimmond et al. 2020). For clarity, in this chapter, we make a distinction between predictions, produced by NWP models, nowcasting systems or statistical processing, and forecasts, generated by forecasters based on the interpretation of predictions (see Fig. 7.1), while recognizing that predictions are increasingly input directly into hazard models. Observations are not only critical inputs to the model prediction process but are also needed by the forecaster and for verification. Sophisticated interactive human-machine software is enabling forecasters to interact more with both observations and predictions and to create digital products based on human interpretation of observations or forecasts (e.g. the Interactive Multi-Sensor Snow and Ice Mapping System, Matson and Wiesnet 1981).

7.2 High-Impact Weather Forecasting

7.2.1 Multiscale Forecasts

High-impact weather involves multiscale processes, so both observations and predictions should capture atmospheric variation from large scale (global) to small scale (e.g. sub-urban). Different sources of forecast information are used to generate products at these different scales. Table 7.1 describes some of their characteristics, categorized by temporal scale. Spatial scale is related to temporal scale, so small-scale features such as convective thunderstorms are only explicitly resolved with fine temporal- and spatial-scale models. In the table, current typical model grid lengths are quoted, but it should be remembered that models can only resolve atmospheric features of several times (typically five to ten times) the grid spacing (Lewis and Toth 2011). As far as possible, models should be seamless across different space/time domains, but computational constraints still require them to use different resolutions for different domain sizes, which may imply using different parameterizations, data assimilation, ensemble perturbations or post-processing. The frequency of model predictions also varies with forecast length. Long-range (seasonal) predictions may be generated once or twice a month, whereas nowcast predictions may be initiated every hour or less Table 7.1.

There is a commensurate wide variation in the observation requirements of models. Generally, observations need to be more accurate and less biased than model predictions and must be quality controlled with respect to the capability of the model to reproduce the observation (e.g. due to its resolution or the processes it represents). It is very much a case of "treasure versus garbage". Fine-scale

Table 7.1 Characteristics of systems for predicting weather and climate at a variety of time scales.

Forecast Ranges	WMO Definitions	Typical Characteristics	Examples	Applications
Climate	2 + years	~300 km grid spacing	Global and Regional Climate Models (GCM/RCM)	Climate change (IPCC) and adaption policy; scenario and hypothesis testing
Climate Normals	30 years averages	Based on "weather station data	30 year "normals"	Climate record, climate change detection, Building/Construction codes; insurance premium setting/risk analysis
Long Range	30 days to 2 years		Global models, e.g. Integrated Forecast System at ECMWF	Interannual weather and seasonal prediction (e.g. El Nino)
Extended Range	10 days to 30 days	~10-20 km; global; processed in different ways (e.g. averaged, ensemble products)	Also statistical regressions with historical events	Hurricane Season and Sub-Seasonal Prediction; Crop choice decision making, disaster risk reduction
Medium Range	3 days to 10 days			Advisories and watches for large scale features (e.g. hurricane tracks, cold air outbreaks),
Short Range	12 h to 3 days	Regional, 3-12 km,	Limited Area Mesoscale Model e.g. HRLAM	Watches and warnings (e.g., refinement of large scale forecasts to prepare to take action messages; winter storms, nor'easters)
Very Short Range	up to 12 h	Regional, 2-3 km,	High Resolution Rapid Refresh	Watches at convective scale (e.g. thunderstorms); crop spraying decision making
Nowcasting	0-2 h	Local, 100m-3 km	Research / demonstration models	Warnings at convective scale (e.g. thunderstorm, tornado, wind shear, urban applications)

variability resolved by a high-resolution model may be "noise" to a lower-resolution model. Noise in the initial model state can grow rapidly and limit accuracy at longer forecast times, so must be filtered out in the observation quality control and assimilation. Variables that change little over the time scale of a forecast (such as sea surface temperature in a short-range forecast) may be set to a fixed value in that forecast but may need to be predicted in a longer forecast.

Local physical influences often drive the details of high-impact weather, so the focus of much current high-impact weather research is in the development of high-resolution prediction, to capture not only the details of the local environment but also the hazard-related atmospheric structures. Accurate resolution of small-scale physical structures, such as orography or the urban fabric, can aid predictability. However, uncertainties in the initial conditions, under-resolved physical processes, inaccuracies in the numerical solutions and the rapid growth of small-scale perturbations ultimately pose limits to predictability that become shorter with decreasing scale. This paradox, of uncertainty increasing as resolution gets finer, can be overcome using ensembles to generate probabilistic forecasts.

In this section, we shall focus on the highest resolutions to illustrate the needs of prediction systems that support severe weather warnings. However, many of these characteristics are relevant also for the other prediction systems listed in Table 7.1.

7.2.2 Forecasters and Decision-Making

7.2.2.1 The Forecaster Process

Given the variety of missions and stakeholders, there is no single process for generating a forecast, but all forecasters use common methods such as recognition of patterns and established rules relating to them, knowledge about instruments, observations, models, products, societal impacts and responses, collaboration with peers in- and outside their organization and constraints of messaging.

At the start of a duty shift, the forecaster is briefed on and reviews the previous forecast to understand the context in which it was made and any special vulnerabilities, such as a first snow event of the season, or an unusual hazard. In reviewing the previous forecast, the forecaster will take account of not just the meteorology but also any extraneous constraints on the forecast environment such as hardware or staffing issues.

The forecaster must understand the overarching nature of their shift and for whom they're generating a forecast as their approach and strategy will differ if issuing sub-seasonal, weekly or aviation forecasts. The forecast goals depend on the needs of the user, and the role of the forecaster is adapted to the message type and to the risk that must be communicated.

The forecaster will then interact with the data to develop a four-dimensional understanding of the weather situation. An analysis of the large-scale pattern sets the context for understanding meteorological structures at smaller scales. This process continues down to the smallest scales in a process called the forecast funnel by Snellman (1982). Longer lead-time forecasts typically depend more on NWP information than shorter lead-time forecasts.

The volume and sophistication of NWP products are increasing rapidly. The forecaster interprets them in the context of past performance in those types of weather, including model climatologies and verification. Even sub-hourly severe local thunderstorm warning decisions will take account of an available mesoscale NWP analysis from a rapid update convection-permitting model (Weisman et al. 1997). The forecaster will use automated guidance including basic weather variables, such as temperature or wind, but may also have access to hazard variables such as visibility, severe wind gusts and winter precipitation amounts.

Forecasters benefit and suffer from the increasing volumes and diversifying types of observations and model data now available to them. As data volumes grow, the forecaster is increasingly reliant on computer systems to organize and present the data in a form that enables easy navigation and interaction so as to avoid data overload causing a decline in forecasting performance (Hoffman et al. 1995). Forecaster workstations must be designed from a human-centric, not system-centric, perspective (Andra et al. 2002; Heizenreder et al. 2015) as ergonomics, human factors, system architecture, bandwidth and speed of presentation are all important for forecaster effectiveness. Automated products can provide valuable guidance, but providing the "answer" is useless without the capability to efficiently interrogate, assess and evaluate the data for decision-making (Joe et al. 2002; Stuart et al. 2007). It is essential that the forecaster is able to maintain a conceptual understanding of the underlying weather processes and to be able to view how the NWP prediction matches with the conceptual model and observations, so as to make sense of potentially conflicting information. For instance, if short-term model guidance fails to produce convection when the observational data shows the necessary ingredients are present, the forecaster will challenge, and may need to abandon, the model guidance.

A forecaster must collaborate with colleagues serving other users in the same or adjacent areas, whether they are located in the next desk or another country. This is especially true when forecasting for widespread weather systems, such as winter storms or tropical cyclones. On smaller scales, the signal from a single instrument may be critical, requiring expert input from technical experts for interpretation. The forecaster must also take account of how their forecast will be used and must be an effective team player within the warning production chain.

As new observations, NWP and prediction systems are introduced or upgraded, the expert forecaster must understand not only the weather but also the capabilities and limitations of each innovation in order to assess and evaluate their efficacy. System and product training must be continually refreshed for this purpose.

7.2.3 Data-Driven Prediction

7.2.3.1 Post-Processed Products

A numerical weather modelling system predicts the mean dynamic and thermodynamic weather variables such as temperature, pressure, wind and moisture using a discretized form of the continuous equations of fluid mechanics on a three-dimensional grid, with unresolved processes (e.g. those occurring in clouds or close to the ground) parameterized in various ways. Warnings of high-impact weather require knowledge of the basic variables at unresolved scales (e.g. wind gusts) and of other variables (e.g. visibility or snow depth). These may be estimated using statistical or empirical post-processing models. Sub-grid wind gusts need to be estimated using statistical relationships. The representation of terrain used in the model is smoothed and may not represent the urban texture/fabric; but the fine details are often important for warnings. For instance, there can be a considerable difference in height between the observation point and nearest model point, and model data should be adjusted to account for these differences by post-processing, either with past observations or using an estimated gradient. In some cases, the model output is adjusted on an hourly basis (Landry et al. 2004). Bias correction is less of an issue for severe storm prediction, as the objective is to model the hazardous phenomena directly (e.g. hail, tornado).

User-relevant warning products require the combination of weather elements in post-processing; for example, blizzard or dust storm warnings require predictions of snow or sand surface conditions combined with surface winds. A variety of statistical, artificial intelligence and analogue techniques are applied, combining model data with historical data and real-time observations to generate these user-oriented products (Burrows and Mooney 2018). For example, "ProbSevere" combines model predictions with highly processed real-time satellite data to create multi-sensor warning guidance products for the prediction of severe weather (Cintineo et al. 2018).

7.2.3.2 Nowcasting

Nowcasting is defined as forecasting a detailed description of the weather, by any method, over a period from the present to 6 hr. ahead (Sun et al. 2014; WMO 2017). Traditionally, nowcasting was focused on severe thunderstorm warnings, but it has evolved to many more applications.

Summer

Nowcasting of summer weather has focused on convective storms and their hazards, including heavy rain, flash floods, tornadoes, hail, damaging winds and lightning strikes, mainly using observation-based identification and extrapolation. These forecasts support warnings of the most immediate hazards where action should be taken immediately to save property and/or life and generally cover the 0–1 h time period (NWS 2021). Automated extrapolation has been based on spatial correlation of two-dimensional radar-derived precipitation maps at different scales (e.g. Bellon and Austin 1978; Rinehart and Garvey 1978) or tracking of thunderstorm features (Dixon and Wiener 1993).

The automated warning of microbursts at US commercial airline airports is probably the most successful of all nowcasts (see example in this chapter). These warnings have eliminated the crashing of jet aircraft, on take-off and landing, caused by microbursts, likely saving hundreds of lives. Controllers and pilots are warned of microbursts based on an automated algorithm that ingests data from the Terminal Doppler Weather Radar located near most US airports (Wilson and Wakimoto 2001).

The biggest challenge in nowcasting is predicting a severe convective storm before it has formed. This requires observation of the 3-D pre-convective environment in the lower troposphere. Unfortunately, this is currently terra incognita in earth system science (Wulfmeyer et al. 2015) despite long-standing evidence that detection of boundary layer convergence lines and of upward motion at the top of the boundary layer are key for predicting the dynamics (Wilson and Schreiber 1986) and that observations of moisture and temperature profiles are needed for predicting clouds and precipitation. High-resolution networks of surface stations (spacing 5–20 km) are valuable for identifying the sharp gradients in wind, temperature and moisture characteristic of the mesoscale boundaries on which storms may develop. Doppler lidar and radar also have the ability to observe convergence boundaries, while geostationary satellites can detect the growth of the boundary layer and subsequent development of clouds at these boundaries (Purdom 1976; Weaver and Purdom 1995). The potential for growth of the incipient storm is dependent on the stability and wind shear of the deep atmosphere, which can be obtained from radiosondes and vertical profilers and from satellite soundings assimilated in NWP models (WMO 2017). However, current observing capabilities lack the high spatial, vertical and temporal resolution profiles of wind, temperature and moisture in the lower troposphere that are needed. Such observations are becoming possible with the new generation of Doppler, Raman and differential absorption lidar systems.

Once a storm has formed, processing of channel differences in geostationary satellite data can be used to identify storm phenomena such as severe convection and overshooting tops. Storm intensity and movement can also be tracked using lightning detection, from both ground-based networks and satellite-based lightning imaging sensors. However, the most valuable observation source is Doppler radar. Within the storm, radar reflectivity information enables identification of developing and decaying areas and of storm movement, while Doppler wind information can pinpoint the development of storm rotation prior to tornado development and of severe up- and down-draughts.

Quantitative hazard information (e.g. from tornadoes, microbursts) is much more difficult to obtain, as it is generally at even finer spatial resolution than operational radars can provide. Severe convective weather warnings are thus often issued on proxy information such as radar-derived storm structure and precipitation intensity.

In order to verify warnings, direct observations of the hazard are required that are typically obtained from human spotters – both professional and volunteer – and from post-event damage surveys. Increasingly, these reports are being obtained through social media with the possibilities of automated processing in real time for operational warning use.

Winter

Winter nowcasting is focused on the prediction of precipitation type, extreme cold, strong winds and poor visibility. Many of these weather variables are poorly observed, making verification of forecasts difficult. For example, in situ snowfall measurements are impacted by wind, and remote sensing by radar is insufficiently precise (WMO 2018; Boudala et al. 2017). Freezing precipitation is particularly difficult to observe and forecast (Strapp et al. 1996). Frost is not observed at routine weather observing stations. Whether snow or rain occurs depends on small changes near 0 °C, where the difference between model terrain height and reality may undermine the skill of the prediction. Blending of in situ observations with high-resolution models is an emerging technique (Huang et al. 2012; Bailey et al. 2014). However, good in situ high-resolution observations are rare, so validation studies of remote sensing techniques (e.g. for snow depth) are also rare. International projects on winter weather nowcasting have documented some of these problems (Isaac et al. 2014b; Kiktev et al. 2017) and have identified the need for observations at high time resolutions (1 min) and for fine-scale models (<2 km).

7.2.3.3 Typhoon/Hurricane Nowcasting

Tropical cyclones (TC), including hurricanes and typhoons, bring significant safety and economic impacts to lives and property particularly in tropical and subtropical coastal areas. The accuracy of TC track forecasts has continuously improved, but prediction of intensity (surface maximum wind and storm size), structure (symmetry and vertical structure), precipitation and associated flooding and storm surge inundation is still a challenge. Emergency and rescue response rely heavily on rapidly updated observations and nowcasts using Doppler Weather Radar and surface rain gauges for frequent updates of precipitation estimates and forecasts for decision-making. Over the open ocean, geostationary and polar orbiting multispectral satellite observations and products are the main data used to monitor, analyse and assimilate into global and regional NWP models. Significant efforts are being made to retrieve ocean surface winds, layered cloud motion vectors, cloud height, rainfall rate, atmospheric stabilities, etc. (EUMETSAT 2021). Recent advances, including the blending of radar-based and satellite-based data, allow precipitation forecasts to be extended up to 6 hours ahead and, over a broader area, to provide longer lead times enabling early preparation and decision-making. There has been

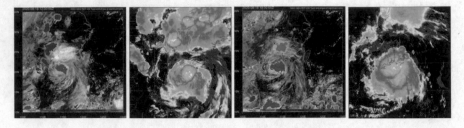

Fig. 7.2 Rapid intensification of Typhoon Higos (Aug 2020) approaching the coast of China over the Northwestern Pacific basin as seen from the "Hot Tower" satellite products. Pink in the middle represents overshooting top by the Rapid Development Thunderstorm product, while light grey hot tower area by the Hot Tower algorithm. There were only 6-hour differences between the two images. (Source: Hong Kong Observatory based on Himawari-8 satellite of the Japan Meteorological Agency)

progress, but there are still significant challenges in detecting rapid intensification (Fig. 7.2; Kaplan et al. 2010; EUMETSAT 2021).

Several existing observation platforms are currently under-utilized in operational TC nowcasts. For example, rapid scan short-wavelength radar, multispectral geostationary satellite imagery, ground-based or spaceborne lightning mapping, dropsondes from reconnaissance flights, aircraft in situ measurements (viz. AMDAR/ACARS upper air winds, temperature, humidity) and Global Positioning System constellation slant-path precipitable water vapour measurements. Studies are required to better utilize this information for rapid analysis of the atmospheric state. Ocean observations (buoys, oil rig AWSs), sea surface wave and current measurements, tide-level measurements, storm surge modelling, hydrological modelling, inundation modelling and their integration still require significant advancement for use in TC disaster nowcasting, warning and protection (Fig. 7.3; WMO RSMC 2021).

These nowcasts should include confidence or calibrated probability information to aid users' risk assessments. Confidence information could be generated efficiently with current computers using an ensemble approach. Probabilities need to be related to the end user's/decision-maker's impact parameters involving the entire value chain.

7.2.4 Numerical Prediction

High-impact weather, related to hazards, occurs mostly on very small scales (e.g. individual convective storms, urban heat islands). Although considerable advances have been made in NWP-based warnings of some high-impact weather events, such as tropical cyclones and disruptive winter weather, detailed high-impact weather forecasts have, until recently, been largely based on observational detection and/or visual confirmation, due to the limitations of operational NWP models in providing accurate predictions at these scales. Skilful probabilistic forecasts are critical to

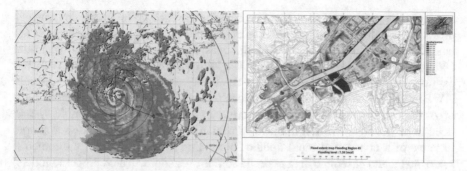

Fig. 7.3 An atmospheric/oceanic observation integrated platform for real-time analysis of the structure and intensity of a TC (left) using latest available radar, scatterometer, automatic weather station, oil rig, buoy, lightning, etc. observations. (Source: Hong Kong Observatory)

provide timely and accurate warnings, requiring access to observations of the dynamics and thermodynamics of the atmosphere at these scales and their assimilation into kilometre (km)-scale ensemble-based numerical weather prediction models. Even though there are many challenges in developing kilometre-scale NWP systems, running them, and post-processing voluminous amounts of output into useful guidance for decision-making, the potential benefits are significant.

7.2.4.1   Kilometre-Scale Numerical Prediction

Kilometre-scale NWP models explicitly represent multiscale processes, including dynamic interactions between scales and organization of different types of high-impact weather. They have a more detailed representation of land surface heterogeneity than coarser-resolution models and use more sophisticated parametrizations of cloud microphysics, boundary layer mixing, turbulent entrainment and radiation. This allows for more realistic and more accurate forecasts of severe weather events. However, these sophisticated schemes are still subject to uncertainty, which needs to be captured in an ensemble prediction system, e.g. by using stochastic parameterization schemes. Improving the scientific foundation and the development of such schemes are on-going research efforts.

One of the most challenging aspects of kilometre-scale NWP is starting the model with an accurate depiction of the atmosphere that includes the representation of fine-scale atmospheric motion including clouds. To explicitly resolve multiscale processes, including deep convection, frequent initialization of the model is critical. This is done by adjusting a very short forecast to match high-resolution observations (in both space and time) of the true state of the atmosphere using data assimilation, enabling frequently updated predictions of high-impact weather events and their associated hazards. A fundamental question is whether a variational, ensemble-based or hybrid data assimilation method yields the best kilometre-scale analyses and forecasts. Current research suggests that hybrid systems may provide the best results, the ensemble-based background error covariances providing the ability to impose balance constraints to create better analyses. At the heart of these considerations is how the considerable uncertainties in the initial conditions can best be represented and how large an ensemble is needed to reliably capture the uncertainty. To maximize the utility and impact of kilometre-scale NWP model and storm-scale observations (e.g. radar and satellite) to users, hundreds of post-processed probabilistic forecast products from the ensemble system need to be generated within minutes of initialization. Figure 7.4 shows a simple timeline of a conceptual rapid update forecasting system based on a kilometre-scale model and ensemble data assimilation. It uses sophisticated process models, starting from frequently updated and perturbed initial states, to generate an ensemble of predictions from which estimates of the probability distribution of future hazards can be made. Successive forecasts should lead to converging advice on the likelihood and severity of the hazard.

This system is conceptually simple but scientifically very challenging. Because the kilometre-scale forecast system aims to produce forecasts from minutes to a few days, it pushes the limits, not only of NWP modelling and advanced data assimilation but also of high-performance computing. Other challenges include lack of high-density observations and optimizing the initial state for multiple space scales.

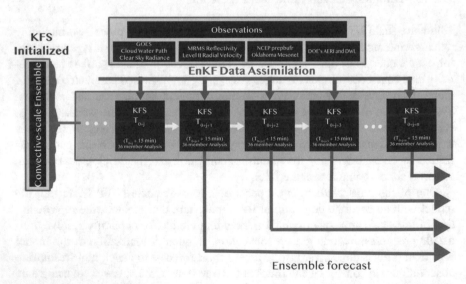

Fig. 7.4 A simple kilometre-scale ensemble data assimilation timeline. This type of frequently updating probabilistic kilometre-scale forecast system (KFS) can assist forecasters with earlier and more accurate communication of hazardous weather threats

7.2.5   *Probabilistic Prediction*

The extreme variability of hazard-related weather requires that forecasts are probabilistic. The basis should always be the use of an ensemble of NWP forecasts that are perturbed in their initial state and/or in some aspects of the model. However, ensembles do not capture the full range of possible outcomes, so there are several post-processing methods used to estimate the probability distribution, and we describe some of them here.

1. Neighbourhood Methods

    Despite the relatively fine horizontal grid spacing employed by kilometre-scale ensembles, probabilistic guidance products are typically not presented at the grid scale due to positional uncertainty. For example, small variations in the location of a small-scale feature, such as a mesocyclone, in different ensemble members, may result in low grid-scale probabilities of feature occurrence within a region, even if every ensemble member has predicted a mesocyclone. These same small displacement errors are responsible for the "double penalty" when applying traditional forecast verification measures to convection-allowing scales. To overcome this, neighbourhood approaches are commonly used for probabilistic forecast product generation (e.g. Schwartz et al. 2010) and for verification (e.g. Ebert 2009; Gilleland et al. 2009).

2. Ensemble Probability of Exceedance and Percentile Products

    The ensemble probability distribution function is used to provide a measure of event likelihood or severity, e.g. the measurable precipitation at a given location, and can provide limited information on event severity as well (e.g. probabilities of precipitation values greater than 200 m^2 s^{-2} imply the potential for a strong mesocyclone). However, specific measures of severity that span the range of ensemble solutions are desirable to forecasters (Novak et al. 2008; Evans et al. 2014). Specific measures of severity can be found using values at a fixed position within the ensemble distribution, represented by a percentile, as opposed to finding the proportion of the ensemble exceeding a specific value. Percentiles that represent "reasonable" best- and worst-case forecast scenarios, such as the tenth and 90th percentiles, are often used to supplement the ensemble maximum (Novak et al. 2014) to avoid overprediction by outliers.

3. Ensemble Statistical and Probability Matched Mean

    The statistical mean of an ensemble is possibly the most familiar ensemble product and provides a more skilful forecast than individual ensemble members when averaged over many forecasts (Leith 1974). The improved skill in the statistical mean comes from smoothing low-confidence events in a forecast while retaining higher-confidence or more frequent features. However, kilometre-scale ensembles are primarily aimed at providing guidance on rare and high-impact events with limited predictability rather than the mean. The localized probability matched mean (PMM) is a post-processing technique that restores characteristic amplitudes of ensemble members to the statistical mean field (Ebert 2001).

4. Pseudo-deterministic Products

While probabilistic guidance products efficiently condense information within the ensemble and provide measures of uncertainty, they provide limited information about the physical processes responsible for the model solutions. This limitation can be overcome by the "postage stamp" plot, which summarizes each ensemble member on a single plot. Postage stamps provide users with deterministic solutions from individual members and all the information available in continuous forecast fields; however, they sacrifice readability, often to the point of being impractical in large ensembles. Alternatively, web-based ensemble viewers can provide a means for rapidly interrogating individual member solutions (Roberts et al. 2019; Schwartz et al. 2019) while preserving output readability.

A second method for displaying deterministic aspects of an ensemble forecast is to extract limited information from each ensemble member on a single plot. These visualizations remove the complexity of full deterministic products, allowing forecasters to rapidly assess ensemble spread in features of interest. The most familiar of these feature-based visualizations is the spaghetti plot (Obermaier and Joy 2014; Rautenhaus et al. 2018), which provides specific contours of a given field for each ensemble member (Sivillo et al. 1997). Spaghetti plots are typically employed to provide information on ensemble spread of features in a continuous field, for example, shortwaves in a 500 hPa geopotential height field or air mass boundaries in a 2 m dew point field. Automated detection of features associated with specific phenomena may be used to produce analogous visualizations for features like frontal boundaries (Hewson and Titley 2010), tropical cyclone tracks (Hamill et al. 2012) or thunderstorm proxies (Schwartz et al. 2015). In particular, kilometre-scale ensembles frequently use feature-based "paintball" plots to display ensemble information of thunderstorm and mesocyclone positions in simulated reflectivity and updraft helicity forecasts, respectively (Schwartz et al. 2015; Roberts et al. 2019; Schwartz et al. 2019).

7.2.6   Forecast Evaluation

Both model predictions and human forecasts must be evaluated regularly to ensure that they have value to the forecast user and to understand the weaknesses that need further research and development. Standard verification techniques are applied to compare the performance of global NWP models, but these have only limited relevance to understanding the prediction of weather-related hazards. Traditional approaches to hazard-related verification have relied on scores such as hit rate and false alarm rate, which relate well to the use of the forecast, but which can be misleading when used to compare different approaches. All evaluation depends on the availability of high-quality observations, and this is perhaps the greatest impediment to verification of hazard-related weather phenomena.

7.3 Observations for High-Impact Weather Monitoring and Prediction

Observations are the heart and language of science and describe structural characteristics of the environment, advancing our understanding of key physical processes governing atmospheric systems track, intensity, structure and impacts. They play a fundamental role in constraining uncertainties in prediction models, directly by sampling the atmospheric initial state and indirectly by providing data for process studies and machine learning approaches to improve the representation of physical processes. They provide datasets for evaluating the performance of models. Observations must be processed to fit the NWP model structure. It is important to know the characteristics of both instrument errors and NWP model-observation representativeness differences.

The current in situ observation network of surface weather stations and upper air soundings was designed for short-range forecasting at synoptic or ~1000 km scales. Forecasting systems still rely heavily on these networks of weather stations operated according to WMO standards. For practical and cost reasons, stations in such networks are generally spaced some tens of kilometres apart for surface data and hundreds of kilometres for sounding stations.

Observation-based predictions have evolved over the years as a pragmatic approach to address the nowcasting of quickly varying small-scale phenomena, such as thunderstorms, and to adjust model predictions that diverge from the observations. Advances in forecast accuracy and forecast range have been demonstrated with high-resolution models using extra data obtained from other agencies, social media/citizen science and mixed technology solutions for specific weather parameters (e.g. precipitation from satellites in remote regions). A vast amount of observations with varying or unknown quality are now available that require new collaborations to bring into effective use. Not only are there challenges in accessing the data on the right time scales, in usable formats and with the required information on error characteristics, but also in optimizing the ways in which high-volume, low-quality data are mixed with low-volume, high-quality data to achieve the best forecasts.

Innovation and adoption of new observations require investment and long-term planning by governments. The development of new technology takes time and needs to be coordinated effectively and efficiently across different mandates and funding sources. In this section, we describe the capabilities and limitations of innovations that will meet the high-resolution observation requirements of high-impact weather, starting with in situ observations (including those derived from social media) and then remote sensing technologies.

7.3.1   In Situ Observations

High-impact weather forecasting requires meteorological observations at spatial and temporal densities significantly higher than available from present National Meteorological and Hydrological Services (NMHS) networks. This has been a prime motivation for investigating the use of third-party data (TPD) which is often collected for purposes other than weather forecasting but nonetheless contains valuable meteorological information. In recent years, the increased reliability and decreased cost of atmospheric sensors, the coming of the "internet of things" and the introduction of machine learning technologies have made available a wealth of new and potentially very useful data on the fine-scale evolution of the atmosphere near the ground. The challenge is how to make this information accessible and usable for the purpose of high-impact weather forecasting. While "big data" and "artificial intelligence" tools and analytics are readily available, accuracy, routine and long-term reliable access to the data, their interpretation and data quality control require scientific and application expertise for usability.

Perhaps the most well-known, and successful, example of third-party data are meteorological observations from commercial aircraft. For several decades, AMDAR (Aircraft Meteorological Data Relay) instruments have been installed by NMHS's on a limited number of aircraft, to provide observations of wind, temperature and humidity at flight level and upon ascent/descent into airports. These data have proven to be of great value for the quality of weather forecasts (Petersen 2016; ECMWF 2020). For cost reasons, relatively few aircraft have been adapted to carry AMDAR instruments. However, more recently, aircraft position messages (Mode-S) have been processed to produce wind and temperature data ~100–1000 times more numerous than AMDAR observations, of comparable quality to radiosonde observations and at a significantly lower cost (WMO 2020a).

A potential new source of high-resolution precipitation data has been demonstrated from cell phone networks (Overeem et al. 2011). Microwave communication signals between cell towers suffer from small time delays which are related to attenuation by precipitation. It has been demonstrated that a network of cell phone towers can provide an accurate and detailed picture of the spatial distribution and amount of precipitation. However, access to these data is a problem as they are considered proprietary by the telecommunications operators.

Observing sensors and platforms are becoming cheaper, more reliable, more widespread and of better quality. Examples include (i) drones equipped with meteorological and/or air quality sensors flying over areas difficult to access, (ii) measurements from wind energy turbines and (iii) near-surface observations from private or charitable (van de Giesen et al. 2014; Kucera 2017) weather stations. These near-surface data can be acquired at affordable cost often by crowdsourcing initiatives such as the Weather Observations Website (2021). Several studies have shown that with careful quality control and bias correction using machine learning techniques, these data (e.g. temperature, pressure and precipitation) can provide significant added value (e.g. Nipen et al. 2020; de Vos et al. 2019; Meier et al. 2017).

For the detection of highly localized severe weather events, the greater density, representativeness and coverage of TPD can be advantageous. Standard meteorological surface weather stations are situated in open fields free from obstacles. This makes them representative of idealized homogeneous surfaces but under-represents heterogeneous surface environments, particularly in urban areas, where the majority of humans live and where high-impact weather forecasts are most needed. A mixture of TPD, from several sources, can provide a useful addition to NMHS ground observing systems (e.g. de Vos et al. 2019; Fenner et al. 2019). Through data sharing, they may be obtained by NMHSs at a fraction of the cost of operating and maintaining their own networks. However, partners, such as internet service providers or wind energy farms, may be reluctant to share data for competitive reasons. For crowdsourced data from smartphones, there are legal, ethical and privacy aspects to consider. Care is needed to strip the acquired data of all but their meteorological information and to anonymize and possibly aggregate them so that the data cannot be traced back to the original provider.

Individual TPD sources are often unable to reach the standards of official meteorological in situ stations, and complex systematic errors need to be removed. However, many studies have indicated that, combined with professional in situ meteorological networks, and after careful quality control, they offer clear added value in the assimilation and post-processing of NWP forecasts. Machine learning algorithms are increasingly proving successful in providing fully automated quality control of TPD; however, they need to be interpreted within the context, scale and purpose of the forecast prediction system.

A long-term challenge will be how to coordinate the acquisition, use and exchange of TPD for meteorological use at a global level. Worldwide uptake of these new data types can be facilitated by creating and fostering a global community of meteorological TPD experts, exchanging experiences and best practices, and requires coordination such as that provided by the World Meteorological Organization to standardize protocols, metadata, formats and mechanisms for exchange of TPD.

Many weather-related elements are now measured from mobile platforms, such as smartphones and cars, and studies are being carried out to assess their value. Pressure data from smartphones have been shown to be of value for weather forecasting (Mass and Madaus 2014; McNicholas and Mass 2018). The mobile nature of these sensors presents some particular challenges to their quality control (Hintz et al. 2019). Lidars and radars are used in vehicle collision avoidance but have yet to be exploited for weather prediction.

Crowdsourcing information about hazardous weather may be provided through common social media (e.g. Twitter, Instagram) or specialized crowdsourcing apps. Given the ubiquity of mobile devices, the data are timely and may be spatially and temporally dense, depending on population density, particularly in existing data sparse regions. These data are used in various ways for warning issuance, nowcasting, verification and providing feedback on the entire high-impact warning chain that cannot be achieved in any other way. Weather-specific apps (e.g. mPing, Elmore et al. 2014) can solicit particular information such as the occurrence of particular weather phenomena (e.g. tornadoes, waterspouts, hail and hail size, storm damage and visibility restrictions) and interactively generate maps or time series products. The frequency and spatial pattern of the reports can provide significant scientific insights. For example, for small-scale hazards such as damaging hail, frequent (5–10 minutes) and high (sub-kilometre)-resolution maps can be produced to understand the evolution of the storm, for damage surveys, and to validate and verify radar and NWP products. The apps are available globally but limited by market penetration into the social media environment. Over time, through peer experience or in-app or on-line training, the quality of the reports should evolve and improve. Statistical or artificial intelligence techniques can be used for quality control. These data greatly expand, complement and supplement reports from trained volunteer spotters which can be used to quality control the reports from the general population.

7.3.2 Ground-Based Remote Sensing

Ground-based remote sensing systems can observe the entire chain of processes leading from land-atmospheric exchange, atmospheric boundary layer (ABL) development, convergence zone formation and evolution and convective initiation to the formation, evolution and decay of clouds, precipitation and other hazards. This capability depends on the remote sensing methodology, e.g. whether passive or active remote sensing is applied and which wavelengths are utilized. For observation of the pre-convective environment, wavelengths from the ultraviolet (UV) up to the infrared (IR) are required. For observation of clouds and precipitation, the microwave spectrum must be used. In passive remote sensing, the emission spectrum of the atmosphere itself or the transmissions of the sun or the moon are used. As the atmospheric variables of interest, such as water vapour and temperature, are indirectly contained in these observations, a retrieval is necessary, which requires a first guess and limits vertical resolution and accuracy. For active remote sensing using sound waves or electromagnetic waves, generally a direct derivation of variables of interest is possible, which intrinsically increases the accuracy as well as the temporal and range resolutions of the results (see, e.g. Wulfmeyer et al. 2015).

7.3.2.1 Land-Atmosphere Exchange

Clear-air observations are required with vertical and temporal resolutions of metres and sub-seconds to resolve atmospheric profiles in the surface layer from the canopy top to a height of about 100 m including turbulence fluctuations. Unfortunately, this is a sore spot of passive and active remote sensing. Recently, the first surface layer scans of wind, temperature and moisture profiles with sufficient resolution and accuracy became possible (Wulfmeyer et al. 2015; Späth et al. 2016) enabling us to study flux-gradient relationships in the surface layer and to make comparisons with current theories such as the Monin-Obukhov similarity theory.

7.3.2.2   Pre-Convective Environment

The simplest way to obtain information about the pre-convective environment is provided by ceilometers and backscatter lidars. These instruments are typically operated in vertically pointing mode and are used for cloud height observations and for volcanic ash monitoring (Adam et al. 2016a). However, a simple backscatter lidar provides only limited information about atmospheric dynamics and thermodynamics. For studies of the pre-convective environment, observations of lower tropospheric wind, temperature and humidity fields with temporal and spatial resolutions of the order of minutes and 100 m are fundamental. Unfortunately, due to a severe lack of availability and coverage with suitable remote sensing systems, this area must be considered as terra incognita in earth system science. As clear-air measurements are required, these observations must be performed with far thermal infrared (FTIR), microwave radiometer (MWR), Global Navigation Satellite Systems (GNSS) and lidar techniques. With respect to thermodynamic profiling, an overview is given in Wulfmeyer et al. (2015). For wind measurements, the operation of Doppler lidar systems or clear-air radar wind profilers is state of the art. If operated in scanning modes, wind profiles can be derived with resolutions of 1 min and 50 m, with an accuracy of 0.5 ms^{-1} in the ABL. The performance of coherent Doppler lidars depends on the presence of aerosol particles in the range of interest so that typically a very high resolution and accuracy are achieved in the atmospheric boundary layer, but this can degrade substantially at greater heights. Scanning Doppler lidars have been developed for wind shear detection at airports (Chan and Lee 2012; Nechaj et al. 2019) and for boundary layer wind profiling.

For vertical measurements of temperature and moisture, passive remote sensing systems such as FTIR and MWR can be applied. However, their vertical resolutions are rather limited: ~1000 m for FTIR and ~2000 m for MWR at 2000 m height, degrading further with greater altitude. For the determination of real profiles of water vapour and temperature with resolutions of minutes and 100 m vertically, Raman lidar and differential absorption lidar (DIAL) can be applied (Turner et al. 2002; Späth et al. 2016; Weckwerth et al. 2016). The new generation of Raman lidars permits temperature measurements throughout the troposphere, day and night (Lange et al. 2018). Water vapour Raman lidar permits measurements up to the lower troposphere during daytime and throughout the troposphere during night-time. Water vapour DIAL measurements have similar performance during daytime and night-time with resolutions of minutes and a few 100 m up to the middle and upper troposphere depending on the atmospheric moisture content. Operational water vapour measurements using DIAL are now possible with low-power, compact systems (Weckwerth et al. 2016).

7.3.2.3 Clouds and Precipitation

Integrated, mainly vertically profiling observations have been deployed for a variety of climate and weather research investigations (Kollias et al. 2007a, b). Wind profilers, aerosol, Doppler and water vapour lidars, radiometers, ceilometers and short wavelength radars (W and Ka band) are maturing technologies that measure within the boundary layer and middle troposphere and have been shown to improve high-impact weather forecasts (Benjamin et al. 2004; Loehnert et al. 2007).

Radars are fundamental tools for the provision of rapidly developing hazardous weather warnings as they observe the precipitation in the atmosphere in three dimensions with spatial and temporal resolutions better than 1 km and 5 minutes, respectively. Radars can "see" precipitation at long distances (250 km or more). They transmit microwave energy into, and receive reflected energy from, the raindrops and other scatterers in the atmosphere – including clouds, airplanes, ground, insects and, if sensitive enough, clouds. The current generation of polarization diversity radars provides greater quality control and hydrometeor identification capabilities. In addition, some highly sensitive modern radars can observe reflections from the clear air (due to insects or Bragg scattering) or due to refractive index fluctuations (Knight and Miller 1998; Wilson et al. 1994; Fabry 2004; Fabry et al. 1997) to retrieve low-level winds and humidity fields that enhance the forecaster's ability to observe pre-cursor signatures of convective initiation and hence potentially extend the lead time for thunderstorm warnings (Wilson and Schreiber 1986). Substantial processing is required to produce precipitation products (Zhang et al. 2016), wind fields (Browning and Wexlar 1968; Sun and Crook 1997) or precipitation types (Park et al. 2009). For data assimilation, radar must be quality controlled to remove features that cannot be represented in NWP models. Hence, partnerships amongst radar specialists, forecasters and assimilation scientists are needed to deliver appropriate application-based quality-controlled radar products. As an example, ground clutter, generally considered a nuisance and often eliminated, has proven useful for monitoring variations in calibration, leading to improvements in the quality of precipitation products (Wolff et al. 2015), and for the retrieval of humidity (Fabry 2004).

Due to the curvature of the earth and beam propagation paths, the radar observing range is limited to near ranges (~50 km) when observing low-level weather phenomena such as tornadoes, wind shear and precipitation type near the ground. Depending on the radar network and largely due to cost, radars generally have spacings of 150–400 km leaving substantial low-level (<1 km altitude) coverage gaps. These are exacerbated by blockage by local obstructions or complex terrain. Dense networks of limited range small radars to sense the lowest levels of the atmosphere have been proposed (Mclaughlin et al. 2009) and have been deployed for demonstration in several urban environments (Cifelli et al. 2018; Misumi et al. 2020; Chandrasekar et al. 2018).

Combining networks of heterogeneous radars, operated by different agencies for different purposes, across multiple countries, and often of mixed technology, can extend and improve the coverage domain. This requires exchanging voluminous quantities of radar data and sharing data quality information, with resulting benefits to NWP assimilation, at reduced cost, increased efficiency of operations and higher quality (Lopez 2011).

Weather radars are the main requirement for nowcasting warnings of convective storms. S-band polarimetric, Doppler, one-degree beam width radars, with good sensitivity are preferred. C-band radars are second in preference in that the unambiguous velocity is less for the same pulse repetition frequency at the cost of higher attenuation. Bragg scattering detection is considerably reduced with C-band radars. X-band radars suffer severe attenuation in regions of high rainfall rates and are thus limited to being deployed in local networks with a spacing of tens of kilometres. Developing, operating, maintaining and sustaining operational radar networks are expensive, and these on-going costs must be considered before initial installation.

Commercial ground-based lightning networks have become ubiquitous. Lightning is a severe weather hazard for fire weather, personal safety and infrastructure. It is associated with convective weather, and so statistical relationships with heavy rain, hail and strong winds have been used to generate precipitation proxy products for convective storms. Lightning has also been assimilated into NWP to improve kilometre-scale predictions (Dixon et al. 2016). The lightning "jump" is a sudden increase in flash rate, associated with the onset of severe weather (Chronis et al. 2015). The causal physical relationships still need to be understood, but they have been used for warnings (Holle et al. 2016).

7.3.3 Satellite Remote Sensing

Weather satellites are the backbone of the global weather observing system (Fig. 7.5). The principal satellite orbits are Geostationary Earth Orbit (GEO) and Low Earth Orbit (LEO) which provide different perspectives of the atmosphere and the earth (WMO 2020). GEO satellites are located at 35,786 kilometres above the earth's surface with an orbit matching the earth's rotation so that the earth and atmosphere can be monitored continuously at the same satellite sub-point. They are the primary source of near-real-time imagery used for nowcasting and the detection of rapidly evolving high-impact environmental phenomena (Goodman et al. 2018, 2019; Schmit et al. 2017, 2018). LEO satellites orbit at about 800 km above the surface, viewing the whole earth twice a day in multiple passes, each at the same local times. These satellites provide the primary source of temperature and humidity profiles of the atmosphere for use in NWP. Together, satellites in the LEO and GEO orbits provide a broad spectrum of atmospheric, land and ocean measurements used in weather forecasting and analysis (Table 7.2).

The new-generation international "GEO-Ring" satellite constellation (Fig. 7.5) provides full disk earth and atmosphere imagery and derived products (e.g. cloud mask, cloud height, cloud phase, precipitable water, stability indices, winds) every 10 minutes and at high frequencies of 1–2.5 min over limited areas. The GEO cloud/moisture-derived atmospheric motion vectors (Fig. 7.6) are widely used in global

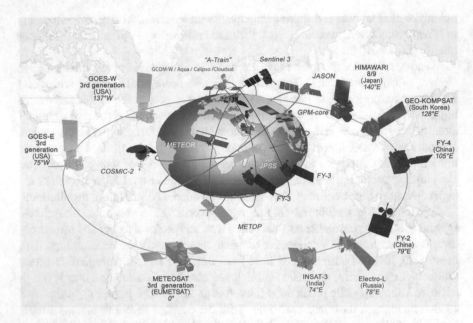

Fig. 7.5 Space-based component of the global observing system. (Source: WMO)

NWP to fill gaps in the global radiosonde network. Information about winds at different levels, areas of wind shear or jet maxima can be identified. Wind vectors are computed using both visible and infrared spectral bands (GOES 2021).

The spectral bands of these new imagers in the visible and infrared portions of the electromagnetic spectrum can be combined in various ways to make decision aids and products for nowcasting and short-range forecasting (e.g. fog, smoke, air mass classification and dust, amongst others). The NOAA GOES and the EUMETSAT MTG (Meteosat Third Generation) also have lightning imagers that provide storm-scale day/night imaging of lightning discharges including their radiant energy, areal extent and propagation.

With the advent of multispectral imagers such as MODIS and the Joint Polar Satellite System (JPSS, Goldberg et al. 2018) VIIRS (visible infrared sensors) LEO satellites are also used increasingly as input to forecaster decision aids. Radiometer and spectrometer instruments in LEO may be active (radars, scatterometers, altimeters, lidars) or passive (multispectral visible (VIS)/near-infrared (NIR)/thermal infrared (TIR) imagers; IR and microwave (MW) sounders). Atmospheric sounding of the vertical temperature and moisture structure of the atmosphere are key contributions for assimilation into NWP. The LEO satellite constellation infrared and passive microwave sounders (Menzel et al. 2018) provide complementary information in clear and cloudy atmospheres as clouds are opaque in the infrared part of the spectrum and largely transparent at microwave frequencies. Operating them together makes it possible to cover a broader range of weather conditions. Infrared sounders have better horizontal and vertical resolution, while microwave sounders, although

Table 7.2 Satellite backbone with specified orbital configuration and measurement approaches

Instruments	Geophysical variables and phenomena
Geostationary core constellation with at least five satellites providing complete earth coverage	
Multispectral VIS/IR imagery with rapid repeat cycles	Cloud amount, type and top height/temperature; wind (through tracking cloud and water vapour features); sea/land surface temperature; precipitation; aerosol content and physical properties; snow cover; vegetation cover; albedo; atmospheric stability; fire properties; volcanic ash; sand and dust storm; convective initiation (combining multispectral imagery with IR sounders data)
IR hyperspectral sounders	Atmospheric temperature; humidity; wind (through tracking cloud and water vapour features); rapidly evolving mesoscale features; sea/land surface temperature; cloud amount and top height/temperature; atmospheric composition (aerosols, ozone, greenhouse gases, trace gases)
Lightning imagers	Total lightning (in particular cloud to cloud), convective initiation and intensity, life cycle of convective systems, NOx production
UV/VIS/NIR sounders	Ozone, trace gases, aerosol, humidity, cloud top height
Sun-synchronous core constellation satellites in three orbital planes (morning, afternoon, early morning)	
IR hyperspectral sounders	Atmospheric temperature and humidity; sea/land surface temperature; cloud amount, water content and top height/temperature; precipitation; atmospheric composition (aerosols, ozone, greenhouse gases, trace gases)
MW sounders	
VIS/IR imagery; realization of a day/night band	Cloud amount, type and top height/temperature; wind (high latitudes, through tracking cloud and water vapour features); sea/land surface temperature; precipitation; aerosol properties; snow and (sea) ice cover; ice-flow distribution; vegetation cover; albedo; atmospheric stability; volcanic ash; sand and dust storm; convective initiation
MW imagery	Sea ice extent and concentration and derived parameters (such as ice motion); total column water vapour; water vapour profile; precipitation; sea surface wind speed [and direction]; cloud liquid water; sea/land surface temperature; soil moisture; terrestrial snow
Scatterometers	Sea surface wind speed and direction; surface stress; sea ice; soil moisture; snow cover extent and SWE
Sun-synchronous satellites at three additional equatorial crossing times, for improved robustness and improved time sampling particularly for monitoring precipitation	
Instruments on other satellites in low earth orbit	
Wide-swath radar altimeters and high-altitude, inclined, high-precision orbit altimeters	Ocean surface topography; sea level; ocean wave height; lake levels; sea and land ice characteristics, snow on sea ice
IR dual-angle view imagers	Sea surface temperature (of climate monitoring quality); aerosols; cloud properties
MW imagery for surface temperature	Sea surface temperature (all-weather)

(continued)

Table 7.2 (continued)

Instruments	Geophysical variables and phenomena
Low-frequency MW imagery	Soil moisture, ocean salinity, sea surface wind, sea ice thickness, snow cover extent and snow water equivalent
MW cross-track upper stratospheric and mesospheric sounders	Atmospheric temperature profiles in stratosphere and mesosphere
UV/VIS/NIR sounders, nadir and limb	Atmospheric composition (ozone, aerosol, reactive gases)
Precipitation radars and cloud radars	Precipitation (liquid and solid), cloud phase, cloud top height, cloud particle distribution and amount and profiles, aerosol, dust, volcanic ash
MW sounder and imagery in inclined orbits	Total column water vapour; precipitation; sea surface wind speed [and direction]; cloud liquid water; sea/land surface temperature; soil moisture
Absolutely calibrated broadband radiometers and TSI and SSI radiometers	Broadband radiative flux; earth radiation budget; total solar irradiance; spectral solar irradiance
GNSS radio occultation (basic constellation)	Atmospheric temperature and humidity; ionospheric electron density, zenith ionospheric total electron content and total precipitable water
Narrow-band or hyperspectral imagers	Ocean colour; vegetation (including burnt areas); aerosol properties; cloud properties; albedo
High-resolution multispectral VIS/IR imagers	Land use, vegetation; flood, landslide monitoring; ice-flow distribution; sea ice extent/concentration, snow cover extent and properties; permafrost
Synthetic aperture radar imagers and altimeters	Sea state, sea surface height, sea ice motion, sea ice classification, ice-flow geometry, ice sheets, soil moisture, floods, permafrost
Gravimetry missions	Ground water, oceanography, ice and snow mass

Source: WIGOS

having lower resolution (~10s of km), can observe the earth's atmosphere and surface day and night even through intervening clouds.

The Global Precipitation Measurement (GPM) mission uses multiple satellites. The core has two primary instruments, a dual-frequency precipitation radar (DPR) and a GPM passive microwave imager. The DPR consists of a Ku-band precipitation radar (KuPR, 13.6 GHz) and a Ka-band precipitation radar (KaPR, 35.5 GHz), both having 5 km spatial resolution at nadir and covering a swath width of 245 km. The DPR is more sensitive than its TRMM predecessor especially in the measurement of light rainfall and snowfall in mid-latitude regions. Rain/snow determination uses the differential attenuation between the Ku band and the Ka band. The GPM microwave imager is a multi-channel, conical-scanning, microwave radiometer that serves as both a precipitation and a radiometric standard for the other GPM international partner satellites. It has 13 microwave channels ranging in frequency from 10 GHz to 183 GHz. The GPM core and its partners are combined with GEO imagers, to create a widely used precipitation product called IMERG (Huffman et al. 2019a, b) that is updated every 30 minutes through temporal morphing of the

21 Oct 2020 13:00Z NOAA/NESDIS/STAR GOES-East ABI DMW over GEOCOLOR

Fig. 7.6 Derived motion wind vectors (DMW) from the GOES East (GOES-16) Advanced Baseline Imager overlaid on a GeoColor false colour RGB image (Miller et al. 2020) at 13 UTC on 21 October 2020. Hurricane Epsilon (29.9°N, 58.8°W) in the central Atlantic was a Category 1 hurricane at this time with maximum sustained winds of 74 kts (85 mph). The wind speed and direction, derived using sequential images, are one of the most important inputs assimilated into the global NWP models, most notably filling gaps in data-sparse areas © NOAA, 2020

instantaneous rainfall fields and is widely used in nowcasting, NWP and flood/land-slide monitoring (Kirschbaum et al. 2017; Kirschbaum and Stanley 2018).

A constellation of satellites can fly in formation to produce synchronized data from several different instruments (Stephens et al. 2018). Current and planned con-stellations and future CubeSat swarms of sensors may greatly augment the capabil-ity of the global observing system and increase the revisit frequency from twice per day to perhaps hourly or better, making these data of potentially great interest and value for nowcasting and regional- to global-scale NWP. Intercalibration of these measurements will be a challenge with each instrument providing a different view geometry and atmospheric path.

Global Positioning System (GPS) radio occultation is another important satellite measurement for NWP data assimilation and is complementary to the infrared and microwave radiances observed by atmospheric sounders. The highly precise radio occultation signal is measured by the Global Navigation Satellite System. It is affected by the density, the moisture content and hence the refractive index of the atmosphere. This alters the propagation path and time of the signal between a GPS satellite and a receiver on a LEO satellite from which the atmospheric temperature and humidity can be retrieved to produce upper-troposphere to lower-stratosphere temperature profiles and lower-troposphere humidity profiles (Menzel et al. 2018).

7.3.4 Aircraft Reconnaissance of Tropical Cyclones

Tropical cyclones (TCs) plague coastal communities around the world, threatening millions of people and causing many billions of dollars in damage to infrastructure – impacts that are increasing as coastal development continues worldwide. These impacts result in severe consequences in all affected ocean basins.

Many platforms are available for observing TCs, including airborne (both manned and unmanned), spaceborne and ground-based (Rogers et al. 2019). Each of these brings advantages and disadvantages to the challenge of observing TCs. For example, spaceborne platforms provide global coverage, but are generally unable to measure structures within the inner core. Aircraft can provide this inner-core information, but their range is limited, and even in the Atlantic basin, only about 35% of TCs are sampled.

The USA has a long history of airborne TC reconnaissance, dating back to the 1940s. Currently, the two main agencies responsible for airborne reconnaissance are the National Oceanic and Atmospheric Administration (NOAA), which operates two WP-3D hurricane-penetrating aircraft and one G-IV high-altitude jet for environmental surveillance, and the Air Force 53rd Weather Reconnaissance Squadron, which operates C-130 J aircraft with capabilities similar to the WP-3Ds. An exciting development in recent years is the proliferation of airborne reconnaissance capabilities in other TC-prone regions of the world. Taiwan has carried out the DOTSTAR (Dropwindsonde Observations for Typhoon Surveillance near the Taiwan Region) programme using a high-altitude ASTRA jet since 2003. Hong Kong Observatory (HKO) began flying reconnaissance missions for TCs over the northern part of the South China Sea in 2011 and continues to do so with a Bombardier Challenger jet aircraft. Japan uses a high-altitude G-II jet as a part of their T-PARC II (Tropical cyclone-Pacific Asian Research Campaign for Improvement of Intensity estimates/ forecasts) project, begun in 2016. The Shanghai Typhoon Institute (STI), in conjunction with HKO, has used a variety of airborne platforms in their Experiment on Typhoon Intensity Change in Coastal Area (EXOTICCA), begun in 2014.

Airborne instruments, both in situ and remote sensing, are used to sample the kinematic and thermodynamic characteristics of the TC inner core and its environment. Conventional instruments include the dropsondes, which provide profiles of temperature, moisture, pressure and winds; airborne Doppler radar, which provides three-dimensional distributions of reflectivity and horizontal and vertical winds in precipitation; flight-level measurements of basic state variables; and stepped frequency microwave radiometer nadir measurements of surface brightness temperatures, which can be used to infer surface wind speed. New technologies are continually being developed, including a variety of low-level and upper-level unmanned aerial systems (UAS; e.g. Braun et al. 2016; Cione et al. 2020; Wick et al. 2020), rocket sondes launched over the top of TCs in the South China Sea (Lei et al. 2017), lidars for the retrieval of kinematic and thermodynamic information when optically thick clouds are not present (Bucci et al. 2018) and dropsondes with infrared sensors to estimate sea surface temperature and provide co-located atmospheric and surface temperature and moisture needed for surface flux estimates (Zhang et al. 2017), to name just a few.

Depending on the platform, measurements are taken in the inner core to provide information vital to operational centres for accurate assessment of TC position and intensity, or they are taken in the environment in data-sparse regions over the ocean to sample features expected to impact the future track of the TC. Typically, missions sampling the inner core are performed every 6–12 h when a TC is a potential threat to land and even more frequently (e.g. every 3 h or less) when landfall is imminent. For environmental sampling, missions may be flown every 12–24 h.

In terms of value to the forecasting community, the main goals of TC airborne data collection (Rogers et al. 2006, 2013) are 1) collect observations that span the TC life cycle in a variety of environments for model initialization and evaluation; 2) develop and refine measurement strategies and technologies that provide improved real-time monitoring of TC intensity, structure and environment; and 3) improve the understanding of physical processes important in track, structure and intensity change for a TC at all stages of its life cycle. When reported in real time and combined with other platforms, e.g. satellites and ground-based sensors, these data can be a powerful tool to provide situational awareness to the forecaster and input to NWP models. Figure 7.7 shows an image combining the near-surface wind field observed by airborne Doppler radar on the WP-3D with

Fig. 7.7 Image in AWIPS-II showing 0.5 km wind speed (shaded, kn) from airborne Doppler radar and GOES-15 infrared image with superimposed cloud-to-ground lightning strikes (white "plus" and "minus" signs) at 1700 UTC in Hurricane Lane (2018). (Image courtesy of Stephanie Stevenson, NHC; lightning data courtesy of Vaisala)

cloud-to-ground lightning detected in Hurricane Lane (2018). It was generated in the Advanced Weather Interactive Processing System (AWIPS) used by forecasters at the National Hurricane Center (NHC) to make real-time assessments and forecasts of TC position, structure and intensity. Such a capability provides an unprecedented opportunity to assess TC inner-core structure in real time and make more informed predictions of intensity changes, at least in the short term (e.g. 6–12 h).

7.4 Bridges: To Forecasts from Observations

7.4.1 Overview

The gap between observationalists and prediction practitioners/forecasters arises because of several pragmatic issues, including the complexity of various individual components of the forecasting system, and the different pace of progress. The specifics of service and warning requirements evolve over time due to advancement in scientific understanding and technology. Solutions require research and development and technology transfer processes combined with long-term implementation strategies and investments. Figure 7.8 provides a simplified schematic showing the gaps, pathways and bridges from observations to weather warnings. In the following text, the numbered sequence corresponds to numbered items in grey boxes in Fig. 7.8.

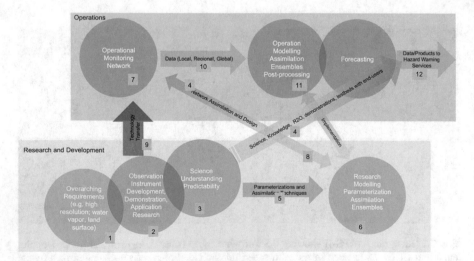

Fig. 7.8 Gaps and pathways bridging observationalists and forecast practitioners amongst research, operations and hazard warnings. See text for details

[1] Existing or anticipated future requirements (such as high-resolution urban observations, hazard impacts for verification, low cost) provide the starting point for the development of new observation technologies. A comprehensive study of gaps in current capability should be informed by user needs, model sensitivity studies and instrument design expertise.

[2] Requirements are translated into instrument concepts, based on knowledge of existing sensor and platform technologies that may have been developed in other fields of science and engineering. Within a dedicated collaboration, interaction between atmospheric scientists and instrument experts should be continuous, with funding for design and development of suitable instruments as new requirements emerge. Collaboration of this sort has contributed to the successful development of satellite remote sensing and would benefit ground-based observation.

[3] New scientific insights arise from new observations and their analysis, and are transferred to operations by early adopters in an ad hoc fashion [4]. Journal, conference and "grey" literature (internet) publications facilitate information exchange. Theoretical or laboratory studies from academia or research institutes lead to improved physical parameterizations and guide NWP development (e.g. sub-grid-scale and physical parameterizations, process rates, urban surface-canopy representations) [5].

[6] NWP models improve their resolution, representation of physics, etc. For models with an operational focus, use of existing available observations [7] remains the focus of data assimilation programmes [8]. Many existing operational observations are still not optimally used or assimilated by operational NWP despite the shortage of high-resolution data (e.g. Raman lidars: see Adam et al. 2016b; Thundathil et al. 2020).

[9] While some technologies are implemented through replacement or upgrade programmes (e.g. radar upgrade to polarization), the value of new technologies (e.g. ADM Aeolus, GOES-R) and network design may need to be demonstrated through improvements to forecast accuracy and cost-benefit analyses using observation system simulation experiments. For these, careful choice of success metrics is needed.

[10] On the global scale, there are existing protocols, standards and agreements for data sharing. For high-resolution weather data, new long-term partnerships, data sharing and quality control protocols need to be put in place, before research-operational prediction/forecast practitioners fully adapt [11], enabling the benefits to reach the hazard warning community [12].

Figure 7.8 also identifies other gaps to bridge both within and beyond the weather forecast system. The most intense partnerships between the observationalist and forecast practitioner occur in the "technology transfer or development" phase during research and development. The biggest gap is in the transfer of new monitoring technologies from research to operations. For operational implementation, developments and investments to reduce risk in terms of cost, support and maintenance cannot be overlooked particularly with limited operating budgets. The success metric for operational monitoring is to ensure the reliability, accuracy and efficient delivery of observations in a timely manner and to protect the historical climate record. In this context, the culture is to maintain the status quo, and innovations in technology or procedures are appropriately approached with great caution. Therefore, the successful innovation will require partnerships between forecasting operations and research observationalists with the knowledge to properly formulate the technical specifications and to demonstrate utility and cost-benefits. Since users of forecasts and warnings only have access to operational data and products, their early involvement in development projects or testbeds is needed to quantify benefits.

7.4.2   New Technology Development

New observing instruments are developed as technology evolves and becomes cost-effective. For example, lidars that measure moisture in the atmosphere are now available, using commercial off-the-shelf equipment, but are not yet deployed in operational weather observing networks. Processing of communication microwave signals delayed by atmospheric moisture or precipitation can retrieve humidity using radar ground clutter, inter-cell tower signals or GPS, respectively. Social media, sensors on mobile phones and on cars (including radars and lidars) and crowdsourcing have provided new opportunities to be exploited to observe the weather, to verify and validate forecasts and to determine hazard impacts.

Assimilation of data is model dependent. Observations are often indirect measures (e.g. polarization radar reflectivity variables are used to estimate precipitation rate, type and winds) and must be converted into model variables, while the scale of the observations must be filtered or processed to match the model resolution. These differences can lead to misunderstandings and misinterpretations as to the quantified impact or verification of new observations or monitoring networks. It is very much a case of "garbage is someone else's treasure" and quality control is an important issue.

Requirements are generally well known and largely unfulfilled as expectations continually increase. Innovation often originates through recognizing and filling gaps between a technology and an application or requirement. Therefore, partnerships between academia, industry and governments to fund technology innovation, development, demonstration and implementation are required to exploit the plethora of opportunities.

7.4.3   Demonstrations, Testbeds and Technology Transfer

The added benefit of new technology is assessed, verified and validated through inter-comparison with existing accepted standards. For example, winds from wind profilers, Doppler radars or lidars, temperature and humidity profiles from satellites, radiometers, DIAL lidars or AMDAR are compared to the radiosonde. New satellite sensors are often first deployed on the ground, then on aircraft and perhaps on demonstrator satellite missions. With the critical role of numerical weather models in high-impact weather forecasting, assimilation studies and demonstrations are now a de facto requirement to bridge the gap between observations and forecasts. Assimilation studies may be in limited field projects, through simulations, at various scales and weather scenarios. These studies take into consideration the errors in the observation and contribute to determining the optimal observation requirements. Good examples include the justification of the ADM Aeolus Doppler Wind Lidar (Reitebuch 2012) and the Global Precipitation Measurement (Hou et al. 2014) satellite missions.

In recent years, the World Weather Research Programme (WWRP) of the WMO has conducted Research Development (RDP) and Forecast Demonstration Projects (FDP) to accelerate and focus progress in key research areas and to demonstrate multiple prediction systems in parallel as part of the research to operations technology transfer process (Keenan et al. 2003; Fig. 7.8). The premise is that advances are made through cooperation and collaboration, taking advantage of experts working on a high-profile project with a very firm deadline. Working on common data across a common domain demonstrates the applicability, strengths and weaknesses and transferability of the technology and allows for proper comparison of the results through verification. The ability to successfully implement and demonstrate provides insights into the maturity of the technology. Demonstration projects are also important in terms of global capacity building as many of the advances in high-impact warnings are technically challenging requiring significant research and operational resources for implementation and not always affordable by all countries (see, e.g. WMO-HIGHWAY 2021). Local relevance and application must be demonstrated, and capacity building training/workshops are a critical step, both in the technology transfer process and in tailoring to different weather regimes and local or national organizations and infrastructures.

Testbeds are on-going long-term programmes and are the next step in bridging the research-operations gap. They are essentially mini-weather services set up to test out new technologies (observations, NWP, products, paradigms) within and external to the weather forecasting system. Particularly valuable are testbeds attached to major observatories such as the Meteorological Observatory Lindenberg (MOL) of DWD in Germany, the Payerne Observatory of MeteoSwiss and the ARM research facilities at the Southern Great Plains (SGP) site. A key element is the iterative aspect where the technology can be developed over time and improved with feedback from users. Another key aspect is the participation of hazard researchers and forecast users with access to pre-operational datasets and products (DFW 2021; HMT-WPC 2021). This allows for the co-development and co-design of the system and products over time, to develop institutional/community partnerships and trust.

The introduction of new technology in the forecast office does not always result in immediate gains, particularly with increased expectations due to their complexities and their consequences (Pliske et al. 1997; Pliske et al. 2004). Introducing new warning services requires forecasters to have more extensive knowledge of the weather, the user, observation limitations, the complexity of the prediction system (NWP, data-driven predictions, system concepts) and its products and effective access to relevant critical information. Expertise and decision-making skills are required to take advantage of innovations (Klein 2000; Andra et al. 2002; Hoffman et al. 2018). It takes time, even for an experienced forecaster, to re-develop the expertise, abilities and skills to adapt, adopt and exploit new innovations. Rapid development of expertise to develop judgment and decision-making (e.g. identification of cues, maintaining situational awareness, consideration of alternative scenarios, managing second-guessing and maintaining self-awareness) has been demonstrated through scenario training and simulation (Klein 1998; WDTD 2021).

Technology transfer or adoption, particularly at the professional forecaster level, is a social diffusive process (Rogers 2003; Fig. 7.9). Innovations or new technology are first discovered by "enthusiasts" within the forecast office or community, even prior to implementation which begins the change process. However, their opinions are not necessarily trusted or followed by many others due to their high risk and technology-biased perspective. Over time, their enthusiasm may infect or diffuse to another group identified as "early adopters" who see the value and worthiness of the new technology and are able to demonstrate its use and effectiveness similar to the role of FDPs (described above). They become recognized, respected and trusted for their opinions. Their ownership of the new technology inspires "early pragmatists" as the easier and more effective way to perform their job functions. Then, the remaining forecasters follow the trend or peer pressure. A small group may remain who are inherently resistant and may need transfer to a different task. One implication is that standard training practices aimed at the "majority" are perhaps misplaced and that the training should initially focus on and be tailored to needs and

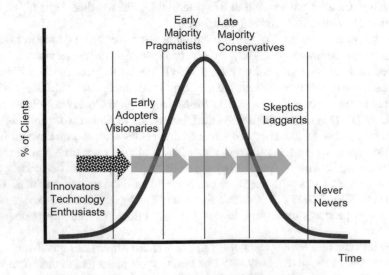

Fig. 7.9 The technology transfer process. (Adapted from Rogers 2003)

personality of the "early adopters" who then develop the training materials for the others.

7.4.4 Strategic Planning: Integrated Observing Network Design

Given the complexity and long implementation time frames, long-term agile technology transfer, strategic plans and frameworks are critical. Integrated meteorological monitoring addresses the observation and measurement of the state and processes of the atmosphere and related geophysical and anthropogenic systems by heterogeneous measurements and technical technologies. Integrated design is the planning and design of interoperable upper air and surface in situ, surface and space remote sensing technologies together with data fusion capabilities, to meet the needs of different application areas. Ideally, the integrated network design achieves cost-efficiencies by considering observation equipment capabilities and capacities (e.g. altitude, spatial and temporal resolution, observation accuracy and uncertainty, etc.), the local weather and climate characteristics, topography (installable or not, representativeness), underlying surface characteristics (geological disaster-prone, ecological protection, etc.) and population distribution.

New technology must be combined with existing equipment in a way that is demonstrated to add value to the global observation network. Separate (non-integrated) networks lead to inefficient use of human and material resources and waste of space and may be environmentally or occupationally hazardous. To resolve these issues and to set priorities for operational deployment of new technologies, the World Meteorological Organization has established a Rolling Review of Requirements (RRR) process to identify and prioritize gaps between observation capability and application requirements, to optimize the large investments that need to be made at global and local scales and to find the optimal layout and balance of observation facilities that meets the requirements of different spatial and temporal resolutions covering all space and time scales.

The RRR process provides the scientific justification for the monitoring network for different user applications in 14 categories that include numerical models, marine, transportation, agriculture, energy, etc. The RRR method analyses the gap between observation capabilities and different application requirements and then designs and implements the network taking into consideration the observation capabilities of different instruments. For example, in order to satisfy the numerical model prediction requirement (initial value and verification), the optimum geographic locations of surface stations are based on NWP assimilation and verification (Riishojgaard 2017). In China, a national surface network design satisfying numerical forecast requirements was completed after 3 years of testing and evaluation. Similarly, the weather radar network was designed taking into account satellite monitoring capabilities, radar coverage, population and economic zone distribution, topographic and geomorphic features and installation and maintenance factors. Network design also needs to respect the existing observations network (e.g. for continuity of the climate record), which may mean reconfiguring existing networks to fill the gaps in sparsely observed areas.

The spatial and temporal resolution of grid data are defined at three levels, basic, objective and ideal, for different application areas. For example, taking temperature profiles for regional numerical model use, the most basic requirement is that its spatial and temporal resolution should reach 1 km/5 h on the ground level, 5 km(horizontal)*0.45 km (vertical)/1 h at the bottom of the troposphere and 25 km (horizontal)*1.5 km (vertical)/1 h at the tropopause. Other application areas have their own standards.

It is necessary to establish a data management system to support the analysis of three-dimensional real-time gridded fields. Requirements, data sharing, data management, quality control and data exchange are issues at the core of cooperatively building an observation network amongst national meteorological services and third-party providers. The WMO is a recognized authority for cooperation in meteorological services and provides substantial guidance in all of these key issues. New technologies require that standards, procedures and requirements are regularly updated.

Basic principles for operation and maintenance of observation stations and data are as follows:

- Consider both the scientific basis and operational characteristics of the instruments, including temporal and spatial resolutions as well as measurement accuracy.
- Categorize observing stations as surface (land and sea surface), upper air and/or space and as contributing to global, regional and/or local needs.
- Meet all domestic laws, regulations and national standards as well as common practice.
- Maintain overall stability/consistency through processing and dynamic adjustments taking into account the needs of operational developments, their categories and their management levels, to realize "multi-purposes in one station".
- Manage all data, incorporating observations and meta-data collected by others. Observation stations contributed by other agencies, industry, volunteers and crowdsourcing need to be included.

7.5 Examples

Box 7.1 Satellite Observations
Steve Goodman

Satellite weather missions provide examples of a mature programmatic/formal process of bridging the observationalist-forecasting practitioner gap. Generally, all missions follow the same process where a nascent idea is first proposed in a competition which in some cases spans multiple sciences. The WMO Coordination Group for Meteorological Satellites (CGMS 2021a) is the primary international partnership body whose main goals are to support operational weather and climate monitoring and forecasting end-to-end in response to user requirements formulated by WMO and other international agencies, and impact and benefits studies are critical to justify the proposal (ESA 2021; JAXA 2021; NASA 2021a; WMO 2021). Workshops are conducted to consider stakeholder needs to identify the next-generation constellation of satellites and instruments (NASA 2021b; CGMS 2021b). For example, with the recent increase and impact of wildfires and hurricanes in a warming climate, high-resolution multispectral VIS/IR imagers, IR hyperspectral sounders and lightning imagers will be the backbone of the future geostationary satellite constellation (GEO-Ring).
  The process includes the following steps:

1. Proposals are evaluated by an expert independent panel of science and technical experts.
2. The evaluation is performed against stated mandates, strategic visions, science importance and impact, technical feasibility, cost (including life cycle costs and organizational impacts) and maturity (both of the science and technology). The evaluation is difficult as it must compare conflicting goals and objectives and different sciences.
3. Several missions may be selected for further study or demonstration or to resolve potential issues. Feasibility and trade-off studies may be initiated.
4. Use of the data or products by practitioners must be demonstrated. Proposed user products are scrutinized by forecasters (or their surrogate), and data are assimilated by appropriately modified prediction systems and their impacts established, thereby ensuring that the observationalist-prediction practitioner gap is bridged. An example is the Atmospheric Dynamics Mission where benefits were quantified well before launch (ESA 1999; Tan and Andersson 2004; Tan et al. 2007).
5. The feedback from the evaluation and impact studies are used to refine the proposals and then re-evaluated and selected for implementation as missions in final competition.

(continued)

Exploitation of satellite data for NWP in the USA is facilitated by the Joint Center for Satellite Data Assimilation (JCSDA 2021). It is a partnership between observationalists and prediction practitioners to advance the ability and shorten the time to use of satellite data, particularly those with a limited lifetime (e.g. science missions), in operational NWP models. Integrated modelling systems replicate operational capability to quantify the expected impacts of new data sources on forecast accuracy. The JCSDA transitions this research to operational and university communities through a robust data infrastructure and open-source software. Successful transitions of advanced satellite data into operations include QuikSCAT winds, MODIS winds, GOES-R winds, Atmospheric Infrared Sounder [AIRS] data and Suomi NPP (National Polar-orbiting Partnership) CrIS (Cross-Track Infrared Sounder) and ATMS (Advanced Technology Microwave Sounder) data.

Box 7.2 Aviation Partnerships

George IsaacJim WilsonPing Wah Peter LiPaul Joe

Aviation has had a long partnership amongst the forecasting community, air traffic management, pilots, airlines, service providers and other aviation stakeholders. It is one of the best examples demonstrating how the information from observations to forecasts are integrated to support end-user planning, strategic and tactical operations. Governance of the airspace is globally coordinated by the World Meteorological Organization and the International Civil Aviation Organization. Regulations and standards are established for observations and products made by national meteorological services, airport authorities or third parties.

Aviation activities are highly weather dependent. Efficiency and safety issues are intertwined as weather can change rapidly. Planes are scheduled for take-off or landing every 30 seconds at some airports but can only take off and land under specific conditions. Aviation hazards include runway surface conditions, wind shear, visibility, crosswinds and the presence of lightning, amongst others. Widespread snowstorms can affect many airports and their alternates for hours or days, as aircraft must de-ice before taking off and runways must be kept clear. Similarly, summer convective systems can have small-scale features (tens of kilometres) that must be avoided, within a broad weather system (hundreds to thousands of kilometres), causing congestion and flight delays. All ground operations at airports stop when thunderstorms are in the vicinity. *En route*, aircraft are sensitive to turbulence, strong winds, volcanic ash, thunderstorms and in-cloud icing due to supercooled liquid and high ice water content.

(continued)

The dependencies within aviation operations are intertwined, and interruptions at a single hub can have a domino effect elsewhere. Airports are rarely closed due to weather, and pilots (the ultimate authority) must exercise their expert judgement regarding weather hazards, in real time, under challenging situations.

Microburst Detection

The implementation of the Terminal Doppler Weather Radar (TDWR 2015) and similar systems around the world (Hong Kong 2021; JMA 2021), and the subsequent elimination of aircraft wind shear accidents, is a prime example of the high-impact "Perfect Warning" and demonstrates the partnerships required to rapidly bridge the various gaps - from initial investigation (no knowledge), to research (including field programmes and analysis), to technology development, implementation, system co-design and implementation, to address a critical end-user hazard. The automated warning of microbursts at many airports is the most successful of all nowcasts and has saved hundreds of lives. Controllers and pilots are now warned of microbursts based on automated alerts based on the TDWR, the low-level wind shear alert system (LLWAS; a network of anemometers positioned around runways) and now Doppler lidars (Chan and Lee 2012; Nechaj et al. 2019).

Initially the reason for the crashes was unknown. A microburst is a small-scale (<4 km) and very-short-lived (<20 min) divergent low-level (<200 m AGL) outflow from a thunderstorm (Fujita 1985). This is an end-user (rather than phenomenon-based) definition based on the inability of airplanes to react and recover from such a small and intense feature and illustrates the need to understand and involve the user community at early stages to determine the requirements of the hazard warning. The research community quickly conducted field programmes to understand microbursts and then to develop techniques to detect and anticipate them (Wilson and Wakimoto 2001). Studies indicated that microbursts occur in both wet (precipitation related) and dry (where precipitation has evaporated before reaching the ground) microbursts. Specialized and specific numerical weather prediction models were developed to test the new understanding. A warning strategy with co-design of products that fit within the culture and technology environment of the aviation industry (tower and cockpit) was developed. Demonstration projects (with engagement of meteorologists and air traffic controllers) were conducted to develop, understand and demonstrate the interpretation of the products, to test the risk and communication modalities. This was followed by the development of the TDWR in universities and industry and the rapid installation of the radars at airports. Intensive education of pilots and controllers about microbursts and what pilots should do when encountering wind shear has completely eliminated wind shear crashes (Serafin et al. 1999). The entire process took less than 20 years. This success story demonstrated the

(continued)

multi-agency support and quick funding by the US National Science Foundation and Federal Aviation Agency and the close working relationship of government, university and private companies, particularly the Lincoln Laboratory, National Center for Atmospheric Research and National Oceanic and Atmospheric Administration.

Icing and De-Icing

Snow and ice can accumulate on aircraft fuselage and wings significantly reducing lift. De-icing fluid is sprayed on aircraft to melt the ice and to prevent accumulation. The type and efficacy of the de-icing fluid is determined by the precipitation conditions. The aircraft has a limited window of time to take-off before the fluid is diluted and becomes ineffective. This is typically 10 to 30 minutes, and so the nowcasting of precipitation conditions is critical for safe and efficient operations in winter. Similar to the microburst story, several accidents led to intense field programmes to better understand the meteorological conditions, the user processes and procedures and to co-design effective products. This led to the implementation of prototype instruments and prediction techniques based on radar or in situ instruments (Rasmussen et al. 2001; Isaac et al. 2014a).

High ice water content in the top of cirrus clouds can also affect en route flight safety. In certain conditions, ice crystals are ingested into aircraft engines causing them to shut down. This resulted in several crashes as a result of which the meteorological research community, aircraft designers and aviation regulators have created partnerships to set new flight regulations and aircraft certification requirements and procedures (Strapp et al. 2016).

Future Aviation

The WMO and ICAO (International Civil Aviation Organization) have partnered to modernize global aviation (Global Aviation Navigation Plan, GANP 2019). A key requirement is the ability to produce highly accurate forecasts for the terminal area with a precision of minutes and hundreds of metres at a lead time of 6 to 12 hours. The Aviation Research and Demonstration Project (AvRDP, from 2015 to 2019) was conducted to develop innovative aviation-specific nowcasting services and to demonstrate their benefits to end users (AvRDP 2019). Eleven international airports participated, covering a variety of climate and technology scenarios. Observation and prediction technologies included advanced cloud radars, satellite, lidar, nowcasting and high-resolution, rapidly updated models and translation of the meteorological information into an Air Traffic Management (ATM) information system. The high-impact weather studied included convection, low visibility, low cloud, dust storms and low-level wind shear (Fig. 7.10).

(continued)

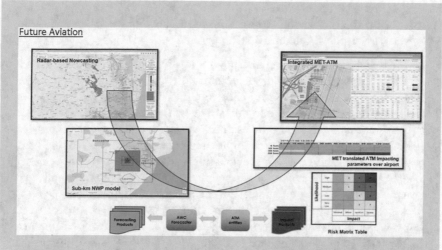

Fig. 7.10 Sample integration of impacting convection nowcast data with Air Traffic Management system. The colour scale is based on an agreed likelihood-impact risk metric

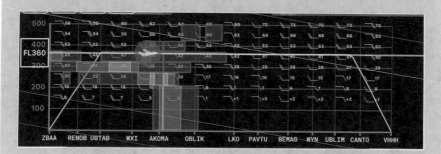

Fig. 7.11 Seamless weather information required to support the whole gate-to-gate flight trajectory: immediate and short-range information during take-off/landing and ascending/descending, combined with regional/global long-term model information during the en route phase. (© Hong Kong Observatory)

A second phase is planned to further demonstrate the concepts of research-to-operations and science-for-services throughout the full value chain through collaboration in the use of advanced aviation meteorological information to seamlessly support safe and efficient gate-to-gate operations (take-off, ascent, cruising, descent, until landing – see Fig. 7.11). Here, "seamless" refers not just to the continuous information across multiple spatial and temporal scales but also across the whole value chain from observations to users' benefits. A long-term collaborative strategic plan provides direction and guidance for both the meteorological and aviation communities (WMO LTP 2019).

Box 7.3 Testbeds, Proving Grounds and Observatories
Nusrat YussoufSteve GoodmanVolker Wulfmeyer

Testbeds and proving grounds are a programmatic bridge between operational forecasters, model developers, social scientists, emergency managers, broadcasters and the private sector to accelerate the transition of novel research ideas and forecast products into operations while ensuring that they receive critical feedback and are co-designed during the development process (Fig. 7.12). Testbeds facilitate the future implementation of new, cutting-edge high-impact weather data or products, improved analysis techniques, better statistical or dynamic models and forecast techniques to improve situational awareness and improve forecaster warning accuracy and lead time. The feedback process is often iterative – incorporating a test-feedback loop between users and developers. Testing and evaluation are conducted with operational forecasters in a quasi-operational environment with the tools and systems the forecasters use in their everyday workflow. NOAA operates a dozen such testbeds and proving grounds (NOAA 2021a), including several successful high-impact weather testbed facilities, e.g. the Joint Hurricane Testbed, Hazardous Weather Testbed, Aviation Weather Testbed and Hydrometeorology Testbed. Satellite observation capabilities are also evaluated at the Joint Center for Satellite Data Assimilation (JCSDA), while derived products are evaluated in various testbeds and proving grounds (Goodman et al. 2012).

Fig. 7.12 NOAA's Hazardous Weather Testbed during the annual Spring Experiment. The Storm Prediction Center is visible through the glass. (Photo Credit: James Murnan/NOAA)

Once a product has been tested with positive results, a project plan is submitted to a formal review committee which then assesses the operational value and identifies any infrastructure, training and funding gaps to ensure a successful implementation into operations. This can take several years. It is highly desired to have a peer-reviewed publication to accompany the new science before the product or algorithm is transitioned to operations.

A specific example is the Hazardous Weather Testbed that jointly conducts the satellite product evaluation and the Experimental Warning Programme Spring Experiment demonstration in the USA. An Annual Guidance Memorandum from the National Weather Service provides a list of products to be demonstrated. Forecaster experiences are shared through weekly seminars (HWT 2021) and satellite application workshops, both nationally (COMET 2019) and internationally (European Severe Storms Laboratory, NWCSAF 2021), and through blogs (NOAA 2021b; ESSL 2021). The organizational structure, goals and objectives of the US and European testbeds are similar and include cross-fertilization as well as international participation from multiple NMHS, researchers and industry practitioners.

Another example is the GEWEX Land-Atmosphere Feedback Observatory (GLAFO), a new project of the Global Land/Atmosphere System Study (GLASS) panel (see http://www.gewex.org/panels/global-landatmosphere-system-study-panel). The scientific goal of the GLAFOs is to understand the land-atmosphere feedback chains that pre-condition the lower atmosphere in different regimes of temperature, soil and snow conditions, vegetation properties and ABL evolutions in the context of large-scale forcing. They will use new instrumentation for high-resolution observations of wind, temperature and moisture profiles. A network of GLAFOs in various climate regions will contribute to process understanding, development of new parameterizations, climate monitoring, model verification and data assimilation.

Box 7.4 Seamless Prediction and Demonstration Project
Paul Joe

ECPASS (Environment Canada Pan Am Science Showcase) was a project demonstrating the multi-facets, benefits and issues of bridging seamless weather, air quality and health prediction (Joe et al. 2018; WMO 2016) associated with the Pan-American Games of 2015 (PA15) in Toronto.

The service requirement was to provide weather, air quality and health warnings at the sporting venues. Existing operational weather warnings are a national responsibility and are provided for areas that are generally 40 km ×

(continued)

40 km in size. These warnings are issued by a single regional forecast office with responsibility for a very large area (1000x1000km), and a single forecaster is responsible for monitoring more than ten radars (including overlapping radars from neighbouring jurisdictions). Air quality warnings are provided at the short-term area/time scale and are issued jointly at the national and provincial level. Health warning responsibilities are issued by the national authority (ECCC), but there are 36 public health units responsible for implementing responses across the province specific to their location and partnership arrangements. In addition, urban services are the direct responsibility of the local municipality in partnership with the various levels of government (Health Ontario 2021).

Venues (such as athletic or sailing facilities) are just a few hundred metres in size and essentially considered as "points" within the context of existing forecast service domains. The venue warnings were a specialized service for PA15 to the public. Unlike other Olympic demonstration projects, services for the conduct of fair or safe competitions that require a higher level of service were not provided (Joe et al. 2010; Golding et al. 2014). As these venue warnings were outside the operational norms of monitoring, production and forecast services, a parallel weather service was set up including separate forecast desks, data management, forecasters and dedicated briefers.

As a single official public warning area may contain several venues, the specificity of the venue warnings could confound or be perceived to be in conflict with the "official operational" warning. Hence, venue warning provision was limited to PA15 officials and to centralized emergency services who were given special training. However, due to the novelty and importance of the warnings, on-site briefers were provided to accurately interpret and translate the high-resolution information. A most important aspect of the presence of on-site briefers was their engagement with the "early adopter" end user that resulted in the development of trust and technology transfer of the state-of-the-art services.

ECPASS provided the opportunity to develop and to demonstrate the concept of the state of the art in seamless weather prediction services. These demonstration projects are opportunities for researchers in different services/disciplines to interact. For example, early research-to-research interaction led to the deployment of black globe temperature sensors in the mesonet, producing high-resolution heat stress prediction products (100 m scale) on a specialized display system, all of which was unprecedented for health warning "technology innovators". This was done through research collaborations as it was outside the requirements and mandate of operational weather monitoring services. While we intended to train health warning users such as long-term care facility operators and hospital admissions (for programming, staff scheduling and other purposes), the time limitations of the diffusion process

(continued)

precluded significant uptake by "early adopters". Follow-up "testbed" programmes are needed to continue the technology transfer/adoption process.

One significant outcome was that discussion and co-design of the heat stress products contributed to the harmonization of heat stress warning standards and policy by the participating health units (Herdt 2017). The current policy requires consecutive days of heat and humidity, while high-resolution predictions provided a pathway to very-short-term heat warnings (6 to 12 hours).

Most of the PA15 venues were located near a big lake (Lake Ontario), and the lake breeze initiates thunderstorms, modifies the air quality and affects the temperature, and so it is a factor (at high resolution) for all the warning services (Mariani et al. 2018). Previous experience with evenly dispersed mesonet stations (typically 10 km spacing) was unsatisfying as the fine structures of high-resolution models could not be evaluated. The mesonet was designed with the urban-lake breeze as a harmonizing focus with stations aligned perpendicular and parallel to the lake geometry and with greater station density near the land-lake boundary for diagnostic and investigative studies.

High-resolution models were configured with parameterizations considering the urban fabric, including buildings of various heights and surfaces (e.g. green, white concrete, black roads). PA15 monitoring stations located on green, rooftop and other urban surfaces were valued for model verification/validation studies. Normally only observations from green sites are acceptable for assimilation or verification in forecast models. Even with 1-minute data, there were not enough observations to verify all the parameterizations. For example, the parameterization of outdoor cooking (barbeques) for air quality, the heat flux from rooftops, temperature variations and wind gusts within urban canyons were identified as missing observations. For the 1-minute wind data, the reporting of maximum wind gust (usually reported as maximum wind in the past hour) needed re-defining.

7.6 Summary

- NWP and nowcasting models provide the foundation of hazardous weather forecasts. Development of higher-resolution, more detailed process models, improved data assimilation, frequent updating and ensemble probability prediction are driving improvements in forecast accuracy.
- The latest generation of kilometre-scale NWP models predicts small-scale weather hazards, such as thunderstorms, embedded in larger-scale weather systems. Optimizing both scales simultaneously is a challenge for NWP research.

- Forecasters use model guidance to formulate scenarios of how the weather will develop, focused on the applications in which the information will be used.
- Observations are the fundamental ingredient for monitoring and prediction of hazardous weather and for verification of forecasts.
- Development of new observational capabilities is a long-term process which needs to be planned a decade or more before the data are required.
- Even in well-observed countries, current observing capabilities are inadequate for the new generation of high-resolution models, so new sources of data are needed, including new instruments, new observing platforms and extraction of weather information from data obtained for non-meteorological purposes.
- There are particular gaps in our capability to observe pre-convective dynamics and thermodynamics of the lower troposphere.
- Prediction models require observations that can be related to model variables, for which there are well-defined performance data, and that can be delivered quickly. Meeting these needs depends on close collaboration between observationalists and forecasters.
- Future forecasting systems will particularly require additional observations of weather variations within urban areas and in areas of complex topography.

References

Adam, M., M. Turp, A. Horseman, C. Ordez, J. Buxmann and J. Sugier, 2016a. From operational ceilometer network to operational Lidar network, Proc. 27 ILRC, S14.06-163. *EPJ Web of Conferences 27007*, https://doi.org/10.1051/epjconf/201611927007

Adam, S., A. Behrendt, T. Schwitalla, E. Hammann and V. Wulfmeyer, 2016b. First assimilation of temperature lidar data into an NWP model: impact on the simulation of the temperature field, inversion strength and PBL depth. *Quart. J. Roy. Meteorol. S.,* **142,** 2882–2896. https://doi.org/10.1002/qj.2875

Andra, D. L., E. M. Quoetone and W. F. Bunting, 2002. Warning Decision Making: The Relative Roles of Conceptual Models, Technology, Strategy, and Forecaster Expertise on 3 May 1999. *Wea. Forecast,* **17,** 559–566, https://doi.org/10.1175/1520-0434(2002)017<0559:WDMTR R>2.0.CO;2.

AvRDP, 2019. Aviation Research Demonstration Project, https://avrdp.hko.gov.hk/, (accessed 28 Feb 2021).

Bailey, M.E., G.A. Isaac, I. Gultepe, I. Heckman and J. Reid, 2014. Adaptive Blending of Model and Observations for Automated Short Range Forecasting: Examples from the Vancouver 2010 Olympic and Paralympic Winter Games. *Pure Appl. Geophys.* 171, 257–276. https://doi.org/10.1007/s00024-012-0553-x

Baklanov, A. , C. S. B. Grimmond, D. Carlson, D. Terblanche, X. Tang, V. Bouchet, B. Lee, G. Langendijk, R. K. Kolli and A. Hovsepyan, 2018. From urban meteorology, climate and environment research to integrated city services. *Urban Clim.,* **23**, 330–341, https://doi.org/10.1016/j.uclim.2017.05.004.

Baklanov, A., M. Lawrence, S. Pandis, A. Mahura, S. Finardi, N. Moussiopoulos, M. Beekmann, P. Laj, L. Gomes, J.-L. Jaffrezo, A. Borbon, I. Coll, V. Gros, J. Sciare, J. Kukkonen, S. Galmarini, F. Giorgi, S. Grimmond, I. Esau, A. Stohl, B. Denby, T. Wagner, T. Butler, U. Baltensperger, P. Builtjes, D. van den Hout, H. D. van der Gon, B. Collins, H. Schluenzen, M. Kulmala, S. Zilitinkevich, R. Sokhi, R. Friedrich, J. Theloke, U. Kummer, L. Jalkinen, T. Halenka,

A. Wiedensholer, J. Pyle and W. B. Rossow, 2010. MEGAPOLI: Concept of multi-scale modelling of megacity impact on air quality and climate. *Adv. Sci. Res.*, **4**, 115–120. https://doi.org/10.5194/asr-4-115-2010

Bellon, A. and G. L. Austin, 1978. The evaluation of two years of real time operation of a short-term precipitation forecasting procedure (SHARP). *J. Appl. Meteorol.*, **17**, 1778–1787.

Benjamin, S. G., B. E. Schwartz, E. J. Szoke, and S. E. Koch, 2004. The value of wind profiler data in U.S. weather forecasting. *Bull. Amer. Meteorol. S.*, **85**, 1871–1886.

Boudala, F.S., G. A. Isaac, P. Filmam, R. Crawford, D. Hudak, and M. Anderson, 2017. Performance of emerging technologies for measuring solid and liquid precipitation in cold climate as compared to the traditional manual gauges. *J. Atmos. Ocean. Tech.*, **34**, 167–185.

Braun, S. A., P. A. Newman and G. M. Heymsfield, 2016. NASA's Hurricane and Severe Storm Sentinel (HS3) Investigation. *Bull. Amer. Meteorol. S.*, **97**, 2085–2102. https://doi.org/10.1175/BAMS-D-15-00186.1

Browning, K. A. and R. Wexler, 1968. The determination of kinematic properties of a wind field using Doppler-Radar. *J. Appl. Met.*, **7**, 105–113.

Bucci L. R., C. O'Handley, G. D. Emmitt, J. A. Zhang, K. Ryan and R. Atlas, 2018. Validation of an Airborne Doppler Wind Lidar in Tropical Cyclones. *Sensors*, **18(12)**, 4288. doi:https://doi.org/10.3390/s18124288

Burrows, W. R. and C. Mooney, 2018. Automated Products for Forecasting Blizzard Conditions in The Arctic, Polar Prediction Matters Blog. *J. Appl. Meteorol.*, **7**, 105–113.

CGMS, 2021a. WMO Coordination Group for Meteorological Satellites, https://www.cgms-info.org/index_.php/cgms/index. (Accessed 28/2/2021)

CGMS, 2021b. 2020 Community Meetings Presentations, https://www.nesdis.noaa.gov/content/2020-community-meetings-presentations. (Accessed 28/2/2021)

Chan, P.W. and Lee, Y.F., 2012. Application of short-range lidar in wind shear alerting. *J. Atmos. Ocean. Tech.*, **29**, 207–220. https://doi.org/10.1175/JTECH-D-11-00086.1.

Chandrasekar, V., H. Chen and B. Philips, 2018. Principles of high-resolution radar network for hazard mitigation and disaster management in an urban environment. *J. Meteorol. S. Japan*, **96A**, 119–131. https://doi.org/10.2151/jmsj.2018-015.

Chronis, T., L. D. Carey, C. J. Schultz, E. V. Schultz, K. M. Calhoun and S. J. Goodman, 2015. Exploring lightning jump characteristics. *Wea. Forecasting*, **30**, 23–37, https://doi.org/10.1175/WAF-D-14-00064.1.

Cifelli, R., V. Chandrasekar, H. Chen and L. E. Johnson, 2018. High resolution radar quantitative precipitation estimation in the San Francisco Bay Area: Rainfall monitoring for the urban environment. *J. Meteorol. S. Japan*, **96A**, 141–155. https://doi.org/10.2151/jmsj.2018-016.

Cintineo, J. L., M. J. Pavolonis, J. M. Sieglaff, D. T. Lindsey, L. Cronce, J. Gerth, B. Rodenkirch, J. Brunner and C. Gravelle, 2018. The NOAA/CIMSS ProbSevere Model: Incorporation of total lightning and validation. *Wea. Forecast.*, **33**, 331–345. https://doi.org/10.1175/WAF-D-17-0099.1.

Cione J. J., G. H. Bryan, R. Dobosy, J. A. Zhang, G. de Boer, A. Aksoy, J. B. Wadler, E. A. Kalina, B. A. Dahl, K. Ryan, J. Neuhaus, E. Dumas, F. D. Marks, A. M. Farber, T. Hock and X. Chen, 2020. Eye of the Storm: Observing Hurricanes with a Small Unmanned Aircraft System. *Bull. Amer. Meteorol. S.*, **101**, E186–E205. https://doi.org/10.1175/BAMS-D-19-0169.1

COMET, 2019. 2019 NWS Satellite User Applications Workshop, https://courses.comet.ucar.edu/course/view.php?id=225&lang=fr). (Accessed 28/2/2021)

de Vos, L.W., H. Leijnse, A. Overeem and R. Uijlenhoet, 2019. Quality Control for Crowdsourced Personal Weather Stations to Enable Operational Rainfall Monitoring. *Geophys. Res. Letters*, **46**, 8820–8829. https://doi.org/10.1029/2019GL083731

DFW, 2021: Dallas Fort Worth Urban Test Bed, http://www.casa.umass.edu/main/research/urban-testbed/. (Accessed 27/2/2021)

Dixon, K., C. F. Mass, G. J. Hakim, and R. H. Holzworth, 2016. The Impact of Lightning Data Assimilation on Deterministic and Ensemble Forecasts of Convective Events. *J. Atmos. Ocean. Tech.*, **33**, 1801–1833. https://doi.org/10.1175/JTECH-D-15-0188.1.

246   P. Joe et al.

Dixon, M. and G. Wiener, 1993. TITAN: Thunderstorm Identification Tracking, Analysis, and Nowcasting: A radar-based methodology. *J. Atmos. Ocean. Tech.*, **10**, 785–797.

Ebert, E. E., 2001. Ability of a Poor Man's Ensemble to Predict the Probability and Distribution of Precipitation. *Mon. Wea. Rev.*, **129**, 2461–2480. https://doi.org/10.1175/1520-0493(2001)12 9<2461:AOAPMS>2.0.CO;2

Ebert, E. E., 2009. Neighborhood Verification: A Strategy for Rewarding Close Forecasts. *Wea. Forecast.*, **24**, 1498–1510. https://doi.org/10.1175/2009WAF2222251.1

ECMWF, 2020. Coordinated response mitigates loss of aircraft-based weather data, https://www.ecmwf.int/en/newsletter/164/news/coordinated-response-mitigates-loss-aircraft-based-weather-data. (Accessed 20/2/2021)

Elmore, K. L., Z. L. Flaming, V. Lakshmanan, B. T. Kanev, V. Farmer, H. D. Reeves and L. P. Rothfusz, 2014. MPING: Crowd-sourcing weather reports for research. *Bull. Amer. Meteorol. S.*, **95(9)**, 1335–1342. https://doi.org/10.1175/BAMS-D-13-00014.1

ESA, 1999. Atmospheric Dynamics Mission, ESA publication SP-1233 (4). https://www.dlr.de/pa/en/Portaldata/33/Resources/dokumente/adm_sp1233_4.pdf (Accessed 28 Mar 2022)

ESA, 2021. The Living Planet Programme, https://www.esa.int/Applications/Observing_the_Earth/The_Living_Planet_Programme/ESA_s_Living_Planet_Programme. (Accessed 28/2/2021)

ESSL, 2021. ESSL Testbed Blog, https://www.essl.org/testbed/blog/?cat=9. (Accessed 28/2/2021).

EUMETSAT, 2021. Support to Nowcasting and Very Short Range Forecasting, http://www.nwc-saf.org/. (Accessed 20/2/2021)

Evans C., D. F. Van Dyke and T. Lericos, 2014. How Do Forecasters Utilize Output from a Convection-Permitting Ensemble Forecast System? Case Study of a High-Impact Precipitation Event. *Wea. Forecast.*, **29**, 466–486. https://doi.org/10.1175/WAF-D-13-00064.1

Fabry, F., 2004. Meteorological Value of Ground Target Measurements by Radar. *J. Atmos. Oceanic Technol.*, **21**, 560–573.

Fabry, F., C. Frush, I. Zawadzki and A. Kilambi, 1997. On the extraction of near-surface index of refraction using radar phase measurements from ground targets. *J. Atmos. Oceanic Technol.*, **14**, 978–987. https://doi.org/10.1175/1520-0426(1997)014<0978.

Fenner, D., A. Holtmann, F. Meier, I. Langer and D. Scherer, 2019. Contrasting changes of urban heat island intensity during hot weather episodes. *Environ. Res. Lett.*, **14(12)**, 124013. https://doi.org/10.1088/1748-9326/ab506b

Fujita, T.T., 1985. The downburst: microburst and macroburst. *SMRP Research Paper 210*, University of Chicago, 122pp. NTIS PB85-148880.

GANP, 2019. *Global Aviation Navigation Plan*, https://www4.icao.int/ganpportal/. (Accessed 20/2/2021)

Gilleland, E., D. Ahijevych, B. G. Brown, B. Casati and E. E. Ebert, 2009. Intercomparison of Spatial Forecast Verification Methods. *Wea. Forecast.*, **24**, 1416–1430. https://doi.org/10.117 5/2009WAF2222269.1

GOES, 2021. *GOES Quick Guide Description*, https://www.star.nesdis.noaa.gov/GOES/docu-ments/QuickGuide_BaselineDerivedMotionWinds.pdf. (Accessed 20/2/2021)

Goldberg, M. D., H. A. Cikanek, L. Zhou and J. Price, 2018. *1.04 – The Joint Polar Satellite System, Comprehensive Remote Sensing*, Academic Press, **1**, 91–118. https://doi.org/10.1016/B978-0-12-409548-9.10314-8.

Golding, B.W., S.P. Ballard, K. Mylne, N. Roberts, A. Saulter, C. Wilson, P. Agnew, L.S. Davis, J. Trice, C. Jones, D. Simonin, Z. Li, C. Pierce, A. Bennett, M. Weeks and S. Moseley, 2014. Forecasting capabilities for the London 2012 Olympics. *Bull. Amer. Meteorol. S.*, **95**, 883–896.

Goodman, S. J., T. J. Schmit, J. Daniels and R. J. Redmon, eds., 2019. *The GOES-R Series: A New Generation of Geostationary Environmental Satellites*, Academic Press, Print and e-book, ISBN-13: 978-0128143278, ISBN-10: 0128143274, 306pp.

Goodman, S. J., T. J. Schmit, J. Daniels, W. Denig and K. Metcalf, 2018. *1.05 – GOES: Past, Present and Future, Comprehensive Remote Sensing*, Academic Press, **1**, 119–149, https://doi.org/10.1016/B978-0-12-409548-9.10315-X.

Goodman, S. J., J. Gurka, M. DeMaria, T. Schmit, A. Mostek, G. Jedlovec, C. Siewert, W. Feltz, J. Gerth, R. Brummer, S. Miller, B. Reed and R. Reynolds, 2012: The GOES-R Proving Ground: Accelerating User Readiness for the Next Generation Geostationary Environmental Satellite System. *Bull. Am. Meteorol. S.*, **93**, 1029–1040. https://doi.org/10.1175/BAMS-D-11-00175.1

Grimmond, S., V. Bouchet, L. Molina, A. Baklanov, J. Tan, K. H. Schluenzen, G. Mills, B. Golding, V. Masson, C. Ren, J. Voogt, S. Miao, H. Lean, B. Heusinkveld, A. Hovespyan, G. Terrugi, P. Parrish and P. Joe, 2020. Integrated Urban Hydrometeorological, Climate and Environmental Services: Concept, Methodology and Key Messages. *J. Urban Climate*, **33**, 100623–100745, https://doi.org/10.1016/j.uclim.2020.100623

Hamill, T. M., M. J. Brennan, B. Brown, M. DeMaria, E. N. Rappaport and Z. Toth, 2012. NOAA's future ensemble-based hurricane forecast products. Bull. Amer. Meteorol. Soc., 93, 209–220. https://doi.org/10.1175/2011BAMS3106.1

Health Ontario, 2021. Heat Warning Information System, http://www.health.gov.on.ca/en/common/ministry/publications/reports/heat_warning_information_system/heat_warning_information_system.aspx#ch1 (Accessed 10/3/2021)

Heizenreder, D., P. Joe, T. Hewson and E. D. Coning, 2015. Development of applications towards a high-impact weather forecast system, in Seamless Prediction of the Earth System: from minutes to months, **WMO -No. 1156,** ISBN 978-92-63-11156-2, https://library.wmo.int/index.php?lvl=notice_display&id=17276#.YM5K-C-94I7

Herdt, A. J., 2017. A multi-index investigation of the spatio-temporal relationships between heat and EMS calls during the 2015 Pan American Games in Toronto, Canada. M.S. thesis, Dept. of Atmospheric Science, Texas Tech University, 87 pp., https://ttu-ir.tdl.org/bitstream/handle/2346/73150/HERDT-THESIS-2017.pdf?sequence=1 (Accessed 10/3/2021)

Hewson, T. D. and H. A. Titley, 2010. Objective identification, typing and tracking of the complete life-cycles of cyclonic features at high spatial resolution. *Meteorol. Appl.,* **17**, 355–381. https://doi.org/10.1002/met.204

Hintz K. S., H. Vedel and E. Kaas, 2019. Collecting and processing of barometric data from smart phones for potential use in numerical weather prediction data assimilation. *Meteorol. Appl.*, 26, 733–746. https://doi.org/10.1002/met.1805.

HMT-WPC, 2021. Hydrometeorological Test Bed, Weather Prediction Center, https://www.wpc.ncep.noaa.gov/hmt/. (Accessed 27/2/2021)

Holle, R. L., N. W. Demetriades and A. Nag, 2016. Objective airport warnings over small areas using NLDN cloud and cloud-to-ground lightning data. *Wea. Forecast.*, **31**, 1061–1069.

Hoffman R. N., V. K. Kumar, S-A. Boukabara, K Ide, F. Yang, and R. Atlas, 2018. Progress in Forecast Skill at Three Leading Global Operational NWP Centers during 2015–17 as Seen in Summary Assessment Metrics (SAMs). *Wea. Forecast.*, **33**, 1661–1679. https://doi.org/10.1175/WAF-D-18-0117.1

Hoffman, R. R., Shadbolt, N., Burton, A. M. and Klein, G. A., 1995. Eliciting knowledge from experts: A methodological analysis. *Organizational Behavior Human Decision Processes*, **62**, 129–158.

Hong Kong, 2021. Windshear and Turbulence Warning Services, https://www.hko.gov.hk/en/aviat/amt/windshear_warning.htm (Accessed 28/2/2021)

Hou, A. Y., R. K. Kakar, S. Neeck, A. Azarbarzin, C. D. Kummerow, M. Kojima, R. Oki, K. Nakamura and T. Iguchi, 2014. The Global Precipitation Measurement Mission. *Bull. Amer. Meteorol. S.*, **95**, 701–722. doi:https://doi.org/10.1175/BAMS-D-13-00164.1.

Huang, L. X., G. A. Isaac and G. Sheng, 2012. Integrating NWP Forecasts and Observation Data to Improve Nowcasting Accuracy. *Wea. Forecast.*, **27**, 938–953.

Huffman, G. J., D. T. Bolvin, D. Braithwaite, K. Hsu, R. Joyce, C. Kidd, E. J. Nelkin, S. Sorooshian, J. Tan and P. Xie, 2019a. *NASA Global Precipitation Measurement (GPM) Integrated Multi-Satellite Retrievals for GPM (IMERG).* NASA Algorithm Theoretical Basis Doc., version 06, 38pp. https://pmm.nasa.gov/sites/default/files/document_files/IMERG_ATBD_V06.pdf.

Huffman, G. J., D. T. Bolvin, E. J. Nelkin and J. Tan, 2019b. *Integrated Multi-Satellite Retrievals for GPM (IMERG)* technical documentation. NASA Tech. Doc., 77pp. https://pmm.nasa.gov/sites/default/files/document_files/IMERG_doc_190909.pdf.

HWT, 2021. *Tales from the Testbed*, https://hwt.nssl.noaa.gov/tales/. (Accessed 28/2/2021)

Isaac, G. A., M. Bailey, F. S. Boudala, W. R. Burrows, S. G. Cober, R. W. Crawford, N. Donaldson, I. Gultepe, B. Hansen, I. Heckman, L. X. Huang, A. Ling, J. Mailhot, J. A. Milbrandt, J. Reid and M. Fournier, 2014a. The Canadian airport nowcasting system (CAN-Now). *Meteorol. Appl.* 21, 3049, https://doi.org/10.1002/met.1342.

Isaac, G. A., P. Joe, J. Mailhot, M. Bailey, S. Blair, F. S. Boudala, M. Brugman, E. Campos, R. L. Carpenter Jr., R. W. Crawford, S. G. Cober, B. Denis, C. Doyle, H. D. Reeves, I. Gultepe, T. Haiden, I. Heckman, L. X. Huang, J. A. Milbrandt, R. Mo, R. M. Rasmussen, T. Smith, R. E. Stewart, D. Wang and L. J. Wilson, 2014b. Science of Nowcasting Olympic Weather for Vancouver 2010 (SNOW-10): A World Weather Research Programme project. *Pure Appl. Geophys.* **171**, 1–24. https://doi.org/10.1007/s00024-012-0579-0

JAXA, 2021. Earth Observation Research Center, https://www.eorc.jaxa.jp/en/. (Accessed 28/2/2021)

JCSDA, 2021. Joint Center for Satellite Data Assimilation, http://jcsda.org. (Accessed 28/2/2021)

JMA, 2021. Marine/Aviation Weather Services, https://www.jma.go.jp/jma/en/Activities/aws.html. (Accessed 28/2/2021)

Joe, P., M. Falla, P. Van Rijn, L. Stamadianos, T. Falla, D. Magosse, L. Ing and J. Dobson, 2002. Radar Data Processing for Severe Weather in the National Radar Project of Canada, SELS, San Antonio, 12–16 August 2002, 221–224.

Joe, P., C. Doyle, A. Wallace, S. G. Cober, B. Scott, G. A. Isaac, T. Smith, J. Mailhot, B. Snyder, S. Belair, Q. Jansen and B. Denis, 2010. Weather Services, Science Advances, and the Vancouver 2010 Olympic and Paralympic Winter Games. *Bull. Amer. Meteorol. S.*, **91**, 31–36. https://doi.org/10.1175/2009BAMS2998.1.

Joe, P., S. Belair, N. B. Bernier, V. Bouchet, J. R. Brook, D. Brunet, W. Burrows, J. P. Charland, A. Dehghan, N. Driedger, C. Duhaime, G. Evans, A.-B. Filion, R. Frenette, J. de Grandpré, I. Gultepe, D. Henderson, A. Herdt, N. Hilker, L. Huang, E. Hung, G. Isaac, C.-H. Jeong, D. Johnston, J. Klaassen, S. Leroyer, H. Lin, M. MacDonald, J. MacPhee, Z. Mariani, T. Munoz, J. Reid, A. Robichaud, Y. Rochon, K. Shairsingh, D. Sills, L. Spacek, C. Stroud, Y. Su, N. Taylor, J. Vanos, J. Voogt, J. M. Wang, T. Wiechers, S. Wren, H. Yang, and T. Yip, 2018: The Environment Canada Pan and ParaPan American Science Showcase Project. *Bull. Amer. Meteorol. S.*, **99**, 139–150. https://doi.org/10.1175/BAMS-D-16-0162.1.

Kaplan J., M. DeMaria and J. A. Knaff, 2010. A Revised Tropical Cyclone Rapid Intensification Index for the Atlantic and Eastern North Pacific Basins. *Wea. Forecast.*, **25**, 220–241.

Keenan, T., P. Joe, J. Wilson, C. Collier, B. Golding, D. Burgess, P. May, C. Pierce, J. Bally, A. Crook, A. Seed, D. Sills, L. Berry, R. Potts, I. Bell, F. Fox, E. Ebert, M. Eilts, K. O'Loughlin, R. Webb, R. Carbone, K. Browning, R. Roberts and C. Mueller, 2003. The Sydney 2000 World Weather Research Programme Forecast Demonstration Project: Overview and current status. *Bull. Amer. Meteorol. S.*, **84**, 1041–1054.

Kiktev, D., P. Joe, G. A. Isaac, A. Montani, I.-L. Frogner, P. Nurmi, B. Bica, J. Milbrandt, M. Tsyrulnikov, E. Astakhova, A. Bundel, S. Blair, M. Pyle, A. Muravyev, G. Rivin, I. Rozinkina, T. Paccagnella, Y. Wang, J. Reid, T. Nipen and K-D. Ahn, 2017. FROST-2014: The Sochi Winter Olympics International Project. *Bull. Amer. Meteorol. S.*, **98**, 1908–1929.

Kirschbaum, D. B., G. J. Huffman, R. F. Adler, S. Braun, K. Garrett, E. Jones, A. McNally, G. Skofronick-Jackson, E. Stocker, H. Wu and B. F. Zaitchik, 2017. NASA's Remotely Sensed Precipitation: A Reservoir for Applications Users. *Bull. Amer. Meteorol. S.*, **98**, 1169–1184. https://doi.org/10.1175/BAMS-D-15-00296.1

Kirschbaum, D. and T. Stanley, 2018. Satellite-Based Assessment of Rainfall-Triggered Landslide Hazard for Situational Awareness. *Earth's Future*, **6**, 505–523. https://doi.org/10.1002/2017EF000715

Klein, G., 1998. Sources of Power, How People Make Decisions, MIT Press, ISBN 0-262-11227-2, 330pp.

Klein, G. 2000. Can information technology reduce expertise? *Proc. Human Performance, Situation Awareness and Automation Conf.*, Savannah, GA, Human Factors and Ergonomics Society, 226.

Knight, C.A. and J. Miller, 1998. Early radar echoes from small, warm cumulus: Bragg and hydrometeor scattering. *J. Atmos. Sci.*, **55**, 2974–2992.

Kollias, P., E. Clothiaux, M. Miller, B. A. Albrecht, G. Stephens and T. Ackerman, 2007a. Millimeter-wavelength radars: New frontier in atmospheric cloud and precipitation research. *Bull. Amer. Meteorol. S.*, **88**, 1608–1624.

Kollias, P., E. E. Clothiaux, M. A. Miller, E. P. Luke, K. L. Johnson, K. P. Moran, K. B. Widener and B. A. Albrecht, 2007b. The Atmospheric Radiation Measurement program cloud profiling radars: Second generation sampling strategies, processing, and cloud data products. *J. Atmos. Ocean. Tech.*, **24**, 1199-1214, https://doi.org/10.1175/JTECH2033.1.

Kucera, P., 2017. Personal communication, https://www.icdp.ucar.edu/core-programs/3dpaws/. (Accessed 28/3/2022)

Landry, C., Oullet, M., Parent, R., Deschenes, J. F. and Verret, R., 2004. Observations and Nowcasting in SCRIBE, 20th Conference on Interactive, Information Processing Systems, 1216 January 2004, Seattle, Washington.

Lange, D., A. Behrendt, and V. Wulfmeyer, 2018. Compact automatic rotational Raman lidar system for continuous day- and nighttime temperature and humidity mapping. *Proc. 20th EGU General Assembly*, EGU2018, 4–13 April, 2018, Vienna, Austria, 9114–9114.

Lei, B. K. Zhao, J. Wang, J. Tang, H. W. Gao, Z. H. Zeng, G. L. Li, J. P. Luo, M. Wu, H. Yu, L. Ye, P. Z. Fang, D. A. Yang, S. Zhang, Y. B. Zou, E. J. Zhou, L. M. Lin, P. Chen, Y. C. Zhong, B. P. Shi, H. L. He, Y. S. Li, X. H. Yang, T. Zhao, X. W. Bao, Z. Liu, G. M. Chen, H. M. Wu, D. Wu, F. Huang, J. Wang, S. Luo and L. Yi, 2017. New technology and experiment of rocket dropsondes for typhoon observation [in Chinese]. Chinese Science Bulletin, 62, 3789–3796.

Leith, C. E., 1974: Theoretical skill of Monte Carlo forecasts. *Mon. Wea. Rev.*, **102**, 409–418.

Lewis P. and G. Toth, 2011. University Corporation for Atmospheric Research/Cooperative Program for Meteorological Education and Training: Ten Common NWP Misconceptions. http://meted.ucar.edu/norlat/tencom. (Accessed 29/3/2022)

Loehnert U., E. van Meijgaard, H.K. Baltink, S. Gro and R. Boers, 2007. Accuracy assessment of an integrated profiling technique for operationally deriving profiles of temperature, humidity, and cloud liquid water. *J. Geophys. Res.*, **112**, D04205. doi: https://doi.org/10.1029/2006JD007379.

Lopez, P., 2011. Direct 4D-Var Assimilation of NCEP Stage IV Radar and Gauge Precipitation Data at ECMWF. *Mon. Wea. Rev.*, **139**, 2098–2116.

Mariani, Z., A. Dehghan, P. Joe and D. M. Sills, 2018. Observations of lake breeze events during the Toronto 2015 Pan-American Games. *Bound. Layer Meteorol.*, **166**, 113–135. https://doi.org/10.1007/s10546-017-0289-3.

Mass, C. F. and L.E. Madaus, 2014. Surface Pressure Observations from Smartphones: A Potential Revolution for High-Resolution Weather Prediction? *Bull. Amer. Meteorol. S.*, **95**, 1343–1349.

Matson, M. and D. R. Wiesnet, 1981. New data base for climate studies. Nature **289**, 451–456.

McLaughlin, D., D. Pepyne, V. Chandrasekar, B. Philips, J. Kurose, M. Zink, K. Droegemeier, S. Cruz-Pol, F. Junyent, J. Brotzge, D. Westbrook, N. Bharadwaj, Y. Wang, E. Lyons, K. Hondl, Y. Liu, E. Knapp, M. Xue, A. Hopf, K. Kloesel, A. DeFonzo, P. Kollias, K. Brewster, R. Contreras, B. Dolan, T. Djaferis, E. Insanic, S. Frasier and F. Carr, 2009. Short wavelength technology and the potential for disturbed networks of small radar systems. *Bull. Amer. Meteorol. S.*, **90**, 1797–1817.

McNicholas, C. and C.F. Mass, 2018. Smartphone Pressure Collection and Bias Correction Using Machine Learning. *J. Atmos. Ocean. Tech.*, **35**, 523–540. https://doi.org/10.1175/JTECH-D-17-0096.1.

Meier, F., D. Fenner, T. Grassmann, M. Otto and D. Shearer, 2017. Crowdsourcing air temperature from citizen weather stations for urban climate research. *Urban Clim.*, **19**, 170–191.

Menzel, W. P., T. J. Schmit, P. Zhang and J. Li, 2018. Satellite-Based Atmospheric Infrared Sounder Development and Applications. *Bull. Amer. Meteorol. S.*, **99**, 583–603. https://doi.org/10.1175/BAMS-D-16-0293.1.

Miller, S.D., D. T. Lindsey, C. J. Seaman and J. E. Solbrig, 2020. GeoColor, A Blending Technique for Satellite Imagery. *J. Atmos. Ocean. Tech.*, **37**, 429–448. https://doi.org/10.1175/JTECH-D-19-0134.1.

Misumi, R., Y. Uji and T. Maesaka, 2020. Feeder interactions between stratiform precipitation and shallow convection observed by X-band polarimetric radar and optical disdrometer. *Atmos. Sci. Letters*, **e1034**. https://doi.org/10.1002/asl.1034

NASA, 2021a. Decadal Survey, https://science.nasa.gov/earth-science/decadal-surveys. (Accessed 28/2/2021)

NASA, 2021b. Geostationary and extended orbit (GEO-XO) Program, https://geo-xo-satellites.wixsite.com/virtual-workshops. (Accessed 28/2/2021)

Nechaj, P., L. Gaál, J. Bartok, O. Vorobyeva, M. Gera, M. Kelemen and V. Polishchuk, 2019. Monitoring of low-level wind shear by ground-based 3D Lidar for increased flight safety, protection of human lives and health. International. *J. Env. Res. Pub. Health*, **16(22)**, 4584. https://doi.org/10.3390/ijerph16224584

Nipen, T., I.A. Seierstad, C. Lussana, J. Kirstianssen and O. Hov, 2020. Adopting Citizen Observations in Operational Weather Prediction. *Bull. Amer. Meteorol. S.*, **101**. https://doi.org/10.1175/BAMS-D-18-0237.1.

NOAA, 2021a. Testbeds and Proving Grounds, https://www.testbeds.noaa.gov. (Accessed 1/4/2021)

NOAA, 2021b. GOES-R & JPSS: The Future of Weather Satellites, https://satelliteliaisonblog.com/. (Accessed 28/2/2021)

Novak, D. R., D. R. Bright and M. J. Brennan, 2008. Operational forecaster uncertainty needs and future roles. *Wea. Forecasting*, **23**, 1069–1084.

Novak, D. R., K. F. Brill and W. A. Hogsett, 2014. Using Percentiles to Communicate Snowfall Uncertainty. *Wea. Forecast.*, **29**, 1259–1265. https://doi.org/10.1175/WAF-D-14-00019.1

NWCSAF, 2021. Nowcasting Satellite Applications Facility, http://www.nwcsaf.org/. (Accessed 28/2/2021)

NWS, 2021. Severe Weather Definitions, https://www.weather.gov/bgm/severedefinitions. (Accessed 20/2/2021)

Obermaier, H. and K. I. Joy, 2014. Future challenges for ensemble visualization, *IEEE Comput. Graph. Aopl.*, **34(3)**, 8–11. DOI: https://doi.org/10.1109/PACIFICVIS.2016.7465251

Overeem, A., H. Leijnse and R. Uijlenhoet, 2011. Measuring urban rainfall using microwave links from commercial cellular communication networks, https://doi.org/10.1029/2010WR010350

Park, H. S., A. V. Ryzhkov, D. S. Zrnic and K. Kim, 2009. The hydrometeor classification algorithm for the polarimetric WSR-88D: Description and application to an MCS. *Wea. Forecast.*, **24**, 730–748. https://doi.org/10.1175/2008WAF2222205.1.

Petersen, R. A., 2016. On the Impact and Benefits of AMDAR Observations in Operational Forecasting—Part I: A Review of the Impact of Automated Aircraft Wind and Temperature Reports. *Bull. Amer. Meteorol. S.*, **97**, 585–602. https://doi.org/10.1175/BAMS-D-14-00055.1

Pliske, R. M., B. Crandall and G. Klein, 2004. *Competence in weather forecasting. Psychological Investigations of Competence in Decision Making*, K. Smith, J. Shanteau, and P. Johnson, Eds., Cambridge University Press, 4068.

Pliske, R., D. Klinger, R. Hutton, B. Crandall, B. Knight and G. Klein, 1997. *Understanding skilled weather forecasting: Implications for training and the design of forecasting tools.* Contractor Rep. AL/HR-CR-1997-003, Material.

Purdom, J., 1976. Some uses of high-resolution GOES imagery in the mesoscale forecasting of convection and its behavior. *Mon. Wea. Rev.*, **104**, 14741483.

Rasmussen, R., M. Dixon, F. Hage, J. Cole, C. Wade, J. Tuttle, S. McGettigan, T. Carty, L. Stevenson, W. Fellner, S. Knight, E. Karplus and N. Rehak, 2001. Weather Support to Deicing Decision

Making (WSDDM): A Winter Weather Nowcasting System. *Bull. Amer. Meteorol. S.*, **82**, 579–596, https://doi.org/10.1175/1520-0477(2001)082<0579:WSTDDM>2.3.CO;2.

Rautenhaus, M., M. Bottinger, S. Siemen, R. Hoffman, R. M. Kirby, M. Mirzargar, N. Röber and R. Westermann, 2018. Visualization in Meteorology - A Survey of Techniques and Tools for Data Analysis Tasks. *IEEE Trans. on Vis. and Comp. Graph*, **24** (12), 3268–3296. https://doi.org/10.1109/TVCG.2017.2779501

Reitebuch, O., 2012. The Spaceborne Wind Lidar Mission ADM-Aeolus. In: Schumann U. (eds) Atmospheric Physics. Research Topics in Aerospace. Springer, Berlin, Heidelberg. https://doi.org/10.1007/978-3-642-30183-4_49

Riishojgaard, L. P., 2017. *Report from the Sixth WMO Workshop on the Impact of Various Observing Systems on NWP.* https://www.cgms-info.org/Agendas/PPT/CGMS-45-WMO-WP-02 (Accessed 27/2/2021).

Rinehart, R. E., and E. T. Garvey, 1978. Three-dimensional storm motion detection by conventional weather radar. *Nature*, **273**, 287289.

Roberts, R. B., I. L. Jirak, A. J. Clark, S. J. Weiss and J. S. Kain, 2019. PostProcessing and Visualization Techniques for Convection-Allowing Ensembles. *Bull. Amer. Meteorol. S.,* **100**, 1245–1258. https://doi.org/10.1175/BAMS-D-18-0041.1

Rogers, E.M., 2003. *Diffusion of Innovations*, 5th edition, Simon and Schuster, Free Press, 576pp.

Rogers, R. F., S. Aberson, M. Black, P. Black, J. Cione, P. Dodge, J. Dunnion, J. Gamache, J. Kaplan, M. Powell, N. Shay, N. Surgi and E. Uhlhorn, 2006. The Intensity Forecasting Experiment (IFEX): A NOAA multiyear field program for improving tropical cyclone intensity forecasts. *Bull. Amer. Meteorol. S..*, **87**, 1523–1537.

Rogers R., P. Reasor and S. Lorsolo, 2013. Airborne Doppler Observations of the Inner-Core Structural Differences between Intensifying and Steady-State Tropical Cyclones. *Mon. Wea. Rev.,* **141**, 2970–2991. https://doi.org/10.1175/MWR-D-12-00357.1

Rogers, D. P., V. V. Tsirkunov, H. Kootval, A. Soares, D. Kull, A. Bogdanova and M. Suwa, 2019. Weather the Change: How to Improve Hydromet Services in Developing Countries? Global Facility for Disaster Reduction and Recovery (GFDRR) and World Bank, Washington, D.C.

Schmit, T. J., P. Griffith, M. M. Gunshor, J. M. Daniels, S. J. Goodman and W. J. Lebair, 2017. A closer look at the ABI on the GOES-R series. *Bull. Amer. Meteorol. S.,* **98**, 681–698. https://doi.org/10.1175/BAMS-D-15-00230.1.

Schmit, T. J., S. S. Lindstrom, J. J. Gerth and M. M. Bunshor, 2018. Applications of the 16 spectral bands on the Advanced Baseline Imager (ABI).. *J. Oper. Meteorol.*, **6**, 33–46, https://doi.org/10.15191/nwajom.2018.0604

Schwartz, C. S., J. S. Kain, S. J. Weiss, M. Xue, D. R. Bright, F. Kong, K. W. Thomas, J. J. Levit, M. C. Coniglio and M. S. Wandishin, 2010. Toward improved convection-allowing ensembles: Model physics sensitivities and optimizing probabilistic guidance with small ensemble membership. *Wea. Forecast.*, **25**, 263–280, https://doi.org/10.1175/2009WAF2222267.1.

Schwartz, C. S., G. S. Romine, M. L. Weisman, R. A. Sobash, K. R. Fossell, K. W. Manning and S. B. Trier, 2015. A real-time convection-allowing ensemble prediction system initialized by mesoscale ensemble Kalman filter analyses. *Wea. Forecasting*, **30**, 1158–1181. https://doi.org/10.1175/WAF-D-15-0013.1.

Schwartz C. S., G. S. Romine, R. A. Sobash, K. R. Fossell and M. L. Weisman, 2019. NCAR's Real-Time Convection-Allowing Ensemble Project. *Bull. Amer. Meteorol. S.,* **100**, 321–343. https://doi.org/10.1175/BAMS-D-17-0297.1

Serafin R.J., J. W. Wilson, J. McCarthy and T. T. Fujita, 1999: Progress in understanding windshear and implications on aviation, Chapter 43, Storms Vol II, editor R. Pielke Sr and R. Pielke Jr. 237–251, New York, Routledge.

Sivillo, J. K., J. E. Ahlquist and Z. Toth, 1997. An Ensemble Forecasting Primer. *Wea. Forecast.*, **12**, 809–818. https://doi.org/10.1175/1520-0434(1997)012<0809:AEFP>2.0.CO;2

Snellman, L. W., 1982. Impact of AFOS on operational forecasting. *Preprints, Ninth Conf. on Weather Forecasting and Analysis*, Seattle, WA, Amer. Meteorol. S., 1316.

Späth, F., A. Behrendt, S. Kumar Muppa, S. Metzendorf, A. Riede and V. Wulfmeyer, 2016. 3-D water vapor field in the atmospheric boundary layer observed with scanning differential absorption lidar. *Atmos. Meas. Tech.,* **9**, 1701–1720.

Stephens, G., D. Winker, J. Pelon, C. Trepte, D. Vane, C. Yuhas, T. LEcuyer and M. Lebsock, 2018. CloudSat and CALIPSO within the A-Train: Ten Years of Actively Observing the Earth System. *Bull. Amer. Meteorol. S.,* **99**, 569–581, https://doi.org/10.1175/BAMS-D-16-0324.1.

Strapp, J. W., A. Korolev, T. Ratvasky, R. Potts, A. Protat, P. May, A. Ackerman, A. Fridlind, P. Minnis, J. Haggerty, J. T. Riley, L. E. Lilie and G. A. Isaac, 2016. The High Ice Water Content Study of Deep Convective Clouds: Report on Science and Technical Plan. **DOT/FAA/ TC-14/31**. See http://www.tc.faa.gov/its/worldpac/techrpt/tc14-31.pdf. (Accessed 21/6/2021)

Strapp, J. W., R. A. Stuart and G. A. Isaac, 1996. Canadian climatology of freezing precipitation, and a detailed study using data from St. John's, Newfoundland. FAA International Conference on Aircraft Inflight Icing, Springfield, Virginia, Volume II. **DOT/FAA/AR-96/81**, 45–55. See http://www.tc.faa.gov/its/worldpac/techrpt/ar96-81-2.pdf. (Accessed 21/6/2021)

Stuart, N.A., D.M. Schultz and G. Klein, 2007. Maintaining the role of humans in the forecast process. *Bull. Amer. Meteorol. S.,* **88**, 1893–1898.

Sun, J. and N. A. Crook, 1997. Dynamical and microphysical retrieval from Doppler radar observations using a cloud model and its adjoint. Part I: Model development and simulated data experiments. *J. Atmos. Sci.,* **54**, 1642–1661. https://doi.org/10.1175/1520-0469(1997)054,164 2:DAMRFD.2.0.CO;2.

Sun, J., M. Xue, J. W. Wilson, I. Zawadzki, S. P. Ballard, J. Onvlee-Hooimeyer, P. Joe, D. M. Barker, P.-W. Li, B. Golding, M. Xu and J. Pinto, 2014. Use of NWP for nowcasting convective precipitation: Recent progress and challenges. *Bull. Amer. Meteorol. S.,* **95**, 409–426. https://doi. org/10.1175/BAMS-D-11-00263.1.

Tan, D. G. H. and E. Andersson, 2004. *Expected benefit of wind profiles from the ADM-Aeolus in a data assimilation system.* Final Report for ESA contract 15342/01/NL/MM, Report, ECMWF. (Accessed 28/2/2021)

Tan, D. G. H., E. Andersson, M. Fisher and L. Isaksen, 2007. Observing system impact assessment using a data assimilation ensemble technique: application to the ADMAeolus wind profiling mission. *Quart. J. Roy. Meteorol. S.,* **133**, 381-390. https://doi.org/10.1002/qj.43.

TDWR, 2015: Terminal Doppler Weather Radar, https://www.faa.gov/air_traffic/weather/tdwr/, (Accessed 28/2/2021)

Thundathil, R., T. Schwitalla, A. Behrendt, S. K. Muppa, S. Adam and V. Wulfmeyer, 2020. Assimilation of lidar water vapour mixing ratio and temperature profiles into a convection-permitting model. *J. Meteor. Soc. Japan,* **98**, 959–986. https://doi.org/10.2151/jmsj.2020-049.

Turner, D. D., R. A. Ferrare, L. A. H. Brasseur, W. F. Feltz and T. P. Tooman, 2002. Automated Retrievals of Water Vapor and Aerosol Profiles from an Operational Raman Lidar. *J. Atmos. Ocean. Tech.,* **19**, 37–50. https://doi.org/10.1175/1520-0426(2002)019<0037:AROWV A>2.0.CO;2

van de Giesen, N., R. Hut and J. Selker, 2014. The Trans-African Hydro-Meteorological Observatory (TAHMO). *Wiley Interdiscip. Rev.: Water,* **1**, 341–348.

WDTD, 2021. Warning Decision Training Division, https://training.weather.gov/wdtd/index.php (Accessed 27/2/2021)

Weaver, J. and J. F. W. Purdom, 1995. An interesting mesoscale storm environment interaction observed just prior to changes in severe storm behavior. *Wea. Forecasting,* **10**, 449–453.

Weckwerth, T. M., K. J. Weber, D. D. Turner and S. M. Spuler, 2016. Validation of a water vapor micropulse differential absorption lidar (DIAL). *J. Atmos. Ocean. Tech.,* **33**, 2353–2372. doi:https://doi.org/10.1175/JTECH-D-16-0119.1

Weisman, M. L., W. C. Skamarock and J. B. Klemp, 1997. The Resolution Dependence of Explicitly Modeled Convective Systems. *Mon. Wea. Rev.,* **125**, 527–548. https://doi.org/10.117 5/1520-0493(1997)125<0527:TRDOEM>2.0.CO;2

Wick, G. A., J. P. Dunion, P. G. Black, J. R. Walker, R. D. Torn, A. C. Kren, A. Aksoy, H. Christophersen, L. Cucurull, B. Dahl, J. M. English, K. Friedman, T. R. Peevey, K. Sellwood,

J. A. Sippel, V. Tallapragada, J. Taylor, H. Wang, R. E. Hood and P. Hall, 2020. NOAA's Sensing Hazards with Operational Unmanned Technology (SHOUT) Experiment Observations and Forecast Impacts. *Bull. Amer. Meteorol. S.*, **101**, E968–E987. https://doi.org/10.1175/BAMS-D-18-0257.1

Wilson, J. W., T. M. Weckwerth, J. Vivekanandan, R. M. Wakimoto and R. W. Russell, 1994. Boundary layer clear-air radar echoes: Origin of echoes and accuracy of derived winds. *J. Atmos. Ocean. Tech.*, **11**, 1184–1206.

Wilson, J.W. and R.M. Wakimoto, 2001. The Discovery of the Downburst: T. T. Fujita's Contribution. *Bull. Amer. Meteorol. S.*, **82**, 49–62. https://doi.org/10.1175/1520-0477(2001)08 2<0049:TDOTDT>2.3.CO;2.

Wilson, J.W. and W.E. Schreiber, 1986. Initiation of convective storms at radar observed boundary layer convergence lines. *Mon. Wea. Rev.*, **114**, 2516-2536.

WMO, 2020. WMO Observing Systems Capability Analysis and Review (OSCAR) database for Space-based Capabilities (OSCAR/Space), https://www.wmo-sat.info/oscar/spacecapabilities (Accessed 22/2/2021)

WMO RSMC, 2021: Regional Specialized Meteorological Centre for Nowcasting, https://swirls.hko.gov.hk/rsmc/deepLearning.html.

WMO-HIGHWAY, 2021. High Impact Weather Lake System (HIGHWAY) Project, https://public.wmo.int/en/projects/high-impact-weather-lake-system-highway-project (Accessed 27/2/2021)

WMO, 2016. Toronto 2015 Pan and Parapan American Games - An Environment and Climate Change Canada Perspective. *WMO Bulletin* **65** (1). https://public.wmo.int/en/resources/bulletin/toronto-2015---pan-and-parapan-american-games-environment-and-climate-change (Accessed 10/3/2021)

WMO, 2017. *Guidelines for Nowcasting Techniques*. **WMO-No. 1198**. Geneva: WMO.

WMO, 2018. WMO Solid Precipitation Intercomparison Experiment (SPICE) (2012-2015), https://library.wmo.int/doc_num.php?explnum_id=5686 (Accessed 20/2/2021)

WMO LTP, 2019. Long Term Plan for Aeronautical Meteorology. WMO AeM Series No.5. https://www.wmo.int/aemp/LTP-AeM.

WMO, 2020a. https://community.wmo.int/activity-areas/aircraft-based-observations/amdar.

WMO, 2021. Vision for the WMO Integrated Global Observing System in 2040, https://library.wmo.int/doc_num.php?explnum_id=10278 (Accessed 28/2/2021)

Wolff, D.B., D. A. Marks and W.A. Petersen, 2015. General Application of the Relative Calibration Adjustment (RCA) Technique for Monitoring and Correcting Radar Reflectivity Calibration. *J. Atmos. Ocean. Tech.*, **32**, 496-506. https://doi.org/10.1175/JTECH-D-13-00185.1.

Weather Observations Website, 2021. https://wow.metoffice.gov.uk, (Accessed 20/2/2021)

Wulfmeyer, V., R. M. Hardesty, D. D. Turner, A. Behrendt, M. P. Cadeddu, P. Di Girolamo, P. Schlssel, J. Van Baelen and F. Zus, 2015. A review of the remote sensing of lower tropospheric thermodynamic profiles and its indispensable role for the understanding and the simulation of water and energy cycles. *Rev. Geophysics*, **53**, 819-895.

Zhang J., K. Howard, C. Langston, B. Kaney, Y. Qi, L. Tang, H. Grams, Y. Wang, S. Cocks, S. Martinaitis, A. Arthur, K. Cooper, J. Brogden and D. Kitzmiller, 2016. Multi-Radar Multi-Sensor (MRMS) Quantitative Precipitation Estimation: Initial Operating Capabilities. *Bull. Amer. Meteorol. S.*, **97**, 621–638. https://doi.org/10.1175/BAMS-D-14-00174.1

Zhang J. A., J. J. Cione, E. A. Kalina, E. W. Uhlhorn, T. Hock and J. A. Smith, 2017. Observations of Infrared Sea Surface Temperature and Air–Sea Interaction in Hurricane Edouard (2014) Using GPS Dropsondes. *J. Ocean. Atmos. Tech.*, **34**, 1333–1349. https://doi.org/10.1175/JTECH-D-16-0211.1

Chapter 8
Pulling It All Together, End-to-End

Brian Golding

Abstract In this concluding chapter, we emphasise the performance of the whole chain. We look at fire warnings as an effective system to see what we can learn from them. We see that each step in the chain is important but that the bridges are at least as important as each of the steps. Partnership is key to the effectiveness of each bridge and is also crucial for the overall chain. Understanding the way in which value propagates up and down the chain can enable improvements to be targeted in the most beneficial places. We conclude with a summary of some attributes of the "perfect" warning chain.

Keywords Warning chain · Fire warning · Partnership · Evaluation · Attribution · Checklist

8.1 Introduction

We started this book looking at the disaster risk reduction environment, of which a warning system forms a part, and the governance of the warning system itself. We then proceeded to look in detail at the individual steps that form a complete warning system, starting with the decision-maker and their need for information in order to take protective decisions and actions, proceeding to how they receive the warning, how a warning is put together, how the impacts of hazards are forecast, how hazards are forecast and how the hazard-generating weather is forecast. At each of these steps, we have looked particularly at how the various professionals involved in creating that information can work in partnership to more effectively deliver the warning service.

B. Golding (✉)
Met Office, Exeter, UK

WMO/WWRP HIWeather project, Geneva, Switzerland
e-mail: brian.golding@metoffice.gov.uk

© The Author(s) 2022
B. Golding (ed.), *Towards the "Perfect" Weather Warning*,
https://doi.org/10.1007/978-3-030-98989-7_8

Now it is time to put it all back together again and to look at the warning chain as a whole.

As we have seen, a successful warning draws on many sources of expertise. In particular, we have emphasised the expertise of the receiver of the warning who is the best source of information on their need at the time of receiving the warning. The expertise of the communicator is quite different from the expertise of the satellite instrument designer or the meteorologist or hydrologist, and the range of expertise needed to translate hazard into impact is extraordinarily wide. Making a successful warning depends on these sources of expertise adding to each other rather than acting independently, in competition or in conflict. We have characterised the connections that achieve this as the bridges of our warning chain and described how they may be built through partnership. At its best, this chain of partnerships transparently communicates a consistent specification of the needs of the receiver to all of the contributing experts and communicates back reliable and trustworthy information about future hazards and impacts together with their uncertainties and appropriate responses, meeting the needs of the receiver, so that trust is created and communicated back up the chain, cementing and strengthening the overall relationship. Thus, while a failure of the warning chain arises from its weakest link, its strength comes from its end-to-end-to-end integration.

Research into the performance of whole warning systems is only just beginning, and so much of this chapter is based on anecdotal evidence. In this chapter, we will:

- Look at a long-established warning chain that is highly effective and consider what we can learn from it.
- Consider how to evaluate the outcomes of a warning system and look at the sensitivity of those outcomes to aspects of system performance.
- Identify some levers that can be used in optimising outcomes.

8.2   An Integrated Warning Chain

As an example of a familiar and long-established warning chain, we consider fire alarms in public buildings. The first consideration is that they exist because the government has legislated that they must and in general has defined such things as how quickly those at risk must be removed from harm. At its simplest, a fire alarm detects smoke, indicating that a fire has started. Although smoke is a hazard in its own right, it is here being used as an indirect sensor of the existence of a fire. A responder hears the alarm, knows that it is a fire alarm, evacuates the building and calls the fire service, who come and put out the fire. If the fire is small enough, a responder may also use local facilities to attack it. A visitor also recognises the alarm and follows accessible instructions or responder directions to evacuate. If it works well, no one is hurt and damage is minimal. The alarm may be communicated by a siren and/or a flashing light and/or audible instructions. It may also communicate directly with the fire and rescue service, perhaps with a manual override available to remove false alarms and tests.

For the responder, the requirement is that everyone should be evacuated safely during the short time that may be available between detection and people being trapped. This requires that regular practices are held so that they recognise the alarm and know what to do when they hear it. It may also require that the building is zoned so that different groups of people take different routes, to avoid congestion, and that the responder should know where the fire is so that proximate zones can be cleared first while avoiding any exits affected by the fire. For the visitor, the requirement is that they, personally, should evacuate safely and be inconvenienced as little as possible. For the fire service, if it is a complex building, they need to know in advance, and have trained for, entry routes, access to water, presence of hazardous materials and so on.

Taking this out of its fire context, we see that governance is essential and that it should define the desired outcome for the good of the community. Each country will define for itself what that requirement should be, though safety of life will always be fundamental. Where fire insurance is widespread, and especially if it funds the fire service, alarms may be mandated as a requirement for cover, in which case minimisation of damage will be part of the requirement. We also see the importance of preparation for responders, who should practice sufficiently frequently to be familiar with their duties. The remainder of those affected need to recognise the alarm and then either be told what to do by the alarm system, have easy access to pre-prepared instructions or be guided by responders. On the face of it, a fire alarm system has the advantage of being fixed. However, in reality, exits get blocked for repairs, and the needs of those in the building will be different each day. A best practice alarm system will direct people only to available exits and ensure responders are trained to provide assistance to those who cannot walk downstairs, for instance. We have also seen that to provide appropriate information to different receivers, the alarm system may provide tailored information.

In all of this, some key partnerships are essential for safety. The three-way relationship among the building manager, the fire service and the local responders is critical in ensuring adequate preparation. The relationships between building manager, alarm system designer and alarm sensor manufacturer determine the structure of an alarm system that will give adequate warning. The relationship between the local responders and visitors will affect the behaviour of visitors in response to the alarm. It is noteworthy that, contrary to traditional practice with natural hazard warnings, the focus here is on designing an alarm system to a specific outcome (defined in legislation), not on installing the latest sensor technology and building a warning system to use it to achieve the best outcome in safety and damage limitation.

8.3   Evaluating the Warning Chain

What makes a good warning and how can we measure it? The first question is not too difficult to answer – a good warning is one that enables negative consequences of hazards to be reduced – and a perfect warning is one that prevents all such consequences that are avoidable. Measuring the value of a warning is much more difficult, since there is no control; every event is different, affecting a different population; and the sample is small. Limited progress has been made using case studies that compare disaster events before and after the introduction of a warning system (e.g. heat wave warnings after 2003 in France – Fouillet et al. 2008). However, such studies make the assumption that each of the cases represents the whole population of events before and after the introduction, respectively, and that nothing else has happened to change people's response to the event, despite the fact that the trauma of the event that precipitated the warning system almost certainly changed behaviour in those affected. Where events are sufficiently frequent, it is possible to compare aggregate statistics before and after introduction of a warning system, as in the HIGHWAY project to provide user-oriented warnings to fishermen on Lake Victoria (Roberts et al. 2021).

Reduction of direct impacts may be estimated by modelling what would have happened without the warning-based decision. For instance, operation of a weir or installation of temporary flood defences will protect a well-defined area from flooding of predictable depth. The cost of repairing the avoided damage can be estimated fairly reliably, at least in countries with an active flood insurance market. Similarly, for the cancellation of a sports event in extreme heat, use of epidemiological analysis can give the expected number of people who would have died or been hospitalised, within defined statistical uncertainties. It is much more difficult to make such estimates when only a fraction of people responds to the warning. For instance, a hurricane evacuation warning may reach 80% of people in an evacuation area but suppose only half of those actually evacuate and the number of evacuees is swollen by those from outside the evacuation area who also choose to evacuate. In this situation, estimating the impact of the warning is much more challenging – undoubtedly depending on the socio-economic characteristics of the affected populations. In these circumstances, it is possible to build a model using historical data and surveys to estimate response inhibitors, but it will have substantial uncertainties and be subject to change after each event.

To ground the results in reality, it is essential to gather data on how people actually respond to warnings routinely. Properly designed and sampled surveys of the at-risk populations after each warning, supported by in-depth interviews, can provide evidence for longitudinal analysis of behavioural responses. There remain issues with the small number of events, with heterogeneous populations and with external changes, especially trends and shocks in the way people behave. Such surveys of how people report their behaviour also need to be supported by evidence of how they actually behaved, and this is increasingly possible through careful analysis of social media (Anderson et al. 2016; Eyre et al. 2020). Again, analysis should be carried out routinely to maximise the sample and to enable longitudinal analysis.

Elucidation of the factors that influence response behaviours can be carried out under test conditions where people say what they would do, provided the results can be related to real-life behaviour. Increasingly, immersive gaming approaches are achieving this, but the results still need checking in real situations.

Once established, the quantitative values of these factors can be used in the models of warning chain effectiveness. However, the factors that determine effectiveness of response are needed not just for the end user but also for the intermediate actors in the warning chain: those who assess the hazard and its impact and who decide how to convert the forecast information into an actionable warning. In some cases, personal prejudices will dominate – including fear of liability, over-optimism and distrust of models – making generalisation difficult.

While it is essential to know the outcomes that result from the issue of warnings, additional information is needed to determine how to make them better. Each chapter in this book has referred to the need to evaluate that step in the warning chain – both the quality of the information going into it and the quality of the information coming out of it – and for that evaluation to be carried out in terms relevant to the decision(s) that the receivers of the warning will need to make. However, the things that are measured are very different at each stage of the warning production process.

Each discipline along the warning chain, as depicted in Fig. 1.3, has developed its own methods of evaluation, geared to optimising its own performance in meeting the needs of those who use its products. These methods are valuable, but a means of relating them to each other and to the outcome of the whole warning chain is needed. Currently, there is no accepted methodology for connecting the quality measures for each step of the warning chain that will link them to the value of the decisions taken by warning receivers. While there are mathematical tools for relating the output of a mathematical model to its inputs, these are only currently deployed for individual models, not to the warning system as a whole.

8.4   Sensitivity of the Outcome

A typical approach to investing in an early warning system is to look at where there are technical options for improvement available, for the relevant organisations to undertake independent cost-benefit analyses on each and for the upgrades to be developed and implemented in isolation. In doing this, there is inevitably double counting of benefits, as each organisation claims the benefits of increased warning effectiveness.

Ideally, we should like to target investment of limited resources where it will be most effective. In short, if a flood warning system reduces the economic loss from floods by £1 m p.a., we want to know how much of that benefit is due to the monitoring network, how much to the weather forecast, how much to the hydrological forecast, how much to the translation into economic damage and how much to the form of the warning communication. We cannot experiment with the real world, so we need to use an alternative approach. Eventually, I hope this will take the form of a systems model, which reliably reproduces the behaviour of a warning chain, and can be used to identify responses to perturbations. Such a model does not yet exist. In the interim, we consider some of the sensitivities of the warning outcomes and how they depend on aspects of the warning system.

8.4.1 Precision

Precision is a characteristic of the underlying forecast information. It can be degraded in the formulation of a warning, but it can't be enhanced. The steps from weather to hazard to impact can either degrade or enhance precision, depending on the nature of the hazard and its impacts.

Precision can be in time, space or intensity. Precision in time may be needed for closing evacuation routes before, or letting relief in after, a disaster. Precision in space can be critical for protecting the right people and evacuating the minimum numbers. Precision in intensity may not matter for the most destructive events but may be critical for deciding whether or not to evacuate or protect from lesser ones.

Precision links to accuracy. Precision is generally of little value if it is not accurate. However, highly precise, but inaccurate, predictions may capture hazards or impacts that would be missed by a lower precision forecast system, and these may inform useful responses through upscaling in time and space.

8.4.2 Timeliness

In general, accuracy, precision and confidence in the available information get better as lead time reduces. However, the protective actions that the receiver will take require time, so there are critical lead times after which some protective actions, such as emptying a reservoir or evacuating a city, can no longer be taken. Receivers often ask for high precision, accuracy and confidence before taking an action and are then restricted in the actions that they can take. Part of an effective partnership is to share what is possible, and what might be possible with a little more investment, so that the opportunities for making a difference are maximised. This is particularly an issue for low-probability, high-impact hazards for which very short lead times are typically available, such as tornadoes. If it takes an hour for the forecast to be processed, interpreted and delivered, the window of opportunity for action has been lost. Speeding that up by 50% could be life-saving.

8.4.3 Accuracy

Accuracy is something that is a particular focus higher up the warning chain. It is a function of spatial and temporal precision and lead time. Highly inaccurate information can kill and cause damage itself, and over time it destroys trust. It is therefore important to post-process forecast information, e.g. by statistical correction, by upscaling in time and space or by subjective interpretation, to a level where accuracy can be demonstrated. Of course, such processing inevitably reduces the information content. If this makes it no longer useful for the intended action, alternative actions may have to be considered until forecast accuracy can be improved.

8.4.4 Reliability and Trust

The reliability of forecast information is an important characteristic in building trust. If it can be demonstrated to the user that a higher level of warning is consistent with a higher level of damage, that higher confidence is associated with greater likelihood of occurrence or greater proximity, the recipient is more likely to take action in the future.

Trust is, perhaps, the most difficult characteristic to invest in. Anecdotally, it is extremely important, and there is increasing documented evidence that this is so. Defining and measuring trust is difficult. As well as depending on the object of the trust, it varies enormously between communities and cultures. Trust may be intrinsic to a product or supplier, but it can also be gained by association. For instance, a respected community leader will engender trust of the community in the information they pass on. In consumer societies, a trusted consumer brand may confer trust on something it associates with, including warning information, especially if there is a perceived link.

8.4.5 Understandability

This is perhaps the most difficult characteristic of a warning message to get right. Describing the location of a hazard precisely without using names that are only recognised locally is extremely difficult. Tying road warnings to intersection identifiers or river flood warnings to local names of reaches or urban flood warnings to street names will immediately lose contact with a substantial proportion of those affected. On the other hand, overcoming those problems by the use of maps will lose those who are unable to read maps. Language is equally problematic, especially in countries with very large numbers of local languages. It is not just that the message may be missed if it is not in a person's first language but that it may still be lost if that language is not used idiomatically.

8.4.6   Reach

The warning chain can be broken at many points. When this happens, the value of the upstream information is lost. The easiest place to break the chain is if the information fails to reach the person making the decision. When everyone watched the evening news bulletin on television or radio, it was relatively easy to reach a large audience. That is no longer the case, with separate audiences for Twitter, Facebook, TikTok, etc. In multi-cultural societies, the media choices of different cultural groups need to be taken into account. And there still remains the challenge of reaching those who lack the ability to engage with any of these media: the frail, blind, deaf and housebound.

Where the decision-maker is a professional with responsibility for taking action to protect others, the importance of reaching them with critical information is much greater. It helps if they have been placed on-call, following an earlier warning. However, they may still be at home, asleep, at the golf course or at a party. Communicating the required information to these people is likely to be more than a Tweet or Facebook post, so making sure that they can access it from a mobile device is critical. In many countries, mobile connectivity is patchy and limited, so backup options need to be available, e.g. through conventional audio telephony.

8.4.7   User-Specific Warnings

We have emphasised a warning chain driven by the receiver of the warning. Yet the warning needs to be delivered to all those at risk and to those tasked with protecting them. Providing different warnings to different groups requires additional work, but can be worthwhile where requirements clearly differ, as between professional responders and the public or between users of specialist websites, such as sports users. The danger is that the same warning, delivered by different routes, in different formats and languages, is perceived to be inconsistent. Developments in technology are beginning to make it possible to tailor warnings automatically using systems like CAP. Genuinely personalised warnings are not currently available, but it is an area that will undoubtedly develop in the future.

8.5   Optimising the Outcome

So how do we design the perfect warning system? Here we distil the key messages from this book. The list given here draws heavily on the Early Warning Checklist produced at the first conference of the International Network for Multi-Hazard Early Warning Systems (IN-MHEWS) that took place from 22 to 23 May 2017 in

Cancún, Mexico (IN-MHEWS 2018), to which reference should be made for more detailed recommendations on how to carry out each step.

1. It is essential to understand the risks that will be faced by the community and how they are perceived by those at risk. Ocean hazards will not affect a land-locked country, but flooding may be a future risk for an arid country. The nature of the risk may also change – from direct risks to rural agriculture, to indirect risks to urban populations depending on vulnerable infrastructure. If a hazard is not considered to pose a risk to a community, then warnings will produce little response unless or until that perception changes.

2. With the community, identify the problem. If flooding is the biggest source of risk, is it river or surface water flooding or both? Is it frequent minor flooding or occasional major flooding? Is it damage to crops or interruption of transport? Warnings may be applied generally, but the design of the warning chain should focus on priority impacts.

3. With the community or organisations concerned, identify candidate mitigation actions that would be acceptable to the user and that would reduce the risk. Identify what information would be needed, by whom, when and where, to enable those actions to be undertaken.

4. Investigate how that information could be assembled – the areas of expertise needed, the organisations with the capability to provide services, those that are trusted to deliver, the existing working partnerships and those that have the connectivity to get the information to those who need it.

5. Identify the limitations to delivery – accuracy, timeliness, precision, format, etc. – and whether those limitations could be removed by investment. Where possible, use quantitative measures of capability.

6. Return to the user and consider whether the available capability would enable a useful and cost-effective mitigation. If there are significant benefits but the investment is unaffordable, or not cost-effective, consider whether the warnings could serve other users and if those benefits would make the service cost-effective or affordable. If not, consider whether there are alternative mitigations that might be more achievable.

7. Plan the selected system in detail, incorporating not just the production and delivery chain but monitoring and evaluation of each step and an assessment of how to recover when a failure occurs in any component.

8. Build a partnership to deliver and manage the warning system.

9. Train the warning producers and users, both before implementation and periodically afterwards.

10. Evaluate every step in the chain continually to ensure good performance. Look for opportunities to invest in further improvement, especially as new research capabilities become available.

11. Periodically – at least every 10 years – repeat the design process to check whether the system is still meeting priority user needs in the most cost-effective manner and to re-design if necessary.

Finally, when you have a system that is working, let all of your stakeholders know how well it is working and how effective it is at saving lives and reducing damage. That will help grow confidence and trust, build the foundations for future investment and encourage others to follow. And remember that every warning that reduces distress is a successful warning.

References

Anderson, J., M. Kogan, M. Bica, L. Palen, K. Anderson, K. Stowe, R. Morss, J. Demuth, H. Lazrus, O. Wilhelmi and J. Henderson, 2016. Far Far Away in Far Rockaway: Responses to Risks and Impacts during Hurricane Sandy through First-Person Social Media Narratives. In *Proc. ISCRAM 2016 conference, Rio de Janeiro, Brazil.*

Eyre, R., F. de Lucia and F. Simini, 2020. Social media usage reveals recovery of small businesses after natural hazard events. *Nature Communications*, **11**, 1629. https://doi.org/10.1038/s41467-020-15405-7

Fouillet, A., G. Rey, V. Wagner, K. Laaidi, P. Empereur-Bissonnet, A. Le Tertre, P. Frayssinet, P. Bessemoulin, F. Laurent, P. De Crouy-Chanel, E. Jougla and D. He'mon. 2008. Has the impact of heat waves on mortality changed in France since the European heat wave of summer 2003? A study of the 2006 heat wave. *Int. J. Epidemiol.* **37**, 309–317. https://doi.org/10.1093/ije/dym253

IN-MHEWS, 2018. *Multi-hazard Early Warning Systems: A Checklist*. WMO. Geneva, Switzerland. 20pp Available at https://www.preventionweb.net/publications/view/57604 (Accessed 25/5/2021)

Roberts, R. D., S. J. Goodman, J. W. Wilson, P. Watkiss, R. Powell, R. A. Petersen, C. Bain, J. Faragher, L. B. Chang'a, J. K. Kapkwomu, P. N. Oloo, J. N. Sebaziga, A. Hartley, T. Donovan, M. Mittermaier, L. Cronce and K. S. Virts, 2021. Taking the HIGHWAY to Save Lives on Lake Victoria. *Bull. Amer. Meteorol. S.* https://doi.org/10.1175/BAMS-D-20-0290.1

Index

© The Author(s) 2022
B. Golding (ed.), *Towards the "Perfect" Weather Warning*,
https://doi.org/10.1007/978-3-030-98989-7